Advances in
CANCER RESEARCH

Volume 113

Advances in
CANCER RESEARCH

Volume 113

Edited by

Kenneth D. Tew
*Professor and Chairman, Department of
Cell and Molecular Pharmacology
John C. West Chair of Cancer Research
Medical University of South Carolina
South Carolina, USA*

Paul B. Fisher
*Professor and Chairman
Department of Human & Molecular Genetics
Thelma Newmeyer Corman Chair in Cancer Research
Director, VCU Institute of Molecular Medicine
Virginia Commonwealth University, School of Medicine
Richmond, Virginia, USA*

AMSTERDAM • BOSTON • HEIDELBERG • LONDON
NEW YORK • OXFORD • PARIS • SAN DIEGO
SAN FRANCISCO • SINGAPORE • SYDNEY • TOKYO
Academic Press is an imprint of Elsevier

Academic Press is an imprint of Elsevier
525 B Street, Suite 1900, San Diego, CA 92101-4495, USA
225 Wyman Street, Waltham, MA 02451, USA
32 Jamestown Road, London, NW1 7BY, UK
Linacre House, Jordan Hill, Oxford OX2 8DP, UK
Radarweg 29, PO Box 211, 1000 AE Amsterdam, The Netherlands

First edition 2012

Copyright © 2012 Elsevier Inc. All rights reserved.

No part of this publication may be reproduced, stored in a retrieval system
or transmitted in any form or by any means electronic, mechanical, photocopying,
recording or otherwise without the prior written permission of the Publisher.

Permissions may be sought directly from Elsevier's Science & Technology Rights
Department in Oxford, UK: phone (+44) (0) 1865 843830; fax (+44) (0) 1865 853333;
email: permissions@elsevier.com. Alternatively you can submit your request online by
visiting the Elsevier website at http://elsevier.com/locate/permissions, and selecting
Obtaining permission to use Elsevier material.

Notice
No responsibility is assumed by the publisher for any injury and/or damage to persons
or property as a matter of products liability, negligence or otherwise, or from any use
or operation of any methods, products, instructions or ideas contained in the material
herein. Because of rapid advances in the medical sciences, in particular, independent
verification of diagnoses and drug dosages should be made.

ISBN: 978-0-12-394280-7
ISSN: 0065-230X

For information on all Academic Press publications visit
our website at www.elsevierdirect.com

Printed and bound in USA

12 13 14   10 9 8 7 6 5 4 3 2 1

Working together to grow
libraries in developing countries

www.elsevier.com | www.bookaid.org | www.sabre.org

ELSEVIER   BOOK AID International   Sabre Foundation

Contents

Contributors to Volume 113   ix

The AID Dilemma: Infection, or Cancer?
Tasuku Honjo, Maki Kobayashi, Nasim Begum, Ai Kotani, Somayeh Sabouri, and Hitoshi Nagaoka

  I. Introduction   2
 II. Two Distinct AID Functions   4
III. DNA Cleavage Mechanism by AID   7
 IV. AID Involvement in Genome Instability and Tumorigenesis   22
  V. Coda   31
     Acknowledgments   32
     References   32

The MicroRNA Regulatory Network in Normal- and HTLV-1-Transformed T Cells
Donna M. D'Agostino, Paola Zanovello, Toshiki Watanabe, and Vincenzo Ciminale

   I. Introduction   46
  II. HTLV-1 and ATLL   49
 III. miRNAs in Normal CD4+ T Cells   53
  IV. Cellular miRNA Expression in HTLV-1-Infected Cell Lines   56
   V. miRNA Profiling in ATLL Samples   63
  VI. Repression of miR-31 Expression in ATLL   67
 VII. Perspectives   70
      Acknowledgments   72
      References   72

The Multifaceted Oncoprotein Tax: Subcellular Localization, Posttranslational Modifications, and NF-κB Activation

Youmna Kfoury, Rihab Nasr, Chloé Journo, Renaud Mahieux, Claudine Pique, and Ali Bazarbachi

- I. The Oncogenic Retrovirus HTLV-I 86
- II. The Viral Oncoprotein Tax: Structural and Functional Domains 88
- III. Tax: A Potent Transactivator and a Deregulator of the Cellular Machinery 89
- IV. The NF-κB Pathway: Generalities 91
- V. Tax: A Powerful Activator of the NF-κB Pathway 93
- VI. Tax Posttranslational Modifications and NF-κB Activation 97
- VII. Tax Posttranslational Modifications and Intracellular Localization 103
- VIII. Conclusion 106
    Acknowledgments 106
    References 106

Lynch or Not Lynch? Is that Always a Question?

Chrystelle Colas, Florence Coulet, Magali Svrcek, Ada Collura, Jean-François Fléjou, Alex Duval, and Richard Hamelin

- I. Introduction 122
- II. What is HNPCC? 123
- III. What is MSI? 128
- IV. What is an MMR Defect? 132
- V. What Is Lynch Syndrome? 133
- VI. Unusual Variants of Lynch Syndrome 143
- VII. Genetic Alterations Responsible for MSI Tumor Progression in Sporadic and Lynch Syndrome Cases 145
- VIII. Current Approaches to the Detection of Lynch Syndrome Patients 146
- IX. Surveillance of Individuals with Lynch Syndrome 149
- X. Conclusions 150
    Acknowledgment 151
    References 151

Activation-Induced Cytidine Deaminase in Antibody Diversification and Chromosome Translocation

Anna Gazumyan, Anne Bothmer, Isaac A. Klein, Michel C. Nussenzweig, and Kevin M. McBride

- I. Introduction 168
- II. Chromosome Translocations 168
- III. Activation-Induced Cytidine Deaminase 171
- IV. Regulation of AID 171
- V. Template Strand Targeting by Exosomes 176
- VI. AID Genome Wide Damage 179

VII. CSR-Specific Cofactors   181
VIII. AID Influences CSR Outcome   182
IX. Conclusions   182
   Acknowledgment   183
   References   183

Opportunities and Challenges in Tumor Angiogenesis Research: Back and Forth Between Bench and Bed

Li Qin, Jennifer L. Bromberg-White, and Chao-Nan Qian

I. Introduction   192
II. Heterogeneity of Tumor Vasculature   193
III. Vessel Co-Option for Tumor Vascular Establishment   196
IV. Microvessels Formed by Tumor Cells and Tumor Stem-Like Cells   198
V. Comprehensive Angiogenesis-Related Signaling Pathways   199
VI. Differential Analyses of Tumor Vasculature   209
VII. Revascularization Within Tumors Following Withdrawal of Antiangiogenic Treatment   212
VIII. Vascular Normalization or Ineffective Targeting of Tumor Vasculature?   214
IX. Highlights in Antiangiogenic Therapy   215
X. Need for a Novel Clinical Evaluation System for Antiangiogenic Therapies   217
XI. Mechanisms of Resistance in Antiangiogenic Therapy   218
XII. Prospects for Future Development   221
   Acknowledgments   223
   References   223

Molecular Logic Underlying Chromosomal Translocations, Random or Non-Random?

Chunru Lin, Liuqing Yang, and Michael G. Rosenfeld

I. Introduction   242
II. Mechanisms of Chromosomal Translocations   250
III. The Role of Epigenetics in Chromosomal Translocations   262
IV. Conclusion   264
   Acknowledgments   267
   References   267

Index   281

Contributors

Numbers in parentheses indicate the pages on which the authors' contributions begin.

Ali Bazarbachi, Department of Internal Medicine, Faculty of Medicine, American University of Beirut, Beirut, Lebanon (85)

Nasim Begum, Department of Immunology and Genomic Medicine, Graduate School of Medicine, Kyoto University, Yoshida Sakyo-ku, Kyoto, Japan (1)

Anne Bothmer, Laboratory of Molecular Immunology, New York, USA (167)

Jennifer L. Bromberg-White, Laboratory of Cancer and Developmental Cell Biology, Van Andel Research Institute, Grand Rapids, Michigan, USA (191)

Vincenzo Ciminale, Department of Surgical Sciences, Oncology and Gastroenterology, University of Padova, Padova, Italy; Istituto Oncologico Veneto-IRCCS, Padova, Italy (45)

Chrystelle Colas, INSERM, UMRS 938, Centre de Recherche Saint-Antoine, Equipe "Instabilité des Microsatellites et Cancers," Paris, France; Université Pierre et Marie Curie-Paris6, Paris, France, and APHP, Laboratoire d'Oncogénétique et d'Angiogénétique, Groupe hospitalier Pitié-Salpétrière, Paris, France (121)

Ada Collura, INSERM, UMRS 938, Centre de Recherche Saint-Antoine, Equipe "Instabilité des Microsatellites et Cancers," Paris, France; Université Pierre et Marie Curie-Paris6, Paris, France (121)

Florence Coulet, Université Pierre et Marie Curie-Paris6, Paris, France; APHP, Laboratoire d'Oncogénétique et d'Angiogénétique, Groupe hospitalier Pitié-Salpétrière, Paris, France (121)

Donna M. D'Agostino, Department of Surgical Sciences, Oncology and Gastroenterology, University of Padova, Padova, Italy; Istituto Oncologico Veneto-IRCCS, Padova, Italy (45)

Alex Duval, INSERM, UMRS 938, Centre de Recherche Saint-Antoine, Equipe "Instabilité des Microsatellites et Cancers," Paris, France; Université Pierre et Marie Curie-Paris6, Paris, France (121)

Jean-François Fléjou, INSERM, UMRS 938, Centre de Recherche Saint-Antoine, Equipe "Instabilité des Microsatellites et Cancers," Paris,

France; Université Pierre et Marie Curie-Paris6, Paris, France, and AP-HP, Hôpital Saint-Antoine, Service d'Anatomie et Cytologie Pathologiques, Paris, France (121)

Anna Gazumyan, Laboratory of Molecular Immunology, New York, USA; Howard Hughes Medical Institute, The Rockefeller University, New York, USA (167)

Richard Hamelin, INSERM, UMRS 938, Centre de Recherche Saint-Antoine, Equipe "Instabilité des Microsatellites et Cancers," Paris, France; Université Pierre et Marie Curie-Paris6, Paris, France (121)

Tasuku Honjo, Department of Immunology and Genomic Medicine, Graduate School of Medicine, Kyoto University, Yoshida Sakyo-ku, Kyoto, Japan (1)

Chloé Journo, Retroviral Oncogenesis Laboratory, INSERM-U758 Human Virology, Lyon Cedex 07, France; Ecole Normale Supérieure de Lyon, Lyon Cedex 07, France (85)

Youmna Kfoury, Department of Internal Medicine, Faculty of Medicine, American University of Beirut, Beirut, Lebanon (85)

Isaac A. Klein, Laboratory of Molecular Immunology, New York, USA (167)

Maki Kobayashi, Department of Immunology and Genomic Medicine, Graduate School of Medicine, Kyoto University, Yoshida Sakyo-ku, Kyoto, Japan (1)

Ai Kotani, Medical Science Division, Tokai University Institute of Innovative Science and Technology, Shimokasuya, Isehara City, Kanagawa, Japan (1)

Chunru Lin, Howard Hughes Medical Institute, University of California, San Diego, School of Medicine, La Jolla, California, USA; Department of Medicine, Division of Endocrinology and Metabolism, University of California, San Diego, School of Medicine, La Jolla, California, USA (241)

Renaud Mahieux, Retroviral Oncogenesis Laboratory, INSERM-U758 Human Virology, Lyon Cedex 07, France; Ecole Normale Supérieure de Lyon, Lyon Cedex 07, France (85)

Kevin M. McBride, Department of Molecular Carcinogenesis, University of Texas MD Anderson Cancer Center, Smithville, Texas, USA (167)

Hitoshi Nagaoka, Department of Molecular Pathobiochemistry, Graduate School of Medicine, Gifu University, Yanagido 1-1, Gifu, Japan (1)

Rihab Nasr, Department of Anatomy, Cell Biology and Physiological Sciences, Faculty of Medicine, American University of Beirut, Beirut, Lebanon (85)

Michel C. Nussenzweig, Laboratory of Molecular Immunology, New York, USA; Howard Hughes Medical Institute, The Rockefeller University, New York, USA (167)

Claudine Pique, INSERM-U1016, CNRS UMR8104, Université Paris Descartes, Paris, France (85)

Chao-Nan Qian, State Key Laboratory on Oncology in South China, Sun Yat-sen University Cancer Center, Guangzhou, Guangdong, PR China; Laboratory of Cancer and Developmental Cell Biology, Van Andel Research Institute, Grand Rapids, Michigan, USA (191)

Li Qin, State Key Laboratory on Oncology in South China, Sun Yat-sen University Cancer Center, Guangzhou, Guangdong, PR China; Division of Pharmacoproteomics, Institute of Pharmacy and Pharmacology, University of South China, Hengyang, Hunan, PR China (191)

Michael G. Rosenfeld, Howard Hughes Medical Institute, University of California, San Diego, School of Medicine, La Jolla, California, USA; Department of Medicine, Division of Endocrinology and Metabolism, University of California, San Diego, School of Medicine, La Jolla, California, USA (241)

Somayeh Sabouri, Department of Immunology and Genomic Medicine, Graduate School of Medicine, Kyoto University, Yoshida Sakyo-ku, Kyoto, Japan (1)

Magali Svrcek, INSERM, UMRS 938, Centre de Recherche Saint-Antoine, Equipe "Instabilité des Microsatellites et Cancers," Paris, France; Université Pierre et Marie Curie-Paris6, Paris, France, and AP-HP, Hôpital Saint-Antoine, Service d'Anatomie et Cytologie Pathologiques, Paris, France (121)

Toshiki Watanabe, Department of Medical Genome Sciences, Laboratory of Tumor Cell Biology, Graduate School of Frontier Sciences, The University of Tokyo, Minato-ku, Tokyo, Japan (45)

Liuqing Yang, Howard Hughes Medical Institute, University of California, San Diego, School of Medicine, La Jolla, California, USA; Department of Medicine, Division of Endocrinology and Metabolism, University of California, San Diego, School of Medicine, La Jolla, California, USA (241)

Paola Zanovello, Department of Surgical Sciences, Oncology and Gastroenterology, University of Padova, Padova, Italy; Istituto Oncologico Veneto-IRCCS, Padova, Italy (45)

The AID Dilemma: Infection, or Cancer?

Tasuku Honjo,* Maki Kobayashi,* Nasim Begum,* Ai Kotani,† Somayeh Sabouri,* and Hitoshi Nagaoka‡

*Department of Immunology and Genomic Medicine, Graduate School of Medicine, Kyoto University, Yoshida Sakyo-ku, Kyoto, Japan
†Medical Science Division, Tokai University Institute of Innovative Science and Technology, Shimokasuya, Isehara City, Kanagawa, Japan
‡Department of Molecular Pathobiochemistry, Graduate School of Medicine, Gifu University, Yanagido 1-1, Gifu, Japan

I. Introduction
II. Two Distinct AID Functions
   A. Phenotypes of AID Deficiency
   B. Phenotypes of AID N-Terminal Mutations
   C. Phenotypes of AID C-Terminal Mutations
III. DNA Cleavage Mechanism by AID
   A. The DNA Deamination Hypothesis
   B. The RNA-Editing Model
   C. AID Target Specificity for DNA Cleavage
IV. AID Involvement in Genome Instability and Tumorigenesis
   A. Evolutionary Consideration of AID-Induced Genome Instability
   B. Evidence for AID's Involvement in Tumorigenesis
   C. Normal and Pathogen-Induced AID Expression
V. Coda
Acknowledgments
References

Activation-induced cytidine deaminase (AID), which is both essential and sufficient for forming antibody memory, is also linked to tumorigenesis. AID is found in many B lymphomas, in myeloid leukemia, and in pathogen-induced tumors such as adult T cell leukemia. Although there is no solid evidence that AID causes human tumors, AID-transgenic and AID-deficient mouse models indicate that AID is both sufficient and required for tumorigenesis. Recently, AID's ability to cleave DNA has been shown to depend on topoisomerase 1 (Top1) and a histone H3K4 epigenetic mark. When the level of Top1 protein is decreased by AID activation, it induces irreversible cleavage in highly transcribed targets. This finding and others led to the idea that there is an evolutionary link between meiotic recombination and class switch recombination, which share H3K4 trimethyl, topoisomerase, the MRN complex, mismatch repair family proteins, and exonuclease 3. As Top1 has recently been shown to be involved in many transcription-associated genome instabilities, it is likely that AID took advantage of basic genome instability or diversification to evolve its mechanism for immune diversity. AID targets are therefore not highly specific to immunoglobulin genes and are relatively abundant, although they have strict requirements for transcription-induced H3K4 trimethyl modification and repetitive sequences prone to forming non-B structures. Inevitably, AID-dependent cleavage takes place in nonimmunoglobulin targets and eventually causes tumors. However, battles against infection are waged in the context of acute emergencies, while tumorigenesis is rather a chronic, long-term process. In the interest of survival, vertebrates must have evolved AID to prevent infection despite its long-term risk of causing tumorigenesis. © 2012 Elsevier Inc.

I. INTRODUCTION

The successful smallpox vaccine introduced by Jenner in 1789 paved the way for future vaccines against a wide variety of bacterial and viral infections. While some of medical science's critical contributions to human healthcare have come through microbiology and immunology, the reasons behind the efficacy of vaccines have long remained a mystery. The question is twofold: how does the immune system recognize specific antigens out of the huge variety of antigens the body is exposed to? And, how does the immune system recognize pathogens as the same antigens previously encountered in a vaccine? These two questions are central to modern immunology. Two contrasting hypotheses to answer the first question were extensively debated from the 1950s through the 1970s, one proposing that we have a large number of genes encoding antigen receptors, and the other suggesting that a limited number of genes are mutated to amplify the antigen receptor repertoire. The subsequent development of recombinant technology contributed to proof that the latter hypothesis, proposed by Burnet, is basically correct. However, the precise mechanism is more complex than originally anticipated.

Vertebrates have two types of antibody diversification mechanisms, each taking place at a different lymphocyte differentiation phase. VDJ recombination creates an enormous repertoire of both T- and B-cell receptors by assembling various combinations of V, D, and J segments into one exon (Bassing *et al.*, 2002; Fugmann *et al.*, 2000; Gellert, 2002). The mechanism is highly regulated, and proceeds in step with the T and B lymphocyte developmental program. RAG1 and RAG2, which mediate VDJ recombination, are biochemically well characterized and appear to have been introduced rather recently in evolution, most likely by a transposon-like element (Kapitonov and Jurka, 2005; Schatz, 2004). However, VDJ recombination does not explain how antigenic antibody memory is generated, since its process is completely independent from antigen stimulation.

A second layer of diversity, somatic hypermutation (SHM), is introduced in the V exon of B lymphocytes by antigen stimulation. Evidence for SHM has accumulated through a series of experiments by Milstein, Weigert, and Cohn, among others. The most striking observation by Weigert *et al.* (1970)

is that 7 out of 19 mouse Vλ amino acid sequences are highly homologous except for several amino acid substitutions, while the remaining 12 Vλ sequences are identical. Direct proof that SHM occurs through genetic modification was obtained through the work of Tonegawa, Weigert, and other researchers, who compared DNA sequences between the germline and rearranged immunoglobulin Vλ genes (Rajewsky, 1996).

Antigenic stimulation of mature B lymphocytes induces yet another genetic alteration, called class switch recombination (CSR), into the immunoglobulin heavy-chain locus. Class-switching phenomena were originally reported by Uhr, Nossal, and Cooper, whose combined observations clearly indicated that B cells that express IgM change their isotype to other classes after antigenic stimulation (Honjo *et al.*, 2002). In 1978, Kataoka and Honjo proposed class switching to be caused by DNA recombination with a looping-out deletion (Honjo and Kataoka, 1978). This genetic alteration was (Honjo and Kataoka, 1978) directly demonstrated by cloning class-switched immunoglobulin loci (Cory *et al.*, 1980; Dunnick *et al.*, 1980; Maki *et al.*, 1980; Rabbitts *et al.*, 1980; Yaoita and Honjo, 1980).

By the end of the 1990s, we had learned that antigen-specific antibody memory is represented by two genetic alterations in the immunoglobulin locus: SHM, which is a point mutation in the V region exon, and CSR, a recombination event that replaces the CH gene in the heavy-chain locus. Many researchers looked for the enzymes or proteins regulating these genetic alterations. The gene inducing SHM was considered to be a mutator gene, and the gene for CSR was expected to encode a recombinase. The two proteins were assumed to be different.

The year 2000 brought the surprising discovery that a single protein, activation-induced cytidine deaminase (AID), regulates both SHM and CSR (Muramatsu *et al.*, 2000; Revy *et al.*, 2000). AID was cloned by subtractive hybridization between stimulated and nonstimulated CH12F3-2A B lymphoma cells, which switch efficiently from IgM to IgA when stimulated (Muramatsu *et al.*, 1999; Nakamura *et al.*, 1996). AID was demonstrated to regulate both SHM and CSR, both in studies of AID-deficient mice and through the identification of AID mutations in hyper-IgM syndrome type II (HIGM II) patients (Revy *et al.*, 2000). Subsequently, using artificial constructs to measure SHM and CSR, it was demonstrated that AID induces SHM and CSR in nonlymphoid cells (Okazaki *et al.*, 2002; Yoshikawa *et al.*, 2002). It therefore became clear that AID is essential and sufficient to induce SHM and CSR.

AID overexpression was subsequently shown to cause tumors in mice, indicating that AID is indeed a mutator. Transgenic mice carrying AID cDNA under the chicken β actin promoter frequently develop T lymphoma and lung microadenoma, and less frequently, B lymphoma, muscle-derived tumors, and hepatoma (Okazaki *et al.*, 2003). Inversely, AID deficiency reduces the frequency of *c-myc/Ig* chromosomal translocations that lead to the plasmacytoma formation associated with IL-6 overexpression (Ramiro et al., 2004). More recently, several lines of evidence have indicated that viruses that can cause tumors also frequently induce AID. These include the Epstein–Barr virus (EBV) (Epeldegui et al., 2007), HTLV-1 (Ishikawa et al., 2011), and hepatitis virus type C (Endo et al., 2007). In addition, *Helicobacter pylori* was shown to induce AID in gastric epithelial cells (Matsumoto et al., 2007). Another interesting association between AID and the Philadelphia chromosome-encoded *Bcr-Abl* kinase is considered to link AID with tumorigenesis or tumor progression (Feldhahn et al., 2007). Although AID's preferred target is the Ig locus, AID obviously attacks other genes as well, and there are many lines of evidence indicating that AID may be involved in tumorigenesis. Thus, although AID is essential for antibody memory generation as a core function of adaptive immunity, AID expression may simultaneously cause tumors. This dilemma regarding AID will be discussed from an evolutionary point of view.

II. TWO DISTINCT AID FUNCTIONS

AID function has been extensively studied using AID mutants. Strikingly, in HIGM II patients, AID mutation sites are found scattered across almost all of its 198 amino acid residues (Durandy et al., 2006; Revy et al., 2000). This finding clearly indicates that mutations of various AID regions affect CSR function. This could be due to the AID protein interacting with multiple proteins at its various regions, or to its structure being so unstable that point mutations in its various regions alter the structure enough to damage AID's function. Subsequent studies have revealed that both may be the case.

A. Phenotypes of AID Deficiency

AID mutations in HIGM II patients are scattered through almost all of AID's regions (Fig. 1 and Table I). Loss-of-function mutations are located not only within the catalytic region, which is well defined by conserved histidine (residue 50), tryptophan (residues 68 and 80), and cysteine (residue 87) residues, but also in the NLS (N-terminus), linker, apobec-like, and nuclear export signal (NES) (C-terminus) regions. In most cases, phenotypes result from defects in CSR alone or in both CSR and SHM, leading to severe immune deficiency. These findings strongly support the idea that AID is essential for these two genetic alterations required for antibody memory.

The phenotypes of AID knockout mice are almost identical to those of HIGM II patients, convincingly demonstrating that AID is essential for both CSR and SHM (Muramatsu et al., 2000). AID-deficient mice showed

Function: essential and sufficient to induce
- Class switch recombination (CSR)
- Somatic hypermutation (SHM)

Expression:
- (Physiological) activated B cells
- (Pathogenic) non-lymphoid cells by HCV and *H. pylori*

Structure:

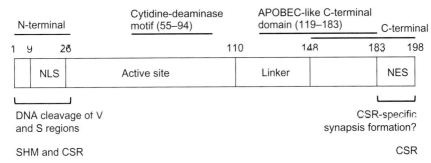

Fig. 1 AID structure and function. Note that the AID N-terminal region is required for DNA cleavage of the V and S regions, and the AID C-terminal region is required for CSR after DNA cleavage. Therefore, AID has two separate functions although it contains only one catalytic center in the middle of the protein.

generalized activation of immune cells that associates with enlarged lymphoid organs. Especially in the gut, the activation of immune cells caused hyperplasia of gut follicular structures such as Peyer's patches and isolated lymphoid follicles scattered throughout the length of the small intestine (Fagarasan *et al.*, 2002). The abnormal bacterial communities in gut mainly caused the activation of immune cells in $Acid^{-/-}$ mice (Fagarasan *et al.*, 2002). Subsequently, it was found that the gut flora in $Acid^{-/-}$ mice was composed mostly of uncultured anerobes, such as segmented filamentous bacteria, which are known to attach firmly to the gut epithelial cells and to cause activation of B and T cells in the gut (Suzuki *et al.*, 2004). Indeed, the treatment of AID-deficient animals with wide spectrum antibiotics resulted in the decrease of anerobic bacteria and simultaneous subsidence of activation of B cells and hyperplasia of germinal center (GC) in all lymphoid tissues. A mutant AID incapable to support diversification of IgAs by SHM (AID^{G23S}) showed very similar phenotype with $Acid^{-/-}$ mice (see later discussion) (Wei *et al.*, 2011). Together, these studies in $Acid^{-/-}$ mice revealed a previously underestimated role of IgAs in regulating the bacterial composition and diversity in the gut. They also showed that gut microbiota critically affects the whole body immune homeostasis and rekindled the study of immunology in the context of infection and immunity (Suzuki *et al.*, 2004).

B. Phenotypes of AID N-Terminal Mutations

AID N-terminal mutations in humans and mice affect both CSR and SHM. Doi *et al.* (2009) showed that in N-terminal-defective AID mouse mutants, DNA cleavage activity was deficient in both the V and S regions, indicating that the AID N-terminal region is indispensable for DNA cleavage. While SHM and CSR are deficient in most N-terminal mutants, they are augmented in others (Shinkura *et al.*, 2004; Shivarov *et al.*, 2008). In either case, CSR and SHM are both affected in the same direction. These results further imply that the cleavage mechanism may be common in the V and S regions. Whether the N-terminal mutation augments or decreases the AID activity, SHM is affected more strongly than CSR. These results are consistent with the observation that AID cleaves DNA at least five times more efficiently in the S region than in the V region (Doi *et al.*, 2009). One possible explanation for the high efficiency of S-region cleavage may be the abundance of repetitive sequences that are prone to form a non-B DNA structure (Dunnick *et al.*, 1993; Nikaido *et al.*, 1981). Thus, in several N-terminal region mutants, SHM is almost completely lost while the CSR activity is only slightly reduced (Shinkura *et al.*, 2004). One such mutant, G23S, is used to specifically abolish SHM and retain CSR *in vivo*, as we will discuss further. Taken together, these results show that AID introduces DNA cleavage in the V and S regions with different efficiencies, and that the target specificity between the V and S regions is not determined by AID itself, but probably by other mechanisms that control the status of the locus.

Studies of G23S knock-in animals show that without SHM, gut microbes expand aberrantly. AID deficiency and GC B cell hyperplasia are observed in the spleen as well as the gut, indicating that SHM is critical for maintaining gut microbial homeostasis (Wei *et al.*, 2011). Without preimmunization, G23S animals are sensitive to cholera toxin, whereas wild-type mice are completely resistant to cholera toxin at the same dose. This increased susceptibility of G23S mice is obvious even after cholera toxin immunization, indicating that in wild-type animals, the SHM that generates the gut immunoglobulin repertoire is spontaneously induced by gut microbes. This observation suggests that generation of a fully competent gut immunoglobulin repertoire depends not only on VDJ recombination, but also on diversification by SHM. How the gut microbes are sensed by the host immune cells is not yet clear. Bacteria might be taken up by M cells in Payer's patches, or by dendritic cells or macrophages situated beneath the epithelial layer. In any case, the results clearly indicate that gut microbes and the generation of the whole-body immune repertoire are mutually dependent. The lymphoid organ hyperplasia in HIGM II patients has been interpreted to be the result of recurrent infection (Quartier *et al.*, 2004). Interestingly, the lymph nodes are not enlarged in HIGM II patients with C-terminal mutations, who retain SHM (Imai *et al.*, 2005). It is therefore important to carefully assess the gut microbiota in HIGM II patients.

C. Phenotypes of AID C-Terminal Mutations

Most AID C-terminal mutations have one of two phenotypes. One accumulates AID in the nucleus because of a defective NES localized at the C-terminus, and the other loses CSR without reducing SHM. A series of mutations in the NES motif (183–198) all cause a specific CSR defect associated with augmented SHM activity and nuclear AID accumulation (Barreto *et al.*, 2003; Ito *et al.*, 2004; Ta *et al.*, 2003). Interestingly, the insertion of 34 amino acids at residue 182 (a clinical mutation called the P20 mutation) specifically causes CSR loss without accumulating the mutant AID protein in the nucleus (Ito *et al.*, 2004). This appears to indicate that the P20 mutation disrupts the domain that interacts with the CSR-specific cofactor. This domain appears to partially overlap the NES motif, as most NES-motif mutants lose CSR. In all of these C-terminal mutants, the DNA cleavage activity is normal or augmented in both the S and V regions (Doi *et al.*, 2009). These results clearly indicate that CSR requires not only DNA cleavage, but also an additional C-terminal-specific AID activity. Although the nature of this C-terminal-specific activity is not clear, we propose that it is involved in synapse formation between the cleaved ends (Doi *et al.*, 2009).

In summary, it is evident that AID has two clear functions (Fig. 1), one involving DNA cleavage of the S and V regions localized in the N-terminus, and the other involving CSR-specific activity in the C-terminus, probably related to synapse formation. The results can be most easily interpreted by the assumption that two separate cofactors interacting at the N-terminus and C-terminus of AID are responsible for DNA cleavage and putative synapse formation, respectively. Cleavage may become more efficient in C-terminal mutants because the cofactor for DNA cleavage monopolizes the catalytic center, assuming that AID's catalytic activity is required for both activities.

III. DNA CLEAVAGE MECHANISM BY AID

While it seems that AID's involvement in tumorigenesis must be closely linked with DNA cleavage, AID's molecular mechanism for DNA cleavage is still controversial. There are two contrasting hypotheses to explain how AID introduces DNA cleavage in the genome (Neuberger *et al.*, 2003; Petersen-Mahrt *et al.*, 2002; Muramatsu *et al.*, 2007).

A. The DNA Deamination Hypothesis

According to the DNA deamination hypothesis, AID first recognizes C bases on DNA and deaminates them to generate U (Di Noia and Neuberger, 2002). The resultant U/G mismatch is recognized by the base excision repair pathway enzymes. The U's are then removed by uracil-DNA glycosylase or uracil nucleotide glycosylase (UNG). The abasic sites thus created are attacked by apurine/apyrimidine (AP) endonuclease, which cleaves phosphodiester bonds at these sites.

1. AID DEAMINATION ACTIVITY

There are several lines of evidence to support the DNA deamination model. First, AID deaminates C to U on single-stranded DNA *in vitro*. Numerous experiments have revealed the properties of the *in vitro* reaction of DNA deamination by AID, including the nucleotide preference of the cleavage sites (Bransteitter *et al.*, 2003; Chaudhuri *et al.*, 2003; Shen *et al.*, 2005). These experiments showed that AID prefers single-stranded DNA, which may be generated *in vivo* by an R-loop or a transient single-strand bubble during transcription. The single-stranded DNA-binding protein RPA was shown to enhance the *in vitro* reaction (Chaudhuri *et al.*, 2004). However, criticism of this *in vitro* assay noted that the enzyme-to-substrate ratio was very high; the molar ratio of the AID added to the DNA far exceeded that seen in the catalytic activity *in vivo*. In addition, the DNA substrate *in vivo* is not naked DNA, but rather a very tight chromatin complex wrapped around histone octamers. Storb's group (Shen *et al.*, 2009) showed that the *in vitro* DNA deamination reaction is substantially reduced if histones are added to the DNA. In addition, it is notable that the *bona fide* RNA-editing enzyme APOBEC1 shows a similar DNA deamination activity *in vitro* (Petersen-Mahrt and Neuberger, 2003).

Another line of evidence consistent with the DNA deamination model is that AID overexpression can introduce C-to-T mutations in the *E. coli* and yeast genomes (Petersen-Mahrt *et al.*, 2002; Poltoratsky *et al.*, 2004). However, under these conditions there is no clear target specificity, in contrast to the clear-cut preference of the immunoglobulin gene in vertebrates. Further, APOBEC1 overexpression shows similar phenotypes (Harris *et al.*, 2002). Therefore, these phenotypes very likely reflect an artificial situation due to overexpression. Further, Shivarov *et al.* (2008) screened AID mutants *in vitro*, and found a series of mutants in which the DNA deamination activity is deficient or entirely lost. They then introduced these mutants into AID deficient B cells and examined whether the CSR activity was reduced. Interestingly, the N51A mutation almost completely abolished the *in vitro* DNA deamination activity but retained the CSR activity, indicating that the *in vitro* DNA deamination activity is not directly related to the CSR activity *in vivo*. The *in vitro* DNA deamination activity is also abolished in the homologous APOBEC1 N51A mutant. However, this mutation does not eliminate APOBEC1's RNA-editing activity, again confirming that APOBEC1's *in vitro* DNA deamination activity is not essential for its RNA-editing activity.

2. UNG INVOLVEMENT IN CSR

Another observation supporting the DNA deamination model is that UNG deficiency reduces CSR and gene conversion (Rada *et al.*, 2002; Saribasak *et al.*, 2006). While UNG deficiency reduces CSR by 80–90%, there is always residual activity (Rada *et al.*, 2002). SHM is enhanced in the absence of UNG, which is puzzling (Rada *et al.*, 2002; Saribasak *et al.*, 2006). Although the DNA deamination model attributes the absence of the SHM phenotype in UNG deficiency to compensation with mismatch repair (MMR) enzymes (Rada *et al.*, 1998), a dual deficiency of Ung and Msh2 reduces SHM by only 30%, with a heavy bias toward C-to-T and G-to-A mutations (Rada *et al.*, 2004). In contrast, a dual deficiency in UNG and Msh2 drastically reduces CSR, indicating that these two proteins are active at different steps of CSR (Honjo *et al.*, 2005). Further, this dissociation of CSR and SHM defects suggests that UNG and Msh2 may not be involved in AID-induced DNA cleavage; if they were, a dual UNG and Msh2 deficiency should affect CSR and SHM at similar levels. Gene conversion (GC), unlike SHM, is sensitive to UNG deficiency, although both SHM and GC give rise to point mutations in the V region. GC and CSR are mechanistically more similar to each other than to SHM; while all three processes require DNA cleavage, GC and CSR also require recombination. It is therefore likely that UNG is involved in a step after DNA cleavage, probably having to do with repairing cleaved ends for recombination.

Since UNG deficiency reduces CSR but not SHM, Begum *et al.* (2004) wondered whether this defect is associated with AID's DNA cleavage activity. They showed that even in the absence of UNG, DNA cleavage remains intact in the S regions. For this assay, IgM hybridomas were generated from UNG-deficient spleen cells after *in vitro* stimulation. Many IgM hybridomas had levels of $S\mu$–region mutations or deletions comparable to those in wild-type spleen cells, clearly indicating that DNA cleavage takes place in UNG-deficient spleen cells. Similarly, in two assays conducted in CH12F3-2 cells, UNG was completely blocked with its specific protein inhibitor Ugi to determine whether DNA cleavage could be induced by either (a) stimulating γH2AX focus formation, or (b) directly labeling the cleaved ends with biotin-labeled dUTP and terminal deoxynucleotide transferase. In both cases, completely blocking

UNG did not reduce the DNA cleavage activity in the activated CH12F3-2 cell S region (Begum et al., 2004). UNG is most likely involved in repairing or pairing DNA after cleavage. In fact, Durandy's group (Kracker et al., 2010) recently showed that UNG-deficient human B cells have much longer microhomology at the switch recombination junction, suggesting that UNG is required to support short heteroduplex regions for recombination, and that only very long microhomology pairing can be stabilized for recombination in the absence of UNG. The effects of UNG deficiency are far more severe in humans than in mice; this probably reflects human CSR's greater dependence on microhomology-mediated endjoining, which uses ligase 3, rather than nonhomologous endjoining, which uses ligase 4 and does not require any homology at the junction.

Begum et al. (2007) made a series of UNG catalytic-center mutants, and found that several mutations that abolish the U removal activity still retain the CSR activity in UNG-deficient spleen B cells (Table II). They also found that the WXXF motif required for UNG's interaction with the HIV Vpr protein is critical for UNG CSR. This WxxF motif is essential for recruiting UNG to the HIV virion to support HIV propagation. Thus, UNG appears to serve as a scaffold for forming protein complexes rather than for U removal (Fig. 2).

Accumulating evidence indicates that UNG has a function other than U removal: (a) host UNG is required for HIV viral genome integration, in which UNG interacts with an integrase and DNA preintegration complex, and UNG's catalytic activity has recently been shown to be indispensable for this interaction (Guenzel et al., 2011); (b) although UNG is essential for vaccinia virus genome replication, its catalytic activity is not required (De Silva and Moss, 2003, 2008); and (c) the histone H3 variant CENP-A is required for chromosome segregation during mitosis. CENP-A assembly on DNA depends on UNG, but not on UNG's catalytic activity (Zeitlin et al., 2011).

It was recently reported that AID induces U accumulation in UNG-deficient spleen cells (Maul et al., 2011). However, highly abundant U amounts are accumulated, as though the MMR proteins are not functional. These results should be quantitatively examined to determine whether the amount of U agrees with a known mutation frequency.

3. THE AP ENDONUCLEASE (APE) REQUIREMENT

Another important criterion for DNA deamination is the requirement of APE for CSR. In mammals, there are two types of APE: APE1 and APE2. In APE2-deficient mice, CSR is not reduced and SHM is enhanced (Sabouri et al., 2009). This is consistent with enzymological studies showing APE2 to be an exonuclease rather than endonuclease (Burkovics et al., 2006). Another report indicates that APE2 deficiency reduces CSR slightly under weak stimulation (Guikema et al., 2007). The significance of this defect is not clear, as

Table I CSR and SHM Activities of Mutant AID

AID Domain[a]	Mutation (AA) mAID	Mutation (AA) hAID	Patients SHM (%)	Patients CSR (%)	In vitro SHM (%)	In vitro CSR (%)	SHM CSR (%) xtAID	SHM CSR (%)	References
N-terminal domain	S3A		ND[b]	ND	207.2	143.6	28×10^{-4}	15.6	Gazumyan et al. (2010)
	S3D		ND	ND	ND	142.3			Gazumyan et al. (2010)
	M6A		ND	ND	1.1	3.3	18.1×10^{-4}	15	Okazaki et al. (2011)
		M6T	(−)[c]	(−)	ND	(−)	ND	ND	Durandy et al. (2006)
	N7A		ND	ND	47.5	93.3			Okazaki et al. (2011)
	R8A		ND	ND	139.2	106.6			Okazaki et al. (2011)
		R8_Y13delinsNfsX19	(−)	(−)	ND	(−)			Durandy et al. (2006)
	R9A		ND	ND	92.8	93.3			Okazaki et al. (2011)
	K10A		ND	ND	37.6	60			Okazaki et al. (2011)
	K10R		ND	ND	60.0±5.2	56.7±17	ND	ND	Ta et al. (2003)
		K10R	ND	ND	53.6	80			Okazaki et al. (2011)
	Y13H		ND	ND	10.7	63.2	0.84±0.12	38±7	Shinkura et al. (2004)
		F15X	0	ND	0.5±0.5	2.7±1.8			Ta et al. (2003)
	V18R		ND	ND	10.7	105.3			Shinkura et al. (2004)
	V18S/R19V		ND	ND	4.8	52.6			Shinkura et al. (2004)
	W20K		ND	ND	16.7	102.6			Shinkura et al. (2004)
		L22X	ND	(−)	ND	(−)			Durandy et al. (2006)
	G23S		ND	ND	10.7	97.4			Shinkura et al. (2004)
		R24W	0.9	ND	0.5±0.5	1.3±0.8			Ta et al. (2003)
		Y31X	ND	(−)	ND	(−)			Durandy et al. (2006)

(*continues*)

Table I (continued)

AID Domain[a]	Mutation (AA) mAID	hAID	Patients SHM (%)	Patients CSR (%)	In vitro SHM (%)	In vitro CSR (%)	SHM CSR (%) wtAID		References
Active site domain	S38A		ND	ND	30.2	36.5	4.3×10^{-4}	17	McBride et al. (2006)
	S38A		ND	ND	43.5	30.6	2.3×10^{-4}	17	McBride et al. (2008)
	S38D		ND	ND	ND	37.7			McBride et al. (2006)
	D45A		ND	ND	2	22	51×10^{-4}	41	Shivarov et al. (2008)
	D45A/F46A		ND	ND	2	7.3			Shivarov et al. (2008)
	D45A/R50A		ND	ND	2	7.3			Shivarov et al. (2008)
	D45A/N51A		ND	ND	3.9	2.4			Shivarov et al. (2008)
	D45A/R50A/N51A		ND	ND	0	2.4			Shivarov et al. (2008)
	F46A		ND	ND	2	43.9			Shivarov et al. (2008)
	G47A		ND	ND	2	12.2			Shivarov et al. (2008)
	H48A		ND	ND	9.8	100			Shivarov et al. (2008)
	L49A		ND	ND	3.9	58.5			Shivarov et al. (2008)
	R50A		ND	ND	29.4	129.3			Shivarov et al. (2008)
	N51A		ND	ND	2	51.2			Shivarov et al. (2008)
	R50A/N51A		ND	ND	0	9.8			Shivarov et al. (2008)
	K52A		ND	ND	19.6	119.5			Shivarov et al. (2008)
	S53A		ND	ND	117.6	85.4			Shivarov et al. (2008)
	G54A		ND	ND	21.6	95.1			Shivarov et al. (2008)
	C55A		ND	ND	37.3	109.8			Shivarov et al. (2008)
		H56Y	0	ND	0.5±0.5	2.1±1.7			Ta et al. (2003)
		H56Y	(−)	(−)	ND	(−)			Durandy et al. (2006)
		H56_E58delinsV	(−)	(−)	ND	(−)			Durandy et al. (2006)
		W68X	0	ND	0.5±0.5	2.0±0.9			Ta et al. (2003)
		W80R	ND	ND	0.5±0.5	3.8±3.2			Ta et al. (2003)
		S83P	ND	ND	ND	(−)			Durandy et al. (2006)
		W84X	ND	ND	ND	ND	ND	ND	Minegishi et al. (2000)

Domain	Mutation							Reference
	S85N	ND	(−)	ND	(−)			Durandy et al. (2006)
	C87R	ND	ND	0	2.1±1.1			Ta et al. (2003)
	C87S	(−)	(−)	ND	(−)			Durandy et al. (2006)
	KSS(H56K,C87S,C90S)	ND	ND	ND	0		16	Doi et al. (2009)
	L106P	ND	ND	0	1.4±0.9			Ta et al. (2003)
Linker domain	A111E	(−)	(−)	ND	(−)			Durandy et al. (2006)
	JP8A (R112C)	ND	ND	0.8±0.75	4.2±3.5			Ta et al. (2003)
	R112H	0–0.34	ND	0.5±0.5	1.8±0.9			Ta et al. (2003)
	L113P	ND	ND	ND	(−)			Durandy et al. (2006)
	M139T	ND	(−)	ND	(−)			Durandy et al. (2006)
	M139V	ND	ND	0.5±0.5	4.6±0.3			Ta et al. (2003)
T140A		ND	ND	87	82.4			McBride et al. (2008)
	D143_E163del20	(−)	(−)	ND	(−)			Durandy et al. (2006)
	C147X	0.76	ND	0	1.7±0.7			Ta et al. (2003)
C-terminal domain	F151S	ND	ND	ND	ND		ND	Revy et al. (2000)
	S169X	ND	(−)	ND	(−)			Durandy et al. (2006)
	L172A	ND	ND	27.7	63.5		26	Doi et al. (2009)
	L172A/G197A	ND	ND	ND	11.5	1.3×10^{-4}	26	Doi et al. (2009)
	R174S	(−)^d	(−)	ND	(−)			Durandy et al. (2006)
	L181delinsCfsX26	(+)^d	(−)	ND	(−)			Durandy et al. (2006)
	L181_P182ins31	(+)	(−)	ND	(−)			Durandy et al. (2006)
	p20(P182ins)	3.4	ND	71.0±23.4	3.2±1.1			Ta et al. (2003)
	JP8B(L183rep)	ND	ND	70.0±8	4.5±2.5			Ta et al. (2003)
	JP8Bdel(L183X)	ND	ND	548	ND	7.7×10^{-4}	ND	Ito et al. (2004)
	JP8Bdel(L183X)	ND	ND	200	3.6		14	Doi et al. (2009)
	L189A	ND	ND	430.8	35.6		26	Doi et al. (2009)

(*continues*)

Table I (continued)

AID Domain[a]	Mutation (AA) mAID	hAID	Patients SHM (%)	Patients CSR (%)	In vitro SHM (%)	In vitro CSR (%)	SHM CSR (%) wtAID		References
	L189X	R190A	ND	ND	100	8.3	11×10^{-4}	30	Barreto et al. (2003)
		JP41(R190X);	ND	ND	18.5	98.1		26	Doi et al. (2009)
		R190X	ND	ND	75.0±15	6.6±2			Ta et al. (2003)
		D191A	65	(−)	ND	ND	2±1.5	ND	Imai et al. (2005)
		A192G	ND	ND	74.6	71.2		26	Doi et al. (2009)
		F193A	ND	ND	100	80.8		26	Doi et al. (2009)
		F193X	ND	ND	592.3	15.4		26	Doi et al. (2009)
		R194A	ND	ND	327.3	ND			Ito et al. (2004)
		T195A	ND	ND	39.2	57.7		26	Doi et al. (2009)
		L196A	ND	ND	27.7	86.5		26	Doi et al. (2009)
		L196X	ND	ND	430.8	21.2		26	Doi et al. (2009)
		G197A	ND	ND	436	ND			Ito et al. (2004)
		L198A	ND	ND	73.9	51.9		26	Doi et al. (2009)
			ND	ND	376.9	21.2		26	Doi et al. (2009)
N/C domain double mutants		S3G/N168S	ND	ND	6.9±2.5	24.1±3			Ta et al. (2003)
		ΔN5JP8Bdel	ND	ND	550.7	ND	ND		Ito et al. (2004)
Linker/C double mutants		JP8A(R112C)/JP8B (L183rep)	1.3	ND	ND	ND	ND		Ta et al. (2003)
		D143_L181delinsAfsX4	(+)	(−)	ND	(−)			Durandy et al. (2006)

All activities shown on this table are relative to wtAID values shown on the right column.
del, deletion; fs, frameshift; ins, insertion; X, stop; ΔN5, N-terminal 5AA deletion; Rep, replacement.
[a]As indicated in Fig. 1.
[b]Not done.
[c]Undetectable.
[d]Detectable.

Table II Rescue of CSR by UNG, SMUG1, and Their Mutants in UNG$^{-/-}$ B Cells

Gene/mutation	Feature/property	CSR
Mouse UNG2	Wild type	(+)[a,e]
D145N	Catalytic inactivation	(+)[a,e]
H268L	Catalytic inactivation	(+)[a,e]
N204V	Catalytic inactivation (does not bind U)	(+)[a,e]
F242S[f]	Unknown/unstable/mito-targeted[g]	(+)[a]
D145N+H268L	Catalytic inactivation	(+)[a]
D145N+N204V	Catalytic inactivation	(−)[a]
N204D	CDG and residual UDG	(−)[a,b]
Y147A	TDG and residual UDG	(−)[a,b]
L272A	Unable to flip Uracil	(+)[c]
L272R	Increased binding to DNA	(+)[c]
R276E	Single strand specific activity	(+)[c]
W231A	Single mutation at Vpr binding site (WxxF)	(+)[b]
W231K	Single mutation at Vpr binding site (WxxF)	(−)[b,d]
F234G	Single mutation at Vpr binding site (WxxF)	(+)[b]
F234Q	Single mutation at Vpr binding site (WxxF)	(+)[b]
NΔ28	Lacks interaction motif for PCNA	(+)[d]
NΔ65	Lacks interaction motif for PCNA and NLS	(+)[b]
NΔ77	Lacks interaction motif for PCNA and NLS	(+)[d]
NΔ86	Lacks interaction motif for PCNA and RPA and NLS	(+)[d]
NΔ86.D145N	Catalytic inactivation	(−)[b]
NΔ86.H268L	Catalytic inactivation	(+)[b]
NΔ86.W231A	Single mutation at Vpr binding site (WxxF)	(−)[d,b]
NΔ86.W231K	Single mutation at Vpr binding site (WxxF)	(−)[d,b]
NΔ86.F234Q	Single mutation at Vpr binding site (WxxF)	(−)[d,b]
NΔ86.F234G	Single mutation at Vpr binding site (WxxF)	(−)[d,b]
NΔ86.W231A + F231G	Double mutation at Vpr binding site (WxxF)	(−)[d,b]
E. coli UNA	Wild type	(+)[b]
D64N	Catalytic inactivation	(−)[b]
H187Q	Catalytic inactivation	(+)[b]
Mouse SMUG1	Wild type	(+)[e,b]
N85A	H$_2$O coordination	(−)[b]
H239L	Stabilization of transition state	(−)[b]
N163D	Substrate binding	(−)[b]
G87Y	Thymine expulsion	(−)[e,b]
G87V	Thymine expulsion	(−)[b]
W144K	WxxF site	(−)[b]

[a]Begum et al. (2004).
[b]Begum et al. (2009).
[c]Muramatsu et al. (2007).
[d]Begum et al. (2007).
[e]Di Noia et al. (2006).
[f]Imai et al. (2003).
[g]Kavli et al. (2005).

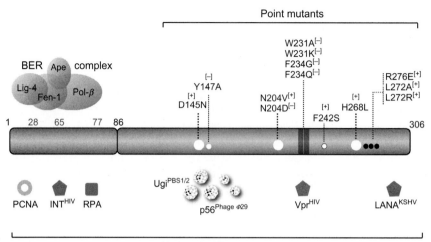

Fig. 2 Structural features of mouse UNG2, showing key mutations and interactive partners that contributed to our understanding of UNG in CSR. Amino acid numbers and the names of individual interacting proteins, along with their origin (upper case) if not cellular, are indicated. Active site residues are shown by wide white circles; two vertical bands denote the WxxF motif position. A mutant's ability [+] or inability [−] to induce or rescue CSR is shown beside its name. *Abbreviations*: UNG, uracil-DNA glycosylase; CDG, cytosine DNA glycosylase; TDG, thymine DNA glycosylase; RPA, replication protein A; PCNA, proliferating cell nuclear antigen; INT, integrase; HIV, human immunodeficiency virus; UGI, uracil-DNA glycosylase inhibitor; Vpr, viral protein R; LANA, latency-associated nuclear antigen; KSHV, Kaposi's sarcoma-associated herpesvirus; CENP-A, centromere protein-A; PBS1/2, phage of *Bacillus subtilis*.

there are no significant differences in serum Ig isotypes in the $APE2^{-/-}$ mice (Ide *et al.*, 2004). Since APE1 is essential for cell survival, $APE1$-knockout mice are lethal at an early stage of embryogenesis (Xanthoudakis *et al.*, 1996). APE1 knockdown in CH12F3-2 cells does not affect CSR (Sabouri *et al.*, 2009). These results clearly indicate that APE 1 and APE2 are not essential for CSR. Taken together, these results indicate that the DNA deamination model has to be reexamined more carefully.

B. The RNA-Editing Model

An alternative hypothesis, called the RNA-editing model, proposes that AID deaminates C bases in RNA, and the edited RNA mediates AID function (Honjo *et al.*, 2005; Muramatsu *et al.*, 2007). Studies on AID mutants suggest that AID has at least two functions: DNA cleavage, and synapsis formation of the cleaved ends. These functions are located at AID's N-terminal and C-terminal regions, respectively. The RNA-editing hypothesis postulates that

AID edits two different types of RNA for these two functions. For DNA cleavage by AID, it is assumed that AID's N-terminal region captures and edits a microRNA precursor associated with the N-terminal cofactor protein, generating a new microRNA that interacts with the 3' UTR of topoisomerase 1 (Top1) mRNA to suppress its translation (Fig. 3). This reduces the level of Top1, preventing the restoration of excessive DNA supercoiling accumulated during active transcription (Kobayashi et al., 2009). This facilitates the formation of unusual DNA structures, that is, non-B-form in the S or V region, resulting in irreversible cleavage by Top1 itself. The lack of restoration of transcription-associated negative DNA supercoiling by Top1 induces the DNA helix to loosen in the rear of the transcriptional machinery (Fig. 3). Under these conditions, thermodynamic calculations suggest that both the S and V regions form a non-B structure. Top1 normally cuts, rotates, and religates DNA to correct the aberrant supercoil accumulated by transcription. While Top1 can cut the non-B DNA form, it cannot rotate around the helix, as it is trapped by being covalently bound to the DNA. It has to be removed by resection enzymes such as Ctip and the MRN complex, which remove Spo11 (Top2) in meiotic recombination (Hartsuiker et al., 2009). This model is proposed based on the series of observations described below.

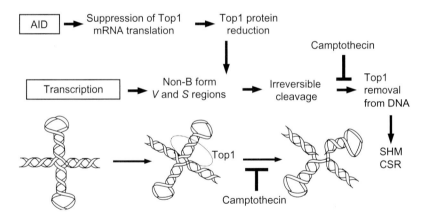

Fig. 3 A hypothetical method for AID-induced DNA cleavage and target transcription. AID suppresses Top1 mRNA translation, thus reducing the level of Top1 protein. Active target DNA transcription causes a negative supercoil accumulation behind the transcription complex. These factors favor the formation of non-B DNA structure in the V and S regions, which contain inverted repeats or repetitive sequences. Top1 introduces irreversible cleavage in a non-B structure due to steric hindrance of its rotation around the DNA helix. This causes Top1 to bond covalently with DNA, and its subsequent removal from the DNA creates a single-stranded break. Camptothecin intercalates in the Top1 and DNA complex and prevents Top1 from being removed from the DNA, thus inhibiting SHM and CSR.

1. AID INDUCES DNA CLEAVAGE BY REDUCING TOP1

AID reduces Top1's translation efficiency by half, decreasing Top1 protein levels by half within 24h, in accordance with Top1's half-life of about 3.7h in CH12F3-2A cells (Kobayashi *et al.*, 2009). Artificial Top1 reduction in CH12F3-2 cells by siRNA augments the CSR and DNA cleavage of *S* regions by AID. Similarly, the SHM activity is enhanced in a mutant plasmacytoma cell line (P388/CPT45) that has much less Top1 than the parental cell line (P388) (Kobayashi *et al.*, 2011). Overexpressing Top1 in P388/CPT45 cells strongly reduces SHM, while SHM is greatly augmented in *Top1* heterozygote B cells, which express about half the normal amount of Top1. It has also been reported that reducing Top1 with antisense oligonucleotides causes general genome instability (Miao *et al.*, 2007; Tuduri *et al.*, 2009). Thus, reducing Top1 augments the DNA cleavage in immunoglobulin genes and other loci. Experiments with the Top1-specific inhibitor camptothecin showed that the DNA cleavage activity in CSR and SHM depends on Top1 (Kobayashi *et al.*, 2009, 2011). A screen of endonuclease inhibitors identified camptothecin as the most potent inhibitor of the DNA cleavage in *S*-region CSR and SHM. Camptothecin traps DNA-bound Top1 by inhibiting Top1's release from the DNA protein complex. This release is essential to initiate CSR and SHM.

2. CSR DEPENDS ON THE ASSOCIATION OF AID'S C-TERMINUS WITH MRNA

AID is proposed to interact with messenger RNA captured by a cofactor that interacts with AID's C-terminus. The edited mRNA is postulated to generate a novel protein responsible for synapsis formation of cleaved ends. This activity is unique to CSR and not required for SHM. Nonaka *et al.* (2009) showed that AID is associated with poly A mRNA, and that this association depends on AID's C-terminal region. The edited mRNA has to be translated, and thus exported from the nucleus to the cytoplasm, in a C-terminal NES motif-dependent manner. In support of this assumption, Doi *et al.* (2003) showed that CSR is sensitive to very brief exposure to protein synthesis inhibitors. When CH12F3-2A cells are treated with cycloheximide or puromycin 1h before AID activation, CSR is completely blocked 3 or 6 h after AID activation, as measured by the formation of $S\mu$–$S\alpha$ recombinant DNA. The requirement of AID nuclear export for CSR has been demonstrated by two types of experiments. First, mutations in the C-terminal NES motif, which cause AID to accumulate in the nucleus, abolish CSR but do not reduce SHM. In addition, a 1-h preincubation with leptomycin B, which specifically inhibits the nuclear export protein CRM1, also blocks CSR (Doi *et al.*, 2003). An association with a CSR-specific cofactor other than the nuclear export cofactor at the C-terminus is supported by the finding that in the P20 mutant, which has an insertion at residue 182, CSR is completely blocked, but there is no effect on nuclear export of AID. SHM is normal in the P20 mutant, clearly indicating that the DNA cleavage activity and switch recombination depend on different regions of AID (Shinkura *et al.*, 2004; Ta *et al.*, 2003). We proposed that the C-terminal domain of AID is responsible for the recombination step after DNA cleavage (Doi *et al.*, 2009). In agreement with this hypothesis, a heterozygote human AID mutant with a C-terminal truncation (R190X) shows unusually longer junctions at Sμ–Sα recombination (Kracker *et al.*, 2010). Interestingly, many repair-related proteins such as DNA-PKcs, UNG, and MMR were reported to interact with the C-terminal domain of AID (Ranjit *et al.*, 2011; Wu *et al.*, 2005). Another group confirmed the independent role of the C-terminal domain from DNA cleavage and proposed its involvement in stabilization of AID (Ellyard *et al.*, 2011). However, this hypothesis does not agree with the fact that the P20 mutant is stable in spite of its loss of the recombination activity. So far, no experimental evidence against the RNA-editing model has been obtained. However, no direct evidence for RNA editing has been obtained, either. To prove that AID edits RNA, it is important to identify the microRNAs and mRNAs that are edited for DNA cleavage and for synapsis formation, respectively.

C. AID Target Specificity for DNA Cleavage

1. NON-IG AID TARGETS

Although AID was originally believed to specifically target immunoglobulin gene V and S regions, it has become evident that AID also cleaves other genes. AID-transgenic mice show tumors developed in various cells, in which mutations are accumulated in non-Ig genes, including proto-oncogenes (Kotani *et al.*, 2005; Okazaki *et al.*, 2003). However, the number of genes AID actually targets is not clear. There have been several attempts to identify AID's target genes by analyzing AID-induced mutations. Schatz and his colleagues showed that AID could introduce SHM in about 30 genes they examined by PCR (Liu *et al.*, 2008). Since this study only included preselected targets, it is uncertain whether AID preferentially targets these genes over the rest of the genome. Indeed, in the Peyer's patch B cells of 6-month-old mice, the mutation frequencies of all the non-Ig genes are less than half those of the J_H loci (Liu *et al.*, 2008). Recently, several groups have tried to identify AID target genes using the ChIP microarray or ChIP sequence method with various markers. Stavnezer's group, using Nbs1 as a DSB marker, reported that Nbs1 ChIP detected hundreds of genes after AID expression (Staszewski *et al.*, 2011). Since the authors did not confirm the mutations in these sites, it is not clear whether all the Nbs1-binding sites were cleaved by AID expression. Curiously, this report showed that Nbs1 accumulated in many nontranscribed loci, even though it has been well established that Ig gene transcription is mandatory for CSR and SHM (Betz et al., 1994; Jung et al., 1993; Peters and Storb, 1996; Zhang et al., 1993). This puzzle will be clarified when Nbs1's exact function in DNA metabolism in the genome is elucidated. Another group carried out ChIP analysis across the whole genome to identify AID-associated targets in mice (Yamane et al., 2011). This study listed about 5900 genes as possible AID targets. However, the order of the targets' AID-binding abundance did not correlate with their mutation frequencies. A physical association of AID and chromatin may or may not indicate loci that are functionally targeted by AID. Nussenzweig and his colleagues used Spt5 as a ChIP marker, assuming that since Spt5 interacts with AID, it is a target locus guiding factor for AID to directly deaminate DNA (Pavri et al., 2010). In this study, approximately 9000 genes were crosslinked with Spt5, which is a component of the transcriptional machinery. Not surprisingly, most of the genes associated with Spt5 are also identified by ChIP against RNA polymerase II. Here again, the physical interaction of Spt5 and AID does not necessarily indicate their functional involvement. It is also not certain whether these targets are actually cleaved or not; the 10 genes with the highest Spt5 binding have mutation frequencies as low as those of the $\beta 2$ microglobulin and β actin genes, which are known to be inert AID targets. We examined the genes functionally affected by Spt5 knockdown, and found that only 196 of them were upregulated, and 16 were downregulated (unpublished). These results suggest that physical Spt5 binding does not confirm its functional involvement in these loci, as has also been shown for RAG1 (Zhang and Swanson, 2008).

Recently, we took a different approach to the same question. We treated the BL2 cell line, which expresses a C-terminally truncated AID mutant fused with the ER domain (JP8del-ER), with tamoxifen for 3 h and directly labeled DNA cleavage sites by a DNA linker containing biotin-labeled dU (Kato et al., 2012). DNA fragments carrying this biotin-dU linker were concentrated by streptavidin beads and sequenced by high-throughput DNA sequencing, to compare the distribution of linker tags between tamoxifen-treated and nontreated BL2 JP8del-ER cells. The same tagged libraries were also analyzed by the promoter microarray method. Statistically significant targets identified by whole-genome sequencing were compared with those identified by microarray data. The expression profiles of candidate targets were examined, and the differential enrichment of target fragments in the two libraries was further confirmed by qPCR (Kato et al., 2012). Some of the targets were further analyzed for mutations. MALAT1 and SNHG3, two unexpected AID target sites identified by these studies, accumulate mutations as efficiently as the V region gene in BL2 JP8Bdel cells, and are known to be chromosomal translocation targets in various tumors (Davis *et al.*, 2003; Levin *et al.*, 2009; Rajaram *et al.*, 2007).

2. MARKERS FOR AID TARGETS

A big question is how AID chooses its Ig and non-Ig targets from among other sites in the genome. It is also not known how AID differentially regulates CSR and SHM. Any model that tries to explain AID-dependent DNA cleavage has to provide the mechanism for the target specificity. To answer this question according to the DNA deamination model, extensive studies were carried out to identify AID-binding cofactor(s) that might account for the target specificity of the AID function. Many proteins have been reported to interact with AID, including RNA polymerase II (Nambu *et al.*, 2003), replication protein A (RPA) (Chaudhuri *et al.*, 2004), protein kinase A (PKA) (Basu *et al.*, 2005; Pasqualucci *et al.*, 2006), DNA-PKcs (Wu *et al.*, 2005), MDM2 (MacDuff *et al.*, 2006), CTNNBL1 (Conticello *et al.*, 2008), Spt5 (Pavri *et al.*, 2010), PTBP2 (Nowak *et al.*, 2011), Spt6 (Jeevan-Raj *et al.*, 2011; Okazaki *et al.*, 2011), and Trim28/KAP1 (Jeevan-Raj *et al.*, 2011; Okazaki *et al.*, 2011). Unfortunately, however, no functional correlation has been shown for any of these proteins to support their role as an AID-specific target. Most of the proteins, such as RNA polymerase II, PKA, Spt5, and PTBP2 interact with many proteins other than AID. PTBP2 is a splicing factor, and Spt5 is one of the transcription elongation factors that associate with RNA polymerase II. RPA, DNA-PKcs, and MDM2 are proteins involved in general DNA repair. CTNNBL1 has since been shown to be dispensable for CSR (Han *et al.*, 2010). Among 10 molecules we found coimmunoprecipitated with AID (Okazaki *et al.*, 2011), Spt6 was the only functionally important one. Spt6 binds to AID's N-terminal region and regulates CSR but not SHM. Since AID's N-terminal region is responsible for SHM and for DNA cleavage in CSR, Spt6's binding to AID may not be directly relevant to the target specificity. In fact, Spt6 was later shown to regulate CSR as a histone chaperon to modify an epigenetic marker of target chromatin, as described below (unpublished data). Interestingly, in an experiment using CD19-cre and Trim28 conditional knockout mice, Trim28/KAP1 regulated CSR without affecting SHM (Jeevan-Raj *et al.*, 2011). However, the AID expression was drastically reduced in a Trim28 conditional knockout using mb1-cre (Okazaki *et al.*, 2011). This discrepancy is most likely due to the insufficient deletion of a floxed gene by CD19-cre compared with mb1-cre (Hobeika *et al.*, 2006).

Generally speaking, the target specificity of known specific recombinations, such as VDJ and meiotic recombination, is determined by the combination of the *cis* (DNA sequence/structure) and *trans* (DNA-binding proteins and the chromatin modification mark of the target locus) elements (Table III). In VDJ recombination, the recombination signal sequence is

Table III Recombination Target Specificity Determinants

Cleaving enzymes	Cis	Trans
Rag1, 2 (VDJ Rb)	RSS (12/23 bp)	H3K4me3
Spo11 (meiotic Rb)	13 mer	H3K4me3
Top1 (CSR, SHM)	Repetitive sequences and/or non-B form	H3K4me3

widely distributed in the genome, but the chromatin modification, for instance the histone3 lysine4 trimethylation (H3K4me3) recognized by the RAG2 PHD domain, is essential to cleave the target accurately (Liu et al., 2007; Matthews et al., 2007). In meiotic recombination, Spo11 cleaves at loosely conserved DNA target sequences that are also recognized by zinc finger-histone methyltransferase (PRDM9) to generate H3K4me3 at the target chromatin (Baudat et al., 2010; Myers et al., 2010; Parvanov et al., 2010; Wahls and Davidson, 2010). Without PRDM9, meiotic recombination is abortive. We have also shown that H3K4me3 at the target region is essential for CSR and SHM (Stanlie et al., 2010). The FACT complex, which is a histone chaperone composed of SSRP1 and Spt16, modulates the trans-histone modification cascade and is essential for CSR (Stanlie et al., 2010). In the absence of FACT, H3K4me3 modifications are defective at the Sμ and Sα regions, and this defect is associated with an S-region cleavage defect, although transcription of the Sμ and Sα regions is not reduced. Similarly, Spt6 and Spt5 are also required for CSR and assist in generating an H3K4me3 mark at the S region (our unpublished data). The RNA-editing model proposes that the target specificity may be determined by transcription-induced changes in cis and trans elements, which are represented by non-B-DNA structure formation and chromatin modification, respectively (Kobayashi et al., 2009, 2011; Stanlie et al., 2010).

IV. AID INVOLVEMENT IN GENOME INSTABILITY AND TUMORIGENESIS

A. Evolutionary Consideration of AID-Induced Genome Instability

1. CSR AND MEIOTIC RECOMBINATION HOMOLOGY

As mentioned above, H3K4me3 histone modification is an essential marker for all known specific recombinations—VDJ, meiotic, and CSR. This finding led us to compare the overall molecular mechanism of meiotic

Table IV Similar Molecules and Mechanisms in Meiotic Recombination and CSR

Function	Meiotic Rb	CSR
Recognized DNA	Hotspot 13-bp	Non-B (?)
Transcription dependency	Yes	Yes
Epigenetic mark	H3K4me3	H3K4me3
Cleaving enzyme	Spo11 (Top2)	Top1
Cleaved ends	Spo11-5'P-DNA	Top1-3'P-DNA
Resection of Top-DNA	CtIP and MRN	CtIP and MRN (?)
Stabilization of DNA structure	Msh4/5, Mlh1/Mlh3	Msh2/6, Mlh1/Pms2 (?)
5'→3' exonuclease	Exo1	Exo1
Endjoining	Crossing over	C-NHEJ or A EJ

recombination with CSR (Table IV). It became clear that the two recombination mechanisms share many aspects and players in addition to the H3K4me3 mark. First, both recombination mechanisms depend on topoisomerase for DNA cleaving—Spo11, a member of the Top2 family, in meiotic recombination, and Top1 in CSR. Second, both depend on the transcription of the target locus. Third, meiotic recombination requires Ctip and the MRN complex to resect covalently bound Spo11 from DNA to initiate recombination. CSR also requires Ctip and the MRN complex, although their precise roles in CSR are not established (Lee-Theilen et al., 2011). Finally, both recombinations depend on MMR proteins. In meiotic recombination, Msh4 and 5 are required for stabilizing the synapsis of strand invasion products during homologous recombination. CSR requires Msh2 and 6, and the DNA deamination model proposes that Msh2 and 6 are involved in the DNA cleavage step by recognizing mismatches generated by AID cytosine deamination. However, Msh2 knockout reduces CSR by shortening the average length of recombination junction microhomology sequences involved, indicating a reduced amount of atypical endjoining relative to nonhomologous endjoining (Schrader et al., 2002). Msh2 deficiency drastically reduces CSR in $S\mu^{-/-}$ mice (Min et al., 2003). These results suggest that Msh2 might be also involved in stabilizing single-stranded pairing at the CSR junction. Without Msh2 stabilization, atypical endjoining might become inefficient, because it depends on single-stranded pairing with many mismatches. Such conservation between meiotic recombination and CSR suggests that CSR evolved by borrowing the basic mechanism for DNA cleavage in meiotic recombination. If so, AID's loose target specificity is probably inevitable, since the target specificity of meiotic recombination is limited but not absolute.

2. TOP1 INVOLVEMENT IN TRANSCRIPTION-RELATED GENOME INSTABILITY

Both CSR and SHM depend completely on target transcription, which is by nature dangerous to genome integrity. Transcription inevitably exposes single-stranded DNA, either by local denaturation or by R-loop formation. In eukaryotes, transcription involves disassembling the tightly packed chromatin structure, exposing naked DNA, and then reassembling the chromatin. Further, transcription creates local distortion of the superhelix, positive in the front and negative in the rear of the transcription machinery. Because of these situations, excessive transcription has been shown to be associated with mutations, recombination, and other genome instability events. Among these, transcription-associated mutagenesis (TAM) has been extensively studied in yeast and has been shown to occur in mammals (Aguilera, 2002).

Several mechanisms have been proposed to explain why excessive transcription is prone to induce mutation, especially when associated with inefficient repair. Yeast genetic studies have shown that replacement mutations depend heavily on error-prone polymerase zeta, which is also involved in SHM (Zan et al., 2001). The nucleotide excision repair pathway, but not the base excision repair pathway, plays an important role in preventing TAM (Morey et al., 2000). Recently, a type of TAM called 2–5 base deletion was shown to depend almost totally on Top1's catalytic activity (Lippert et al., 2011; Takahashi et al., 2011). Top1's covalent association with the enhanced transcription target was demonstrated by chromatin immunoprecipitation. In addition, 2–5 base deletion takes place at hot spots with di- or trinucleotide repeats. More recently, triplet contraction/expansion, which is associated with many human genetic diseases including Huntington's disease, was also shown to depend on Top1 (Hubert et al., 2011). It is well known that tandem repeats of triplets form a non-B structure when transcribed (Bacolla et al., 2006). Here again, DNA cleavage leading to genome instability appears to share several features—transcription, non-B structure, and irreversible cleavage by Top1—with CSR and SHM. Thus, Top1's DNA cleavage mechanism in TAM and triplet contraction may be evolutionarily related to that in CSR and SHM.

B. Evidence for AID's Involvement in Tumorigenesis

1. AID IN MURINE TUMORS

The first evidence of AID's involvement in tumorigenesis was obtained through transgenic mouse studies. Ubiquitous AID expression in mice causes several types of tumors, most frequently in T cells, in association with mutations in many genes, including *T cell receptors*, *CD4*, *CD5*, *c-myc*, *P53*, and *Pim1* (Kotani *et al.*, 2005; Morisawa *et al.*, 2008; Okazaki *et al.*, 2003). These findings clearly indicate that AID can cause tumors by targeting several different genes, even in T cells, when strongly and continuously expressed. AID deficiency was shown to block *c-myc-Ig* translocation, which is associated with plasmacytomagenesis in mice (Ramiro *et al.*, 2004). Lymphomagenesis studies *in vivo* indicated that AID is required for a typical *c-myc-Ig* translocation-associated B lymphomagenesis (Kotani *et al.*, 2007; Takizawa *et al.*, 2008). AID-deficient Bcl-xL transgenic mice develop delayed atypical plasma cell tumors with the unusual Ig/myc chromosomal rearrangements (Kovalchuk *et al.*, 2007). *C-myc* transgenic animals develop pre-B lymphomas or B lymphomas without SHM. Interestingly, AID deficiency reduces the incidence of B lymphomas, but not pre-B lymphomas (Kotani *et al.*, 2007). AID deficiency prevents GC- and post-GC-derived lymphoma, but not marginal zone lymphoma development, in Iμ Bcl6 transgenic mice. These results indicate that AID is involved in tumorigenesis in mature activated B cells.

2. AID IN GC-DERIVED B-CELL LYMPHOMAS

AID is expressed physiologically in GC B cells (Muramatsu *et al.*, 2000; Yang *et al.*, 2005). Thus, after AID was discovered, initial studies on its expression in human hematological malignancy were performed in GC-derived human B-cell lymphomas, such as diffuse large B-cell lymphoma (DLBCL), follicular B-cell lymphoma (FL), and Burkitt lymphoma (BL). The majority, though not all, of these lymphomas express AID constitutively (Faili *et al.*, 2002; Greeve *et al.*, 2003; Hardianti *et al.*, 2004a,b; Muto *et al.*, 2000; Pasqualucci *et al.*, 2001; Smit *et al.*, 2003). These lymphomas frequently carry chromosomal translocations involving IgH, such as IgH/bcl6, IgH/bcl2, and IgH/c-myc; some of these are thought to stem from illegitimate CSR (Kuppers and Dalla-Favera, 2001). These lymphomas also frequently display hypermutations in oncogenes such as *Bcl6*, *c-myc*, *Pim1*, *RhoH/TTF1*, and *Pax5* that are similar to the SHM in Ig genes (Gordon *et al.*, 2003; Migliazza *et al.*, 1995; Pasqualucci *et al.*, 2001; Shen *et al.*, 1998).

AID is consistently expressed in neoplastic cells in nodular lymphocyte-predominant Hodgkin's lymphoma (LPHL), but infrequently in those of classical Hodgkin's lymphoma (CHL) (Greiner *et al.*, 2005; Mottok *et al.*, 2005). This finding is consistent with the idea that lymphocytic and histiocytic (L&H) cells, the neoplastic cells in LPHL, carry highly mutated Ig genes with ongoing SHM, and are therefore probably derived from GC B cells (Braeuninger *et al.*, 1997). In contrast, Hodgkin and Reed-Sternberg cells, the neoplastic cells in CHL, carry highly mutated Ig genes without ongoing SHM, and are probably derived from pre-apoptotic GC B cells, in which

B-cell-specific gene expression is downregulated (Kanzler et al., 1996; Kuppers et al., 2002; Marafioti et al., 2000). An intriguing possibility is that AID expressed in L&H cells plays a role in transforming LPHL, an indolent subtype of Hodgkin's lymphoma, into aggressive DLBCL by introducing additional transforming mutations (Greiner et al., 2005; Mottok et al., 2005).

3. AID IN NON-GC-DERIVED B-CELL LYMPHOMAS

AID expression has also been extensively studied in human B-cell chronic lymphocytic leukemia (B-CLL), which is a non-GC-derived B-cell lymphoma and the most common leukemia in the western world (Albesiano et al., 2003; Cerutti et al., 2002; Heintel et al., 2004; McCarthy et al., 2003; Oppezzo et al., 2003). B-CLL cells were originally considered to be the neoplastic counterpart of naïve B cells, which do not undergo SHM (Hamblin, 2002). However, this view changed when it was demonstrated that approximately 50% of B-CLL cases express mutated IgV_H genes (Schroeder and Dighiero, 1994). It is currently well accepted that there are at least two distinct B-CLL subsets, one with unmutated IgV_H genes (UM B-CLL) and the other with mutated IgV_H genes (M B-CLL); importantly, M B-CLL cases have a better prognosis than UM B-CLL cases (Damle et al., 1999). Gene expression profiles suggest that both B-CLL subtypes resemble memory B cells but not naïve B cells (Klein et al., 2001; Rosenwald et al., 2001). Surprisingly, AID is constitutively expressed in UM B-CLL but not M B-CLL cells (Albesiano et al., 2003; Cerutti et al., 2002; Heintel et al., 2004; McCarthy et al., 2003; Oppezzo et al., 2003), suggesting that the SHM machinery in UM B-CLL cells is defective or inactivated. In contrast, B-CLL cells that constitutively express AID undergo active CSR without any stimulation (Cerutti et al., 2002; Oppezzo et al., 2003), indicating a dissociation between SHM and CSR in CLL cells. Since AID is differentially expressed in the two subsets of CLL, even though their gene expression profiles are almost identical (Klein et al., 2001; Rosenwald et al., 2001), it is tempting to think that AID is involved in the poor prognosis of UM B-CLL. In addition, several AID splicing variants are expressed in UM B-CLL cells (Marantidou et al., 2010; McCarthy et al., 2003; Oppezzo et al., 2003). It will be interesting to learn whether these splicing variants are involved in UM B-CLL pathogenesis.

Recently, the study of AID expression in human B-cell leukemia or lymphoma has extended beyond GC-derived B-cell lymphomas and CLL to other kinds of human B-lineage leukemias and lymphomas. Several studies showed that AID is constitutively expressed in some cases of MALT lymphoma, which derives from marginal zone B cells in mucosa-associated lymphoid tissue; in immunocytoma, which is derived from plasma cells, and in hairy cell leukemia, which is derived from memory B cells, but not in multiple myeloma, which is derived from plasma cells (Forconi et al., 2004; Greeve et al., 2003;

Smit *et al.*, 2003). At present, the expression of AID in mantle cell lymphoma, which is derived from naïve B cells or an intermediate cell type between naïve and GC cells (Kolar *et al.*, 2007), is controversial (Babbage *et al.*, 2004; Greeve *et al.*, 2003; Klapper *et al.*, 2006; Smit *et al.*, 2003).

Of note, AID is constitutively expressed in lymphoid crisis in chronic myelogous leukemia (CML) and in Ph1+ pre-B acute lymphoblastic leukemia (ALL), both of which carry a t(9;22) translocation (Feldhahn *et al.*, 2007; Klemm *et al.*, 2009). These studies indicate that AID can be expressed not only in GC-derived B-cell lymphomas, but also in leukemias and lymphomas derived from B cells at various stages of differentiation.

4. AID'S POTENTIAL IMPORTANCE IN HUMAN B-CELL MALIGNANCIES

The presence of AID in a variety of human B-cell malignancies supports the assumption that AID plays a critical role in their initiation or progression. However, the AID levels expressed in these malignancies do not always correlate with the SHM level in the IgV gene or in oncogenes (Heintel *et al.*, 2004; McCarthy *et al.*, 2003; Pasqualucci *et al.*, 2004; Smit *et al.*, 2003). Moreover, AID expression is not always associated with ongoing mutations (Lossos *et al.*, 2004; Pasqualucci *et al.*, 2001, 2004; Smit *et al.*, 2003), although a clear association is observed in LPHL (Greiner *et al.*, 2005; Hardianti *et al.*, 2004b; Mottok *et al.*, 2005). The dissociation between AID expression and SHM activity can be explained at least in part by two possibilities: (1) the SHM machinery or AID itself is functionally impaired in tumor cells; or (2) AID may introduce mutations at earlier stages of the disease, but in later stages, AID may be shut off.

Nevertheless, AID expression is correlated with a poor prognosis in several human B-cell lymphomas and leukemias. As mentioned above, AID is associated with UM B-CLL, which has a poorer prognosis than M B-CLL (Heintel *et al.*, 2004; McCarthy *et al.*, 2003). High AID expression is restricted to a tumoral cell subpopulation with ongoing CSR and associated with aberrant SHM in the *c-myc*, *Pax5*, and *RhoH* genes. It is also associated with the transformation into a more aggressive lymphoma, which supports a potential role of AID as a new prognostic marker for B-CLL (Palacios *et al.*, 2010; Reiniger *et al.*, 2006).

In addition, significantly higher AID expression has been observed in a subgroup of DLBCL cases with a significantly poorer prognosis (Lossos *et al.*, 2004; Pasqualucci *et al.*, 2004). DLBCLs are categorized into three subgroups based on their gene expression profiles: the GC B-cell-like (GCB), activated B-cell-like (ABC), and type III DLBCL subgroups. The ABC subgroup has a poorer survival rate (Alizadeh *et al.*, 2000; Rosenwald *et al.*,

2002) and significantly higher AID levels than the other subgroups, although AID is highly expressed in all three subgroups (Lossos et al., 2004; Pasqualucci et al., 2001). Similarly, the association between AID and a poor prognosis was examined in primary cutaneous large B-cell lymphomas (PCLBCLs), which are divided into two main groups: primary cutaneous follicle center lymphoma (PCFCL), which is indolent, and PCLBCL, leg type (PCLBCL-leg), which has an intermediate prognosis (Willemze et al., 2005). Aberrant SHM in the *Bcl6*, *Pax5*, *RhoH*, and/or *c-myc* genes was observed in cases of both PCFCL and PCLBCL-leg, and the AID expression level was significantly higher in PCLBCL-leg than in PCFCL (Dijkman et al., 2006). This observation is consistent with the fact that PCFCL and PCLBCL-leg gene expression profiles are similar to those of GCB and ABC DLBCL, respectively (Hoefnagel et al., 2005).

Finally, two out of seven FL cases with clinical and histological progression showed elevated AID expression and the selective outgrowth of AID-expressing clones during the transformation into DLBCL, suggesting that AID is involved in the transformation from indolent FL to aggressive DLBCL (Smit et al., 2003). These observations suggest that AID may play a role in tumor evolution, and that AID may be useful as a new prognostic marker in certain human B-cell malignancies. In Ph1+ ALL and in the lymphoid blast crisis of CML that carries t(9;22), it is hypothesized that Bcr-abl kinase, which is derived from t(9.22), activates AID, and that AID subsequently introduces mutations into various oncogenes, resulting in clonal evolution. There is considerable interest in treatment with Imatinib, an abl kinase inhibitor. The first molecularly targeted Imatinib therapy dramatically improved the prognosis of Ph1+ ALL and CML. However, *bcr-abl* kinase domain mutations cause resistance to Imatinib, especially in the T315I leukemic clone, which is resistant to all abl kinase inhibitors. AID is involved in generating the T315I mutation in *bcr-abl* (Feldhahn et al., 2007; Gruber et al., 2010; Klemm et al., 2009). Splicing variants have also been identified in Ph1+ ALL (Iacobucci et al., 2010). Further analyses of the molecular mechanisms underlying these mutations and their clinical relevance are required.

C. Normal and Pathogen-Induced AID Expression

1. PATHOGEN-INDUCED AID EXPRESSION

Association studies of AID expressed in various pathogen-induced tumors have clarified its involvement in tumorigenesis. In humans, AID is abundantly expressed in EBV-positive BL cells (Table V). This is probably due to the expression of LMP1 on the cell surface, which mimics continuous CD40

Table V Pathogen-Induced Human Malignant Tumors

Pathogen	Type of cancer	Origin	AID expression	Inducer
HTLV-1	ATL		+	Tax[a]
EBV	BL	GC B cell	+	LMP1[b]
	CHL	GC B cell	−	
	Gastric cancer	Gastric epithelial cells	?	
	Nasopharyngeal carcinoma	Epithelial cells	?	
HCV	Hepatoma	Hepatocyte	+	TGF-beta signaling[c]
	DLBCL	GC B cell	+	NF-κB signaling[d]
H. pylori	Gastric cancer	Gastric epithelial cells	+	NF-κB signaling[e]
	MALT lymphoma	Marginal zone B cells	?	
Papilloma virus	Cervical cancer	Cervical epithelial cells	?	

ATL, acute T-cell lymphoma/leukemia; BL, Burkitt lymphoma; CHL, classical Hodgkin's lymphoma; DLBCL, diffuse large B-cell lymphoma; GC B cell, germinal center B cell
[a]Ishikawa *et al.* (2011).
[b]Epeldegui *et al.* (2007).
[c]Kou *et al.* (2007).
[d]Machida *et al.* (2004).
[e]Matsumoto *et al.* (2007).

signaling stimulation. More recently, many tumor-causing viruses have been shown to induce AID expression in both B and non-B cells. AID is frequently expressed in adult T cell leukemia cells caused by HTLV-1 infection (Ishikawa *et al.*, 2011). The HTLV-1 Tax oncogene, when overexpressed in T cells, activates the *Aicda* gene in a CREB- and NF-κB-dependent manner (Ishikawa *et al.*, 2011; Nakamura *et al.*, 2011). Tumorigenesis occurs in transgenic mice for Tax or bZIP, which are HTLV-1 oncogenes; it will be interesting to test whether AID is required for this T-cell tumor (Satou *et al.*, 2011; Yamazaki *et al.*, 2009). Hepatitis C virus (HCV) also induces AID. HCV hepatocyte and B cell infection induces AID via NF-κB signaling (Endo *et al.*, 2007; Machida *et al.*, 2004). Increasing epidemiological evidence has highlighted the close correlation between HCV infection and B-NHL (de Sanjose *et al.*, 2008; Turner *et al.*, 2003). Thus, it is tempting to hypothesize that the upregulation of AID in HCV infected B cells is at least partly responsible for HCV related DLBCL lymphomagenesis. Further, *H. pylori*, a gastric cancer-causing agent, also induce AID in the gastric epithelium (Matsumoto *et al.*, 2007). Kim *et al.* (2007) reported that 73 of 186 sporadic gastric cancers examined expressed AID. The majority of pathogens known to induce tumors are connected with AID expression. AID expression has also been reported in colitis-associated colorectal cancers (Endo et al., 2008), in epithelial breast cancer cell lines (Babbage et al., 2006), in hepatoma (Kou et al., 2007), and in cholangiocarcinoma (Komori et al., 2008). It is not known whether the papilloma virus, which causes cervical cancer, induces AID; this has not yet been studied.

2. REGULATION OF AID EXPRESSION

It is important to understand how AID is regulated. How is AID restricted to B cells? How can AID be expressed in infected non-B cells? Tran et al. (2010) extensively analyzed the promoter *Aicda* in CH12F3-2 cells. Evolutionary conservation reveals four candidate regions for regulatory elements within and surrounding the *Aicda* locus. The R1 region located 5′ to the transcription initiation site contains Sp and HoxC4-Oct binding elements, which have relatively weak activity (Park et al., 2009; Tran et al., 2010; Yadav et al., 2006). The R2 region located in intron 1 contains the Pax5 and E2A elements, which are the most important B cell-specific transactivation motifs. The most striking feature in the R2 region is the presence of negative regulatory elements that may be responsible for suppressing the *Aicda* expression in non-B cells. Transgenic analysis showed that the R3 region located further downstream of the *Aicda* locus has transactivation activity (Crouch et al., 2007). However, in an *in vitro* luciferase assay system, the locus did not show clear activity (Tran et al., 2010). It was recently reported that BATF, which may directly regulate *Aicda* expression, binds to the R3 region (Ise et al., 2011). The R4 region, located further upstream of the R1 region, contains multiple elements that respond to exogenous stimulation, such as NF-κB, STAT6 (which responds to IL-4), and Smad3/4 (Tran et al., 2010). Estrogen has been reported to enhance *Aicda* expression, but whether it regulates *Aicda* directly or indirectly is still contradictory (Mai et al., 2010; Pauklin et al., 2009). Further, recent studies have shown another layer of AID expression regulation by microRNA-155 (Dorsett et al., 2008; Teng et al., 2008).

Qin et al. used genetic marking to examine cell lineages that had experienced AID expression. Transgenic *Aicda-cre* BAC mice were crossed with Rosa26 reporter mice (R26R or Rosa-tdRFP) (Kwon et al., 2008; Qin et al., 2011) to create a system that marks any cell that has expressed AID. This sensitive method substantially marked not only B-lineage cells, but also T-lineage cells. In particular, $CD4^+$ T cells were consistently marked and accumulated with age; by 18 months of age, up to 25% of these cells were marked as having experienced AID expression (Qin et al., 2011). These cells exhibited an effector-memory phenotype, producing IL-10 and IFN-γ. Marked cells were generated by adoptively transferring naïve $CD4^+$ T cells into a T cell-deficient host, suggesting that specific environmental stimuli may be involved in AID expression in T cells. However, because AID-experienced $CD4^+$ T cells can be generated in a B cell-deficient background, a germinal center environment is not essential (Qin *et al.*, 2011). An AID-experienced population was observed in the natural killer cell fraction, as well (Qin *et al.*, 2011). These results indicate that AID can be expressed in non-B lineage cells even under normal conditions. Thus, it is reasonable to imagine that chronic infection may induce AID expression even in nonlymphoid cells.

V. CODA

An evolutionary consideration of the AID mechanism reveals that AID shares its DNA cleavage mechanism with several types of genome instability mechanisms such as TAM and triplet contraction/expansion, which depend on Top1 nicking activity. These mechanisms also share a dependency on transcription, and probably on non-B structural formation induced by excessive negative supercoil. This indicates that AID does not necessarily uniquely target Ig genes. Many genes that form transcription-induced non-B structures could be candidates for AID targeting, but this requirement is not sufficient, because the AID-induced DNA cleavage also depends on chromatin H3K4me3 modification. In fact, even this mark may not be sufficient and other modifications may be required. This dilemma which has arisen when AID was introduced in vertebrates appears to be resolved at least in part by dual restriction of AID target; non-B structure and chromatin modification; while AID probably took advantage of transcription-induced genome instability to amplify immunoglobulin at the somatic level, it was almost impossible to limit its target. Therefore, AID expression was limited rather strictly to B cells. Nonetheless, AID expression is not absolutely regulated; recent studies clearly indicate that many pathogens can activate AID in non-B cells.

However, AID activation alone is not sufficient for tumorigenesis. This is seen in the fact that tumors do not develop until at least a few months after transgenic AID-induced tumorigenesis; in fact, most viral or pathogenic tumorigenesis takes years. It is possible that AID introduces mutations in various target genes, including oncogenes, while virus infection makes target cells immortal and induces AID activation. The immortalization gradually accumulates AID-induced mutations, and eventually tumor cells are selected and quickly expand. It is therefore reasonable for vertebrates to take advantage of AID-enabled immune diversification in spite of the tumorigenesis risks. Since SHM is acutely required to prevent infection, we have to accept the long-term risk of tumorigenesis.

ACKNOWLEDGMENTS

We thank Drs. S. Fagarasan, K. Kinoshita, M. Muramatsu, R. Shinkura, and I. Okazaki for critical reading of this chapter, and Ms. Y. Shiraki for preparing the manuscript. This work was supported by Grant-in-Aid 17002015 for Specially Promoted Research from the Ministry of Education, Culture, Sports, Science, and Technology of Japan.

REFERENCES

Aguilera, A. (2002). The connection between transcription and genomic instability. *EMBO J.* **21**, 195–201.

Albesiano, E., Messmer, B. T., Damle, R. N., Allen, S. L., Rai, K. R., and Chiorazzi, N. (2003). Activation-induced cytidine deaminase in chronic lymphocytic leukemia B cells: expression as multiple forms in a dynamic, variably sized fraction of the clone. *Blood* **102**, 3333–3339.

Alizadeh, A. A., Eisen, M. B., Davis, R. E., Ma, C., Lossos, I. S., Rosenwald, A., Boldrick, J. C., Sabet, H., Tran, T., Yu, X., Powell, J. I., Yang, L., *et al.* (2000). Distinct types of diffuse large B-cell lymphoma identified by gene expression profiling. *Nature* **403**, 503–511.

Babbage, G., Garand, R., Robillard, N., Zojer, N., Stevenson, F. K., and Sahota, S. S. (2004). Mantle cell lymphoma with t(11;14) and unmutated or mutated VH genes expresses AID and undergoes isotype switch events. *Blood* **103**, 2795–2798.

Babbage, G., Ottensmeier, C. H., Blaydes, J., Stevenson, F. K., and Sahota, S. S. (2006). Immunoglobulin heavy chain locus events and expression of activation-induced cytidine deaminase in epithelial breast cancer cell lines. *Cancer Res.* **66**, 3996–4000.

Bacolla, A., Wojciechowska, M., Kosmider, B., Larson, J. E., and Wells, R. D. (2006). The involvement of non-B DNA structures in gross chromosomal rearrangements. *DNA Repair (Amst.)* **5**, 1161–1170.

Barreto, V., Reina-San-Martin, B., Ramiro, A. R., McBride, K. M., and Nussenzweig, M. C. (2003). C-terminal deletion of AID uncouples class switch recombination from somatic hypermutation and gene conversion. *Mol. Cell* **12**, 501–508.

Bassing, C. H., Swat, W., and Alt, F. W. (2002). The mechanism and regulation of chromosomal V(D)J recombination. *Cell* **109**(Suppl.), S45–S55.

Basu, U., Chaudhuri, J., Alpert, C., Dutt, S., Ranganath, S., Li, G., Schrum, J. P., Manis, J. P., and Alt, F. W. (2005). The AID antibody diversification enzyme is regulated by protein kinase A phosphorylation. *Nature* **438**, 508–511.

Baudat, F., Buard, J., Grey, C., Fledel-Alon, A., Ober, C., Przeworski, M., Coop, G., and de Massy, B. (2010). PRDM9 is a major determinant of meiotic recombination hotspots in humans and mice. *Science* **327**, 836–840.

Begum, N. A., Kinoshita, K., Kakazu, N., Muramatsu, M., Nagaoka, H., Shinkura, R., Biniszkiewicz, D., Boyer, L. A., Jaenisch, R., and Honjo, T. (2004). Uracil DNA glycosylase activity is dispensable for immunoglobulin class switch. *Science* **305**, 1160–1163.

Begum, N. A., Izumi, N., Nishikori, M., Nagaoka, H., Shinkura, R., and Honjo, T. (2007). Requirement of non-canonical activity of uracil DNA glycosylase for class switch recombination. *J. Biol. Chem.* **282**, 731–742.

Begum, N. A., Stanlie, A., Doi, T., Sasaki, Y., Jin, H. W., Kim, Y. S., Nagaoka, H., and Honjo, T. (2009). Further evidence for involvement of a noncanonical function of uracil DNA glycosylase in class switch recombination. *Proc. Natl. Acad. Sci. U S A* **106**, 2752–2757.

Betz, A. G., Milstein, C., Gonzalez-Fernandez, A., Pannell, R., Larson, T., and Neuberger, M. S. (1994). Elements regulating somatic hypermutation of an immunoglobulin kappa gene: critical role for the intron enhancer/matrix attachment region. *Cell* **77,** 239–248.

Braeuninger, A., Kuppers, R., Strickler, J. G., Wacker, H. H., Rajewsky, K., and Hansmann, M. L. (1997). Hodgkin and Reed-Sternberg cells in lymphocyte predominant Hodgkin disease represent clonal populations of germinal center-derived tumor B cells. *Proc. Natl. Acad. Sci. USA* **94,** 9337–9342.

Bransteitter, R., Pham, P., Scharff, M. D., and Goodman, M. F. (2003). Activation-induced cytidine deaminase deaminates deoxycytidine on single-stranded DNA but requires the action of RNase. *Proc. Natl. Acad. Sci. USA* **100,** 4102–4107.

Burkovics, P., Szukacsov, V., Unk, I., and Haracska, L. (2006). Human Ape2 protein has a 3′–5′ exonuclease activity that acts preferentially on mismatched base pairs. *Nucleic Acids Res.* **34,** 2508–2515.

Cerutti, A., Zan, H., Kim, E. C., Shah, S., Schattner, E. J., Schaffer, A., and Casali, P. (2002). Ongoing *in vivo* immunoglobulin class switch DNA recombination in chronic lymphocytic leukemia B cells. *J. Immunol.* **169,** 6594–6603.

Chaudhuri, J., Tian, M., Khuong, C., Chua, K., Pinaud, E., and Alt, F. W. (2003). Transcription-targeted DNA deamination by the AID antibody diversification enzyme. *Nature* **422,** 726–730.

Chaudhuri, J., Khuong, C., and Alt, F. W. (2004). Replication protein A interacts with AID to promote deamination of somatic hypermutation targets. *Nature* **430,** 992–998.

Conticello, S. G., Ganesh, K., Xue, K., Lu, M., Rada, C., and Neuberger, M. S. (2008). Interaction between antibody-diversification enzyme AID and spliceosome-associated factor CTNNBL1. *Mol. Cell* **31,** 474–484.

Cory, S., Jackson, J., and Adams, J. M. (1980). Deletions in the constant region locus can account for switches in immunoglobulin heavy chain expression. *Nature* **285,** 450–456.

Crouch, E. E., Li, Z., Takizawa, M., Fichtner-Feigl, S., Gourzi, P., Montano, C., Feigenbaum, L., Wilson, P., Janz, S., Papavasiliou, F. N., and Casellas, R. (2007). Regulation of AID expression in the immune response. *J. Exp. Med.* **204,** 1145–1156.

Damle, R. N., Wasil, T., Fais, F., Ghiotto, F., Valetto, A., Allen, S. L., Buchbinder, A., Budman, D., Dittmar, K., Kolitz, J., Lichtman, S. M., Schulman, P., *et al.* (1999). Ig V gene mutation status and CD38 expression as novel prognostic indicators in chronic lymphocytic leukemia. *Blood* **94,** 1840–1847.

Davis, I. J., Hsi, B. L., Arroyo, J. D., Vargas, S. O., Yeh, Y. A., Motyckova, G., Valencia, P., Perez-Atayde, A. R., Argani, P., Ladanyi, M., Fletcher, J. A., and Fisher, D. E. (2003). Cloning of an Alpha-TFEB fusion in renal tumors harboring the t(6;11)(p21;q13) chromosome translocation. *Proc. Natl. Acad. Sci. USA* **100,** 6051–6056.

de Sanjose, S., Benavente, Y., Vajdic, C. M., Engels, E. A., Morton, L. M., Bracci, P. M., Spinelli, J. J., Zheng, T., Zhang, Y., Franceschi, S., Talamini, R., Holly, E. A., *et al.* (2008). Hepatitis C and non-Hodgkin lymphoma among 4784 cases and 6269 controls from the International Lymphoma Epidemiology Consortium. *Clin. Gastroenterol. Hepatol.* **6,** 451–458.

De Silva, F. S., and Moss, B. (2003). Vaccinia virus uracil DNA glycosylase has an essential role in DNA synthesis that is independent of its glycosylase activity: catalytic site mutations reduce virulence but not virus replication in cultured cells. *J. Virol.* **77,** 159–166.

De Silva, F. S., and Moss, B. (2008). Effects of vaccinia virus uracil DNA glycosylase catalytic site and deoxyuridine triphosphatase deletion mutations individually and together on replication in active and quiescent cells and pathogenesis in mice. *Virol. J.* **5,** 145.

Di Noia, J., and Neuberger, M. S. (2002). Altering the pathway of immunoglobulin hypermutation by inhibiting uracil-DNA glycosylase. *Nature* **419,** 43–48.

Di Noia, J. M., Rada, C., and Neuberger, M. S. (2006). SMUG1 is able to excise uracil from immunoglobulin genes: insight into mutation versus repair. *Embo J.* **25,** 585–595.

Dijkman, R., Tensen, C. P., Buettner, M., Niedobitek, G., Willemze, R., and Vermeer, M. H. (2006). Primary cutaneous follicle center lymphoma and primary cutaneous large B-cell lymphoma, leg type, are both targeted by aberrant somatic hypermutation but demonstrate differential expression of AID. *Blood* **107**, 4926–4929.

Doi, T., Kinoshita, K., Ikegawa, M., Muramatsu, M., and Honjo, T. (2003). De novo protein synthesis is required for the activation-induced cytidine deaminase function in class-switch recombination. *Proc. Natl. Acad. Sci. USA* **100**, 2634–2638.

Doi, T., Kato, L., Ito, S., Shinkura, R., Wei, M., Nagaoka, H., Wang, J., and Honjo, T. (2009). The C-terminal region of activation-induced cytidine deaminase is responsible for a recombination function other than DNA cleavage in class switch recombination. *Proc. Natl. Acad. Sci. USA* **106**, 2758–2763.

Dorsett, Y., McBride, K. M., Jankovic, M., Gazumyan, A., Thai, T. H., Robbiani, D. F., Di Virgilio, M., Reina San-Martin, B., Heidkamp, G., Schwickert, T. A., Eisenreich, T., Rajewsky, K., *et al.* (2008). MicroRNA-155 suppresses activation-induced cytidine deaminase-mediated Myc-Igh translocation. *Immunity* **28**, 630–638.

Dunnick, W., Rabbitts, T. H., and Milstein, C. (1980). An immunoglobulin deletion mutant with implications for the heavy-chain switch and RNA splicing. *Nature* **286**, 669–675.

Dunnick, W., Hertz, G. Z., Scappino, L., and Gritzmacher, C. (1993). DNA sequences at immunoglobulin switch region recombination sites. *Nucleic Acids Res.* **21**, 365–372.

Durandy, A., Peron, S., Taubenheim, N., and Fischer, A. (2006). Activation-induced cytidine deaminase: structure-function relationship as based on the study of mutants. *Hum. Mutat.* **27**, 1185–1191.

Ellyard, J. I., Benk, A. S., Taylor, B., Rada, C., and Neuberger, M. S. (2011). The dependence of Ig class-switching on the nuclear export sequence of AID likely reflects interaction with factors additional to Crm1 exportin. *Eur. J. Immunol.* **41**, 485–490.

Endo, Y., Marusawa, H., Kinoshita, K., Morisawa, T., Sakurai, T., Okazaki, I. M., Watashi, K., Shimotohno, K., Honjo, T., and Chiba, T. (2007). Expression of activation-induced cytidine deaminase in human hepatocytes via NF-kappaB signaling. *Oncogene* **26**, 5587–5595.

Endo, Y., Marusawa, H., Kou, T., Nakase, H., Fujii, S., Fujimori, T., Kinoshita, K., Honjo, T., and Chiba, T. (2008). Activation-induced cytidine deaminase links between inflammation and the development of colitis-associated colorectal cancers. *Gastroenterology* **135**, 889–898, 898, e881–883.

Epeldegui, M., Hung, Y. P., McQuay, A., Ambinder, R. F., and Martinez-Maza, O. (2007). Infection of human B cells with Epstein-Barr virus results in the expression of somatic hypermutation-inducing molecules and in the accrual of oncogene mutations. *Mol. Immunol.* **44**, 934–942.

Fagarasan, S., Muramatsu, M., Suzuki, K., Nagaoka, H., Hiai, H., and Honjo, T. (2002). Critical roles of activation-induced cytidine deaminase in the homeostasis of gut flora. *Science* **298**, 1424–1427.

Faili, A., Aoufouchi, S., Gueranger, Q., Zober, C., Leon, A., Bertocci, B., Weill, J. C., and Reynaud, C. A. (2002). AID-dependent somatic hypermutation occurs as a DNA single-strand event in the BL2 cell line. *Nat. Immunol.* **3**, 815–821.

Feldhahn, N., Henke, N., Melchior, K., Duy, C., Soh, B. N., Klein, F., von Levetzow, G., Giebel, B., Li, A., Hofmann, W. K., Jumaa, H., and Muschen, M. (2007). Activation-induced cytidine deaminase acts as a mutator in BCR-ABL1-transformed acute lymphoblastic leukemia cells. *J. Exp. Med.* **204**, 1157–1166.

Forconi, F., Sahota, S. S., Raspadori, D., Ippoliti, M., Babbage, G., Lauria, F., and Stevenson, F. K. (2004). Hairy cell leukemia: at the crossroad of somatic mutation and isotype switch. *Blood* **104**, 3312–3317.

Fugmann, S. D., Villey, I. J., Ptaszek, L. M., and Schatz, D. G. (2000). Identification of two catalytic residues in RAG1 that define a single active site within the RAG1/RAG2 protein complex. *Mol. Cell* **5**, 97–107.

Gazumyan, A., Timachova, K., Yuen, G., Siden, E., Di Virgilio, M., Woo, E. M., Chait, B. T., San-Martin, B. R., Nussenzweig, M. C., and McBride, K. M. (2010). Amino terminal phosphorylation of activation-induced cytidine deaminase suppresses c-myc-IgH translocation. *Mol. Cell. Biol.* **31**, 442–449.

Gellert, M. (2002). V(D)J recombination: RAG proteins, repair factors, and regulation. *Annu. Rev. Biochem.* **71**, 101–132.

Gordon, M. S., Kanegai, C. M., Doerr, J. R., and Wall, R. (2003). Somatic hypermutation of the B cell receptor genes B29 (Igbeta, CD79b) and mb1 (Igalpha, CD79a). *Proc. Natl. Acad. Sci. USA* **100**, 4126–4131.

Greeve, J., Philipsen, A., Krause, K., Klapper, W., Heidorn, K., Castle, B. E., Janda, J., Marcu, K. B., and Parwaresch, R. (2003). Expression of activation-induced cytidine deaminase in human B-cell non-Hodgkin lymphomas. *Blood* **101**, 3574–3580.

Greiner, A., Tobollik, S., Buettner, M., Jungnickel, B., Herrmann, K., Kremmer, E., and Niedobitek, G. (2005). Differential expression of activation-induced cytidine deaminase (AID) in nodular lymphocyte-predominant and classical Hodgkin lymphoma. *J. Pathol.* **205**, 541–547.

Gruber, T. A., Chang, M. S., Sposto, R., and Muschen, M. (2010). Activation-induced cytidine deaminase accelerates clonal evolution in BCR-ABL1-driven B-cell lineage acute lymphoblastic leukemia. *Cancer Res.* **70**, 7411–7420.

Guenzel, C. A., Herate, C., Le Rouzic, E., Maidou-Peindara, P., Sadler, H. A., Rouyez, M. C., Mansky, L. M., and Benichou, S. (2011). Recruitment of the nuclear form of uracil DNA glycosylase into virus particles participates in the full infectivity of HIV-1. *J. virol.*

Guikema, J. E., Linehan, E. K., Tsuchimoto, D., Nakabeppu, Y., Strauss, P. R., Stavnezer, J., and Schrader, C. E. (2007). APE1- and APE2-dependent DNA breaks in immunoglobulin class switch recombination. *J. Exp. Med.* **204**, 3017–3026.

Hamblin, T. (2002). Chronic lymphocytic leukaemia: one disease or two? *Ann. Hematol.* **81**, 299–303.

Han, L., Masani, S., and Yu, K. (2010). Cutting edge: CTNNBL1 is dispensable for Ig class switch recombination. *J. Immunol.* **185**, 1379–1381.

Hardianti, M. S., Tatsumi, E., Syampurnawati, M., Furuta, K., Saigo, K., Kawano, S., Kumagai, S., Nakamura, F., and Matsuo, Y. (2004a). Expression of activation-induced cytidine deaminase (AID) in Burkitt lymphoma cells: rare AID-negative cell lines with the unmutated rearranged VH gene. *Leuk. Lymphoma* **45**, 155–160.

Hardianti, M. S., Tatsumi, E., Syampurnawati, M., Furuta, K., Saigo, K., Nakamachi, Y., Kumagai, S., Ohno, H., Tanabe, S., Uchida, M., and Yasuda, N. (2004b). Activation-induced cytidine deaminase expression in follicular lymphoma: association between AID expression and ongoing mutation in FL. *Leukemia* **18**, 826–831.

Harris, R. S., Petersen-Mahrt, S. K., and Neuberger, M. S. (2002). RNA editing enzyme APOBEC1 and some of its homologs can act as DNA mutators. *Mol. Cell* **10**, 1247–1253.

Hartsuiker, E., Mizuno, K., Molnar, M., Kohli, J., Ohta, K., and Carr, A. (2009). Ctp1CtIP and Rad32Mre11 nuclease activity are required for Rec12Spo11 removal, but Rec12Spo11 removal is dispensable for other MRN-dependent meiotic functions. *Mol. Cell. Biol.* **29**, 1671–1681.

Heintel, D., Kroemer, E., Kienle, D., Schwarzinger, I., Gleiss, A., Schwarzmeier, J., Marculescu, R., Le, T., Mannhalter, C., Gaiger, A., Stilgenbauer, S., Dohner, H., et al. (2004). High expression of activation-induced cytidine deaminase (AID) mRNA is associated with unmutated IGVH gene status and unfavourable cytogenetic aberrations in patients with chronic lymphocytic leukaemia. *Leukemia* **18**, 756–762.

Hobeika, E., Thiemann, S., Storch, B., Jumaa, H., Nielsen, P. J., Pelanda, R., and Reth, M. (2006). Testing gene function early in the B cell lineage in mb1-cre mice. *Proc. Natl. Acad. Sci. USA* **103**, 13789–13794.

Hoefnagel, J. J., Dijkman, R., Basso, K., Jansen, P. M., Hallermann, C., Willemze, R., Tensen, C. P., and Vermeer, M. H. (2005). Distinct types of primary cutaneous large B-cell lymphoma identified by gene expression profiling. *Blood* **105**, 3671–3678.

Honjo, T., and Kataoka, T. (1978). Organization of immunoglobulin heavy chain genes and allelic deletion model. *Proc. Natl. Acad. Sci. USA* **75**, 2140–2144.

Honjo, T., Kinoshita, K., and Muramatsu, M. (2002). Molecular mechanism of class switch recombination: linkage with somatic hypermutation. *Annu. Rev. Immunol.* **20**, 165–196.

Honjo, T., Nagaoka, H., Shinkura, R., and Muramatsu, M. (2005). AID to overcome the limitations of genomic information. *Nat. Immunol.* **6**, 655–661.

Hubert, L., Jr., Lin, Y., Dion, V., and Wilson, J. H. (2011). Topoisomerase 1 and single-strand break repair modulate transcription-induced CAG repeat contraction in human cells. *Mol. Cell. Biol.* **31**, 3105–3112.

Iacobucci, I., Lonetti, A., Messa, F., Ferrari, A., Cilloni, D., Soverini, S., Paoloni, F., Arruga, F., Ottaviani, E., Chiaretti, S., Messina, M., Vignetti, M., et al. (2010). Different isoforms of the B-cell mutator activation-induced cytidine deaminase are aberrantly expressed in BCR-ABL1-positive acute lymphoblastic leukemia patients. *Leukemia* **24**, 66–73.

Ide, Y., Tsuchimoto, D., Tominaga, Y., Nakashima, M., Watanabe, T., Sakumi, K., Ohno, M., and Nakabeppu, Y. (2004). Growth retardation and dyslymphopoiesis accompanied by G2/M arrest in APEX2-null mice. *Blood* **104**, 4097–4103.

Imai, K., Slupphaug, G., Lee, W. I., Revy, P., Nonoyama, S., Catalan, N., Yel, L., Forveille, M., Kavli, B., Krokan, H. E., Ochs, H. D., Fischer, A., and Durandy, A. (2003). Human uracil-DNA glycosylase deficiency associated with profoundly impaired immunoglobulin class-switch recombination. *Nat. Immunol.* **4**, 1023–1028.

Imai, K., Zhu, Y., Revy, P., Morio, T., Mizutani, S., Fischer, A., Nonoyama, S., and Durandy, A. (2005). Analysis of class switch recombination and somatic hypermutation in patients affected with autosomal dominant hyper-IgM syndrome type 2. *Clin. Immunol.* **115**, 277–285.

Ise, W., Kohyama, M., Schraml, B. U., Zhang, T., Schwer, B., Basu, U., Alt, F. W., Tang, J., Oltz, E. M., Murphy, T. L., and Murphy, K. M. (2011). The transcription factor BATF controls the global regulators of class-switch recombination in both B cells and T cells. *Nat. Immunol.* **12**, 536–543.

Ishikawa, C., Nakachi, S., Senba, M., Sugai, M., and Mori, N. (2011). Activation of AID by human T-cell leukemia virus Tax oncoprotein and the possible role of its constitutive expression in ATL genesis. *Carcinogenesis* **32**, 110–119.

Ito, S., Nagaoka, H., Shinkura, R., Begum, N., Muramatsu, M., Nakata, M., and Honjo, T. (2004). Activation-induced cytidine deaminase shuttles between nucleus and cytoplasm like apolipoprotein B mRNA editing catalytic polypeptide 1. *Proc. Natl. Acad. Sci. USA* **101**, 1975–1980.

Jeevan-Raj, B. P., Robert, I., Heyer, V., Page, A., Wang, J. H., Cammas, F., Alt, F. W., Losson, R., and Reina-San-Martin, B. (2011). Epigenetic tethering of AID to the donor switch region during immunoglobulin class switch recombination. *J. Exp. Med.* **208**, 1649–1660.

Jung, S., Rajewsky, K., and Radbruch, A. (1993). Shutdown of class switch recombination by deletion of a switch region control element. *Science* **259**, 984–987.

Kanzler, H., Kuppers, R., Hansmann, M. L., and Rajewsky, K. (1996). Hodgkin and Reed-Sternberg cells in Hodgkin's disease represent the outgrowth of a dominant tumor clone derived from (crippled) germinal center B cells. *J. Exp. Med.* **184**, 1495–1505.

Kapitonov, V. V., and Jurka, J. (2005). RAG1 core and V(D)J recombination signal sequences were derived from Transib transposons. *PLoS Biol.* **3**, e181.

Kato, L., Begum, N. A., Burroughs, A. M., Doi, T., Kawai, J., Daub, C. O., Kawaguchi, T., Matsuda, F., Hayashizaki, Y., Honjo, T. (2012). Nonimmunoglobulin target loci of activation-induced cytidine deaminase (AID) share unique features with immunoglobulin genes. *Proc. Natl. Acad. Sci. USA* In press.

Kavli, B., Andersen, S., Otterlei, M., Liabakk, N. B., Imai, K., Fischer, A., Durandy, A., Krokan, H. E., and Slupphaug, G. (2005). B cells from hyper-IgM patients carrying UNG mutations lack ability to remove uracil from ssDNA and have elevated genomic uracil. *J. Exp. Med.* **201**, 2011–2021.

Kim, C. J., Song, J. H., Cho, Y. G., Cao, Z., Kim, S. Y., Nam, S. W., Lee, J. Y., and Park, W. S. (2007). Activation-induced cytidine deaminase expression in gastric cancer. *Tumour Biol.* **28**, 333–339.

Klapper, W., Szczepanowski, M., Heidorn, K., Muschen, M., Liedtke, S., Sotnikova, A., Andersen, N. S., Greeve, J., and Parwaresch, R. (2006). Immunoglobulin class-switch recombination occurs in mantle cell lymphomas. *J. Pathol.* **209**, 250–257.

Klein, U., Tu, Y., Stolovitzky, G. A., Mattioli, M., Cattoretti, G., Husson, H., Freedman, A., Inghirami, G., Cro, L., Baldini, L., Neri, A., Califano, A., *et al.* (2001). Gene expression profiling of B cell chronic lymphocytic leukemia reveals a homogeneous phenotype related to memory B cells. *J. Exp. Med.* **194**, 1625–1638.

Klemm, L., Duy, C., Iacobucci, I., Kuchen, S., von Levetzow, G., Feldhahn, N., Henke, N., Li, Z., Hoffmann, T. K., Kim, Y. M., Hofmann, W. K., Jumaa, H., *et al.* (2009). The B cell mutator AID promotes B lymphoid blast crisis and drug resistance in chronic myeloid leukemia. *Cancer Cell* **16**, 232–245.

Kobayashi, M., Aida, M., Nagaoka, H., Begum, N. A., Kitawaki, Y., Nakata, M., Stanlie, A., Doi, T., Kato, L., Okazaki, I. M., Shinkura, R., Muramatsu, M., *et al.* (2009). AID-induced decrease in topoisomerase 1 induces DNA structural alteration and DNA cleavage for class switch recombination. *Proc. Natl. Acad. Sci. USA* **106**, 22375–22380.

Kobayashi, M., Sabouri, Z., Sabouri, S., Kitawaki, Y., Pommier, Y., Abe, T., Kiyonari, H., and Honjo, T. (2011). Decrease in topoisomerase I is responsible for AID-dependent somatic hypermutation. *Proc. Natl. Acad. Sci. USA* **108**, 19305–19310.

Kolar, G. R., Mehta, D., Pelayo, R., and Capra, J. D. (2007). A novel human B cell subpopulation representing the initial germinal center population to express AID. *Blood* **109**, 2545–2552.

Komori, J., Marusawa, H., Machimoto, T., Endo, Y., Kinoshita, K., Kou, T., Haga, H., Ikai, I., Uemoto, S., and Chiba, T. (2008). Activation-induced cytidine deaminase links bile duct inflammation to human cholangiocarcinoma. *Hepatology* **47**, 888–896.

Kotani, A., Okazaki, I. M., Muramatsu, M., Kinoshita, K., Begum, N. A., Nakajima, T., Saito, H., and Honjo, T. (2005). A target selection of somatic hypermutations is regulated similarly between T and B cells upon activation-induced cytidine deaminase expression. *Proc. Natl. Acad. Sci. USA* **102**, 4506–4511.

Kotani, A., Kakazu, N., Tsuruyama, T., Okazaki, I. M., Muramatsu, M., Kinoshita, K., Nagaoka, H., Yabe, D., and Honjo, T. (2007). Activation-induced cytidine deaminase (AID) promotes B cell lymphomagenesis in Emu-cmyc transgenic mice. *Proc. Natl. Acad. Sci. USA* **104**, 1616–1620.

Kou, T., Marusawa, H., Kinoshita, K., Endo, Y., Okazaki, I. M., Ueda, Y., Kodama, Y., Haga, H., Ikai, I., and Chiba, T. (2007). Expression of activation-induced cytidine deaminase in human hepatocytes during hepatocarcinogenesis. *Int. J. Cancer* **120**, 469–476.

Kovalchuk, A. L., duBois, W., Mushinski, E., McNeil, N. E., Hirt, C., Qi, C. F., Li, Z., Janz, S., Honjo, T., Muramatsu, M., Ried, T., Behrens, T., *et al.* (2007). AID-deficient Bcl-xL transgenic mice develop delayed atypical plasma cell tumors with unusual Ig/Myc chromosomal rearrangements. *J. Exp. Med.* **204**, 2989–3001.

Kracker, S., Imai, K., Gardes, P., Ochs, H. D., Fischer, A., and Durandy, A. H. (2010). Impaired induction of DNA lesions during immunoglobulin class-switch recombination in humans influences end-joining repair. *Proc. Natl. Acad. Sci. USA* **107**, 22225–22230.

Kuppers, R., and Dalla-Favera, R. (2001). Mechanisms of chromosomal translocations in B cell lymphomas. *Oncogene* **20**, 5580–5594.

Kuppers, R., Schwering, I., Brauninger, A., Rajewsky, K., and Hansmann, M. L. (2002). Biology of Hodgkin's lymphoma. *Ann. Oncol.* **13**(Suppl. 1), 11–18.

Kwon, K., Hutter, C., Sun, Q., Bilic, I., Cobaleda, C., Malin, S., and Busslinger, M. (2008). Instructive role of the transcription factor E2A in early B lymphopoiesis and germinal center B cell development. *Immunity* **28**, 751–762.

Lee-Theilen, M., Matthews, A. J., Kelly, D., Zheng, S., and Chaudhuri, J. (2011). CtIP promotes microhomology-mediated alternative end joining during class-switch recombination. *Nat. Struct. Mol. Biol.* **18**, 75–79.

Levin, J. Z., Berger, M. F., Adiconis, X., Rogov, P., Melnikov, A., Fennell, T., Nusbaum, C., Garraway, L. A., and Gnirke, A. (2009). Targeted next-generation sequencing of a cancer transcriptome enhances detection of sequence variants and novel fusion transcripts. *Genome Biol.* **10**, R115.

Lippert, M. J., Kim, N., Cho, J. E., Larson, R. P., Schoenly, N. E., O'Shea, S. H., and Jinks-Robertson, S. (2011). Role for topoisomerase 1 in transcription-associated mutagenesis in yeast. *Proc. Natl. Acad. Sci. USA* **108**, 698–703.

Liu, Y., Subrahmanyam, R., Chakraborty, T., Sen, R., and Desiderio, S. (2007). A plant homeodomain in RAG-2 that binds hypermethylated lysine 4 of histone H3 is necessary for efficient antigen-receptor-gene rearrangement. *Immunity* **27**, 561–571.

Liu, M., Duke, J. L., Richter, D. J., Vinuesa, C. G., Goodnow, C. C., Kleinstein, S. H., and Schatz, D. G. (2008). Two levels of protection for the B cell genome during somatic hypermutation. *Nature* **451**, 841–845.

Lossos, I. S., Levy, R., and Alizadeh, A. A. (2004). AID is expressed in germinal center B-cell-like and activated B-cell-like diffuse large-cell lymphomas and is not correlated with intraclonal heterogeneity. *Leukemia* **18**, 1775–1779.

MacDuff, D. A., Neuberger, M. S., and Harris, R. S. (2006). MDM2 can interact with the C-terminus of AID but it is inessential for antibody diversification in DT40 B cells. *Mol. Immunol.* **43**, 1099–1108.

Machida, K., Cheng, K. T., Sung, V. M., Shimodaira, S., Lindsay, K. L., Levine, A. M., Lai, M. Y., and Lai, M. M. (2004). Hepatitis C virus induces a mutator phenotype: enhanced mutations of immunoglobulin and protooncogenes. *Proc. Natl. Acad. Sci. USA* **101**, 4262–4267.

Mai, T., Zan, H., Zhang, J., Hawkins, J. S., Xu, Z., and Casali, P. (2010). Estrogen receptors bind to and activate the HOXC4/HoxC4 promoter to potentiate HoxC4-mediated activation-induced cytosine deaminase induction, immunoglobulin class switch DNA recombination, and somatic hypermutation. *J. Biol. Chem.* **285**, 37797–37810.

Maki, R., Traunecker, A., Sakano, H., Roeder, W., and Tonegawa, S. (1980). Exon shuffling generates an immunoglobulin heavy chain gene. *Proc. Natl. Acad. Sci. USA* **77**, 2138–2142.

Marafioti, T., Hummel, M., Foss, H. D., Laumen, H., Korbjuhn, P., Anagnostopoulos, I., Lammert, H., Demel, G., Theil, J., Wirth, T., and Stein, H. (2000). Hodgkin and Reed-Sternberg cells represent an expansion of a single clone originating from a germinal center B-cell with functional immunoglobulin gene rearrangements but defective immunoglobulin transcription. *Blood* **95**, 1443–1450.

Marantidou, F., Dagklis, A., Stalika, E., Korkolopoulou, P., Saetta, A., Anagnostopoulos, A., Laoutaris, N., Stamatopoulos, K., Belessi, C., Scouras, Z., and Patsouris, E. (2010). Activation-induced cytidine deaminase splicing patterns in chronic lymphocytic leukemia. *Blood Cells Mol. Dis.* **44**, 262–267.

Matsumoto, Y., Marusawa, H., Kinoshita, K., Endo, Y., Kou, T., Morisawa, T., Azuma, T., Okazaki, I. M., Honjo, T., and Chiba, T. (2007). Helicobacter pylori infection triggers aberrant expression of activation-induced cytidine deaminase in gastric epithelium. *Nat. Med.* **13**, 470–476.

Matthews, A. G., Kuo, A. J., Ramon-Maiques, S., Han, S., Champagne, K. S., Ivanov, D., Gallardo, M., Carney, D., Cheung, P., Ciccone, D. N., Walter, K. L., Utz, P. J., *et al.* (2007). RAG2 PHD finger couples histone H3 lysine 4 trimethylation with V(D)J recombination. *Nature* **450,** 1106–1110.

Maul, R. W., Saribasak, H., Martomo, S. A., McClure, R. L., Yang, W., Vaisman, A., Gramlich, H. S., Schatz, D. G., Woodgate, R., Wilson, D. M., 3rd, and Gearhart, P. J. (2011). Uracil residues dependent on the deaminase AID in immunoglobulin gene variable and switch regions. *Nat. Immunol.* **12,** 70–76.

McBride, K. M., Gazumyan, A., Woo, E. M., Barreto, V. M., Robbiani, D. F., Chait, B. T., and Nussenzweig, M. C. (2006). Regulation of hypermutation by activation-induced cytidine deaminase phosphorylation. *Proc. Natl. Acad. Sci. USA* **103,** 8798–8803.

McBride, K. M., Gazumyan, A., Woo, E. M., Schwickert, T. A., Chait, B. T., and Nussenzweig, M. C. (2008). Regulation of class switch recombination and somatic mutation by AID phosphorylation. *J. Exp. Med.* **205,** 2585–2594.

McCarthy, H., Wierda, W. G., Barron, L. L., Cromwell, C. C., Wang, J., Coombes, K. R., Rangel, R., Elenitoba-Johnson, K. S., Keating, M. J., and Abruzzo, L. V. (2003). High expression of activation-induced cytidine deaminase (AID) and splice variants is a distinctive feature of poor-prognosis chronic lymphocytic leukemia. *Blood* **101,** 4903–4908.

Miao, Z. H., Player, A., Shankavaram, U., Wang, Y. H., Zimonjic, D. B., Lorenzi, P. L., Liao, Z. Y., Liu, H., Shimura, T., Zhang, H. L., Meng, L. H., Zhang, Y. W., *et al.* (2007). Nonclassic functions of human topoisomerase I: genome-wide and pharmacologic analyses. *Cancer Res.* **67,** 8752–8761.

Migliazza, A., Martinotti, S., Chen, W., Fusco, C., Ye, B. H., Knowles, D. M., Offit, K., Chaganti, R. S., and Dalla-Favera, R. (1995). Frequent somatic hypermutation of the 5′ noncoding region of the BCL6 gene in B-cell lymphoma. *Proc. Natl. Acad. Sci. USA* **92,** 12520–12524.

Min, I. M., Schrader, C. E., Vardo, J., Luby, T. M., D'Avirro, N., Stavnezer, J., and Selsing, E. (2003). The Smu tandem repeat region is critical for Ig isotype switching in the absence of Msh2. *Immunity* **19,** 515–524.

Minegishi, Y., Lavoie, A., Cunningham-Rundles, C., Bedard, P. M., Hebert, J., Cote, L., Dan, K., Sedlak, D., Buckley, R. H., Fischer, A., Durandy, A., and Conley, M. E. (2000). Mutations in activation-induced cytidine deaminase in patients with hyper IgM syndrome. *Clin. Immunol.* **97,** 203–210.

Morey, N. J., Greene, C. N., and Jinks-Robertson, S. (2000). Genetic analysis of transcription-associated mutation in Saccharomyces cerevisiae. *Genetics* **154,** 109–120.

Morisawa, T., Marusawa, H., Ueda, Y., Iwai, A., Okazaki, I. M., Honjo, T., and Chiba, T. (2008). Organ-specific profiles of genetic changes in cancers caused by activation-induced cytidine deaminase expression. *Int. J. Cancer* **123,** 2735–2740.

Mottok, A., Hansmann, M. L., and Brauninger, A. (2005). Activation induced cytidine deaminase expression in lymphocyte predominant Hodgkin lymphoma. *J. Clin. Pathol.* **58,** 1002–1004.

Muramatsu, M., Sankaranand, V. S., Anant, S., Sugai, M., Kinoshita, K., Davidson, N. O., and Honjo, T. (1999). Specific expression of activation-induced cytidine deaminase (AID), a novel member of the RNA-editing deaminase family in germinal center B cells. *J. Biol. Chem.* **274,** 18470–18476.

Muramatsu, M., Kinoshita, K., Fagarasan, S., Yamada, S., Shinkai, Y., and Honjo, T. (2000). Class switch recombination and hypermutation require activation-induced cytidine deaminase (AID), a potential RNA editing enzyme. *Cell* **102,** 553–563.

Muramatsu, M., Nagaoka, H., Shinkura, R., Begum, N. A., and Honjo, T. (2007). Discovery of activation-induced cytidine deaminase, the engraver of antibody memory. *Adv. Immunol.* **94,** 1–36.

Muto, T., Muramatsu, M., Taniwaki, M., Kinoshita, K., and Honjo, T. (2000). Isolation, tissue distribution, and chromosomal localization of the human activation-induced cytidine deaminase (AID) gene. *Genomics* **68**, 85–88.

Myers, S., Bowden, R., Tumian, A., Bontrop, R. E., Freeman, C., MacFie, T. S., McVean, G., and Donnelly, P. (2010). Drive against hotspot motifs in primates implicates the PRDM9 gene in meiotic recombination. *Science* **327**, 876–879.

Nakamura, M., Kondo, S., Sugai, M., Nazarea, M., Imamura, S., and Honjo, T. (1996). High frequency class switching of an IgM+ B lymphoma clone CH12F3 to IgA+ cells. *Int. Immunol.* **8**, 193–201.

Nakamura, M., Sugita, K., Sawada, Y., Yoshiki, R., Hino, R., and Tokura, Y. (2011). High levels of activation-induced cytidine deaminase expression in adult T-cell leukaemia/lymphoma. *Br. J. Dermatol.* **165**, 437–439.

Nambu, Y., Sugai, M., Gonda, H., Lee, C. G., Katakai, T., Agata, Y., Yokota, Y., and Shimizu, A. (2003). Transcription-coupled events associating with immunoglobulin switch region chromatin. *Science* **302**, 2137–2140.

Neuberger, M. S., Harris, R. S., Di Noia, J., and Petersen-Mahrt, S. K. (2003). Immunity through DNA deamination. *Trends Biochem. Sci.* **28**, 305–312.

Nikaido, T., Nakai, S., and Honjo, T. (1981). Switch region of immunoglobulin Cmu gene is composed of simple tandem repetitive sequences. *Nature* **292**, 845–848.

Nonaka, T., Doi, T., Toyoshima, T., Muramatsu, M., Honjo, T., and Kinoshita, K. (2009). Carboxy-terminal domain of AID required for its mRNA complex formation *in vivo*. *Proc. Natl. Acad. Sci. USA* **106**, 2747–2751.

Nowak, U., Matthews, A. J., Zheng, S., and Chaudhuri, J. (2011). The splicing regulator PTBP2 interacts with the cytidine deaminase AID and promotes binding of AID to switch-region DNA. *Nat. Immunol.* **12**, 160–166.

Okazaki, I. M., Kinoshita, K., Muramatsu, M., Yoshikawa, K., and Honjo, T. (2002). The AID enzyme induces class switch recombination in fibroblasts. *Nature* **416**, 340–345.

Okazaki, I. M., Hiai, H., Kakazu, N., Yamada, S., Muramatsu, M., Kinoshita, K., and Honjo, T. (2003). Constitutive expression of AID leads to tumorigenesis. *J. Exp. Med.* **197**, 1173–1181.

Okazaki, I. M., Okawa, K., Kobayashi, M., Yoshikawa, K., Kawamoto, S., Nagaoka, H., Shinkura, R., Kitawaki, Y., Taniguchi, H., Natsume, T., Iemura, S., and Honjo, T. (2011). Histone chaperone Spt6 is required for class switch recombination but not somatic hypermutation. *Proc. Natl. Acad. Sci. USA* **108**, 7920–7925.

Oppezzo, P., Vuillier, F., Vasconcelos, Y., Dumas, G., Magnac, C., Payelle-Brogard, B., Pritsch, O., and Dighiero, G. (2003). Chronic lymphocytic leukemia B cells expressing AID display dissociation between class switch recombination and somatic hypermutation. *Blood* **101**, 4029–4032.

Palacios, F., Moreno, P., Morande, P., Abreu, C., Correa, A., Porro, V., Landoni, A. I., Gabus, R., Giordano, M., Dighiero, G., Pritsch, O., and Oppezzo, P. (2010). High expression of AID and active class switch recombination might account for a more aggressive disease in unmutated CLL patients: link with an activated microenvironment in CLL disease. *Blood* **115**, 4488–4496.

Park, S. R., Zan, H., Pal, Z., Zhang, J., Al-Qahtani, A., Pone, E. J., Xu, Z., Mai, T., and Casali, P. (2009). HoxC4 binds to the promoter of the cytidine deaminase AID gene to induce AID expression, class-switch DNA recombination and somatic hypermutation. *Nat. Immunol.* **10**, 540–550.

Parvanov, E. D., Petkov, P. M., and Paigen, K. (2010). Prdm9 controls activation of mammalian recombination hotspots. *Science* **327**, 835.

Pasqualucci, L., Neumeister, P., Goossens, T., Nanjangud, G., Chaganti, R. S., Kuppers, R., and Dalla-Favera, R. (2001). Hypermutation of multiple proto-oncogenes in B-cell diffuse large-cell lymphomas. *Nature* **412**, 341–346.

Pasqualucci, L., Guglielmino, R., Houldsworth, J., Mohr, J., Aoufouchi, S., Polakiewicz, R., Chaganti, R. S., and Dalla-Favera, R. (2004). Expression of the AID protein in normal and neoplastic B cells. *Blood* **104**, 3318–3325.

Pasqualucci, L., Kitaura, Y., Gu, H., and Dalla-Favera, R. (2006). PKA-mediated phosphorylation regulates the function of activation-induced deaminase (AID) in B cells. *Proc. Natl. Acad. Sci. USA* **103**, 395–400.

Pauklin, S., Sernandez, I. V., Bachmann, G., Ramiro, A. R., and Petersen-Mahrt, S. K. (2009). Estrogen directly activates AID transcription and function. *J. Exp. Med.* **206**, 99–111.

Pavri, R., Gazumyan, A., Jankovic, M., Di Virgilio, M., Klein, I., Ansarah-Sobrinho, C., Resch, W., Yamane, A., Reina San-Martin, B., Barreto, V., Nieland, T. J., Root, D. E., *et al.* (2010). Activation-induced cytidine deaminase targets DNA at sites of RNA polymerase II stalling by interaction with Spt5. *Cell* **143**, 122–133.

Peters, A., and Storb, U. (1996). Somatic hypermutation of immunoglobulin genes is linked to transcription initiation. *Immunity* **4**, 57–65.

Petersen-Mahrt, S. K., and Neuberger, M. S. (2003). *In vitro* deamination of cytosine to uracil in single-stranded DNA by apolipoprotein B editing complex catalytic subunit 1 (APOBEC1). *J. Biol. Chem.* **278**, 19583–19586.

Petersen-Mahrt, S. K., Harris, R. S., and Neuberger, M. S. (2002). AID mutates E. coli suggesting a DNA deamination mechanism for antibody diversification. *Nature* **418**, 99–103.

Poltoratsky, V. P., Wilson, S. H., Kunkel, T. A., and Pavlov, Y. I. (2004). Recombinogenic phenotype of human activation-induced cytosine deaminase. *J. Immunol.* **172**, 4308–4313.

Qin, H., Suzuki, K., Nakata, M., Chikuma, S., Izumi, N., Thi Huong, L., Maruya, M., Fagarasan, S., Busslinger, M., Honjo, T., and Nagaoka, H. (2011). Activation-Induced Cytidine Deaminase Expression in CD4 T Cells is Associated with a Unique IL-10-Producing Subset that Increases with Age. *PLoS One.* **6**, e29141.

Quartier, P., Bustamante, J., Sanal, O., Plebani, A., Debre, M., Deville, A., Litzman, J., Levy, J., Fermand, J. P., Lane, P., Horneff, G., Aksu, G., *et al.* (2004). Clinical, immunologic and genetic analysis of 29 patients with autosomal recessive hyper-IgM syndrome due to activation-induced cytidine deaminase deficiency. *Clin. Immunol.* **110**, 22–29.

Rabbitts, T. H., Forster, A., Dunnick, W., and Bentley, D. L. (1980). The role of gene deletion in the immunoglobulin heavy chain switch. *Nature* **283**, 351–356.

Rada, C., Ehrenstein, M. R., Neuberger, M. S., and Milstein, C. (1998). Hot spot focusing of somatic hypermutation in MSH2-deficient mice suggests two stages of mutational targeting. *Immunity* **9**, 135–141.

Rada, C., Williams, G. T., Nilsen, H., Barnes, D. E., Lindahl, T., and Neuberger, M. S. (2002). Immunoglobulin isotype switching is inhibited and somatic hypermutation perturbed in UNG-deficient mice. *Curr. Biol.* **12**, 1748–1755.

Rada, C., Di Noia, J. M., and Neuberger, M. S. (2004). Mismatch recognition and uracil excision provide complementary paths to both Ig switching and the A/T-focused phase of somatic mutation. *Mol. Cell* **16**, 163–171.

Rajaram, V., Knezevich, S., Bove, K. E., Perry, A., and Pfeifer, J. D. (2007). DNA sequence of the translocation breakpoints in undifferentiated embryonal sarcoma arising in mesenchymal hamartoma of the liver harboring the t(11;19)(q11;q13.4) translocation. *Genes Chromosomes Cancer* **46**, 508–513.

Rajewsky, K. (1996). Clonal selection and learning in the antibody system. *Nature* **381**, 751–758.

Ramiro, A. R., Jankovic, M., Eisenreich, T., Difilippantonio, S., Chen-Kiang, S., Muramatsu, M., Honjo, T., Nussenzweig, A., and Nussenzweig, M. C. (2004). AID is required for c-myc/IgH chromosome translocations *in vivo. Cell* **118**, 431–438.

Ranjit, S., Khair, L., Linehan, E. K., Ucher, A. J., Chakrabarti, M., Schrader, C. E., and Stavnezer, J. (2011). AID binds cooperatively with UNG and Msh2-Msh6 to Ig switch regions dependent upon the AID C terminus. *J. Immunol.* **187**, 2464–2475.

Reiniger, L., Bodor, C., Bognar, A., Balogh, Z., Csomor, J., Szepesi, A., Kopper, L., and Matolcsy, A. (2006). Richter's and prolymphocytic transformation of chronic lymphocytic leukemia are associated with high mRNA expression of activation-induced cytidine deaminase and aberrant somatic hypermutation. *Leukemia* **20**, 1089–1095.

Revy, P., Muto, T., Levy, Y., Geissmann, F., Plebani, A., Sanal, O., Catalan, N., Forveille, M., Dufourcq-Labelouse, R., Gennery, A., Tezcan, I., Ersoy, F., *et al.* (2000). Activation-induced cytidine deaminase (AID) deficiency causes the autosomal recessive form of the Hyper-IgM syndrome (HIGM2). *Cell* **102**, 565–575.

Rosenwald, A., Alizadeh, A. A., Widhopf, G., Simon, R., Davis, R. E., Yu, X., Yang, L., Pickeral, O. K., Rassenti, L. Z., Powell, J., Botstein, D., Byrd, J. C., *et al.* (2001). Relation of gene expression phenotype to immunoglobulin mutation genotype in B cell chronic lymphocytic leukemia. *J. Exp. Med.* **194**, 1639–1647.

Rosenwald, A., Wright, G., Chan, W. C., Connors, J. M., Campo, E., Fisher, R. I., Gascoyne, R. D., Muller-Hermelink, H. K., Smeland, E. B., Giltnane, J. M., Hurt, E. M., Zhao, H., *et al.* (2002). The use of molecular profiling to predict survival after chemotherapy for diffuse large-B-cell lymphoma. *N. Engl. J. Med.* **346**, 1937–1947.

Sabouri, Z., Okazaki, I. M., Shinkura, R., Begum, N., Nagaoka, H., Tsuchimoto, D., Nakabeppu, Y., and Honjo, T. (2009). Apex2 is required for efficient somatic hypermutation but not for class switch recombination of immunoglobulin genes. *Int. Immunol.* **21**, 947–955.

Saribasak, H., Saribasak, N. N., Ipek, F. M., Ellwart, J. W., Arakawa, H., and Buerstedde, J. M. (2006). Uracil DNA glycosylase disruption blocks Ig gene conversion and induces transition mutations. *J. Immunol.* **176**, 365–371.

Satou, Y., Yasunaga, J., Zhao, T., Yoshida, M., Miyazato, P., Takai, K., Shimizu, K., Ohshima, K., Green, P. L., Ohkura, N., Yamaguchi, T., Ono, M., *et al.* (2011). HTLV-1 bZIP factor induces T-cell lymphoma and systemic inflammation *in vivo*. *PLoS Pathog.* **7**, e1001274.

Schatz, D. G. (2004). Antigen receptor genes and the evolution of a recombinase. *Semin. Immunol.* **16**, 245–256.

Schrader, C. E., Vardo, J., and Stavnezer, J. (2002). Role for mismatch repair proteins Msh2, Mlh1, and Pms2 in immunoglobulin class switching shown by sequence analysis of recombination junctions. *J. Exp. Med.* **195**, 367–373.

Schroeder, H. W., Jr., and Dighiero, G. (1994). The pathogenesis of chronic lymphocytic leukemia: analysis of the antibody repertoire. *Immunol. Today* **15**, 288–294.

Shen, H. M., Peters, A., Baron, B., Zhu, X., and Storb, U. (1998). Mutation of BCL-6 gene in normal B cells by the process of somatic hypermutation of Ig genes. *Science* **280**, 1750–1752.

Shen, H. M., Ratnam, S., and Storb, U. (2005). Targeting of the activation-induced cytosine deaminase is strongly influenced by the sequence and structure of the targeted DNA. *Mol. Cell. Biol.* **25**, 10815–10821.

Shen, H. M., Poirier, M. G., Allen, M. J., North, J., Lal, R., Widom, J., and Storb, U. (2009). The activation-induced cytidine deaminase (AID) efficiently targets DNA in nucleosomes but only during transcription. *J. Exp. Med.* **206**, 1057–1071.

Shinkura, R., Ito, S., Begum, N. A., Nagaoka, H., Muramatsu, M., Kinoshita, K., Sakakibara, Y., Hijikata, H., and Honjo, T. (2004). Separate domains of AID are required for somatic hypermutation and class-switch recombination. *Nat. Immunol.* **5**, 707–712.

Shivarov, V., Shinkura, R., and Honjo, T. (2008). Dissociation of *in vitro* DNA deamination activity and physiological functions of AID mutants. *Proc. Natl. Acad. Sci. USA* **105**, 15866–15871.

Smit, L. A., Bende, R. J., Aten, J., Guikema, J. E., Aarts, W. M., and van Noesel, C. J. (2003). Expression of activation-induced cytidine deaminase is confined to B-cell non-Hodgkin's lymphomas of germinal-center phenotype. *Cancer Res.* **63**, 3894–3898.

Stanlie, A., Aida, M., Muramatsu, M., Honjo, T., and Begum, N. A. (2010). Histone3 lysine4 trimethylation regulated by the facilitates chromatin transcription complex is critical for DNA cleavage in class switch recombination. *Proc. Natl. Acad. Sci. USA* **107**, 22190–22195.

Staszewski, O., Baker, R. E., Ucher, A. J., Martier, R., Stavnezer, J., and Guikema, J. E. (2011). Activation-induced cytidine deaminase induces reproducible DNA breaks at many non-Ig Loci in activated B cells. *Mol. Cell* **41**, 232–242.

Suzuki, K., Meek, B., Doi, Y., Muramatsu, M., Chiba, T., Honjo, T., and Fagarasan, S. (2004). Aberrant expansion of segmented filamentous bacteria in IgA-deficient gut. *Proc. Natl. Acad. Sci. USA* **101**, 1981–1986.

Ta, V. T., Nagaoka, H., Catalan, N., Durandy, A., Fischer, A., Imai, K., Nonoyama, S., Tashiro, J., Ikegawa, M., Ito, S., Kinoshita, K., Muramatsu, M., et al. (2003). AID mutant analyses indicate requirement for class-switch-specific cofactors. *Nat. Immunol.* **4**, 843–848.

Takahashi, T., Burguiere-Slezak, G., Van der Kemp, P. A., and Boiteux, S. (2011). Topoisomerase 1 provokes the formation of short deletions in repeated sequences upon high transcription in Saccharomyces cerevisiae. *Proc. Natl. Acad. Sci. USA* **108**, 692–697.

Takizawa, M., Tolarova, H., Li, Z., Dubois, W., Lim, S., Callen, E., Franco, S., Mosaico, M., Feigenbaum, L., Alt, F. W., Nussenzweig, A., Potter, M., et al. (2008). AID expression levels determine the extent of cMyc oncogenic translocations and the incidence of B cell tumor development. *J. Exp. Med.* **205**, 1949–1957.

Teng, G., Hakimpour, P., Landgraf, P., Rice, A., Tuschl, T., Casellas, R., and Papavasiliou, F. N. (2008). MicroRNA-155 is a negative regulator of activation-induced cytidine deaminase. *Immunity* **28**, 621–629.

Tran, T. H., Nakata, M., Suzuki, K., Begum, N. A., Shinkura, R., Fagarasan, S., Honjo, T., and Nagaoka, H. (2010). B cell-specific and stimulation-responsive enhancers derepress Aicda by overcoming the effects of silencers. *Nat. Immunol.* **11**, 148–154.

Tuduri, S., Crabbe, L., Conti, C., Tourriere, H., Holtgreve-Grez, H., Jauch, A., Pantesco, V., De Vos, J., Thomas, A., Theillet, C., Pommier, Y., Tazi, J., et al. (2009). Topoisomerase I suppresses genomic instability by preventing interference between replication and transcription. *Nat. Cell Biol.* **11**, 1315–1324.

Turner, N. C., Dusheiko, G., and Jones, A. (2003). Hepatitis C and B-cell lymphoma. *Ann. Oncol.* **14**, 1341–1345.

Wahls, W. P., and Davidson, M. K. (2010). Discrete DNA sites regulate global distribution of meiotic recombination. *Trends Genet.* **26**, 202–208.

Wei, M., Shinkura, R., Doi, Y., Maruya, M., Fagarasan, S., and Honjo, T. (2011). Mice carrying a knock-in mutation of Aicda resulting in a defect in somatic hypermutation have impaired gut homeostasis and compromised mucosal defense. *Nat. Immunol.* **12**, 264–270.

Weigert, M. G., Cesari, I. M., Yonkovich, S. J., and Cohn, M. (1970). Variability in the lambda light chain sequences of mouse antibody. *Nature* **228**, 1045–1047.

Willemze, R., Jaffe, E. S., Burg, G., Cerroni, L., Berti, E., Swerdlow, S. H., Ralfkiaer, E., Chimenti, S., Diaz-Perez, J. L., Duncan, L. M., Grange, F., Harris, N. L., et al. (2005). WHO-EORTC classification for cutaneous lymphomas. *Blood* **105**, 3768–3785.

Wu, X., Geraldes, P., Platt, J. L., and Cascalho, M. (2005). The double-edged sword of activation-induced cytidine deaminase. *J. Immunol.* **174**, 934–941.

Xanthoudakis, S., Smeyne, R. J., Wallace, J. D., and Curran, T. (1996). The redox/DNA repair protein, Ref-1, is essential for early embryonic development in mice. *Proc. Natl. Acad. Sci. USA* **93**, 8919–8923.

Yadav, A., Olaru, A., Saltis, M., Setren, A., Cerny, J., and Livak, F. (2006). Identification of a ubiquitously active promoter of the murine activation-induced cytidine deaminase (AICDA) gene. *Mol. Immunol.* **43**, 529–541.

Yamane, A., Resch, W., Kuo, N., Kuchen, S., Li, Z., Sun, H. W., Robbiani, D. F., McBride, K., Nussenzweig, M. C., and Casellas, R. (2011). Deep-sequencing identification of the genomic targets of the cytidine deaminase AID and its cofactor RPA in B lymphocytes. *Nat. Immunol.* **12**, 62–69.

Yamazaki, J., Mizukami, T., Takizawa, K., Kuramitsu, M., Momose, H., Masumi, A., Ami, Y., Hasegawa, H., Hall, W. W., Tsujimoto, H., Hamaguchi, I., and Yamaguchi, K. (2009). Identification of cancer stem cells in a Tax-transgenic (Tax-Tg) mouse model of adult T-cell leukemia/lymphoma. *Blood* **114**, 2709–2720.

Yang, G., Obiakor, H., Sinha, R. K., Newman, B. A., Hood, B. L., Conrads, T. P., Veenstra, T. D., and Mage, R. G. (2005). Activation-induced deaminase cloning, localization, and protein extraction from young VH-mutant rabbit appendix. *Proc. Natl. Acad. Sci. USA* **102**, 17083–17088.

Yaoita, Y., and Honjo, T. (1980). Deletion of immunoglobulin heavy chain genes from expressed allelic chromosome. *Nature* **286**, 850–853.

Yoshikawa, K., Okazaki, I. M., Eto, T., Kinoshita, K., Muramatsu, M., Nagaoka, H., and Honjo, T. (2002). AID enzyme-induced hypermutation in an actively transcribed gene in fibroblasts. *Science* **296**, 2033–2036.

Zan, H., Komori, A., Li, Z., Cerutti, A., Schaffer, A., Flajnik, M. F., Diaz, M., and Casali, P. (2001). The translesion DNA polymerase zeta plays a major role in Ig and bcl-6 somatic hypermutation. *Immunity* **14**, 643–653.

Zeitlin, S. G., Chapados, B. R., Baker, N. M., Tai, C., Slupphaug, G., and Wang, J. Y. (2011). Uracil DNA N-glycosylase promotes assembly of human centromere protein A. *PLoS One* **6**, e17151.

Zhang, M., and Swanson, P. C. (2008). V(D)J recombinase binding and cleavage of cryptic recombination signal sequences identified from lymphoid malignancies. *J. Biol. Chem.* **283**, 6717–6727.

Zhang, J., Bottaro, A., Li, S., Stewart, V., and Alt, F. W. (1993). A selective defect in IgG2b switching as a result of targeted mutation of the I gamma 2b promoter and exon. *EMBO J.* **12**, 3529–3537.

The MicroRNA Regulatory Network in Normal- and HTLV-1-Transformed T Cells

Donna M. D'Agostino,[*,†] Paola Zanovello,[*,†] Toshiki Watanabe,[‡] and Vincenzo Ciminale[*,†]

*Department of Surgical Sciences, Oncology and Gastroenterology, University of Padova, Padova, Italy
†Istituto Oncologico Veneto-IRCCS, Padova, Italy
‡Department of Medical Genome Sciences, Laboratory of Tumor Cell Biology, Graduate School of Frontier Sciences, The University of Tokyo, Minato-ku, Tokyo, Japan

I. Introduction
  A. miRNA Biogenesis
  B. Consequences of miRNA–mRNA Interactions
II. HTLV-1 and ATLL
  A. HTLV-1 Genome Organization and Expression
  B. HTLV-1 Transmission and Persistence
  C. ATLL and Other HTLV-1-Associated Diseases
III. miRNAs in Normal CD4+ T Cells
  A. miRNA Profiles in T Cell Development
  B. miRNAs in Activated T Cells
  C. miRNAs in Tregs
IV. Cellular miRNA Expression in HTLV-1-Infected Cell Lines
V. miRNA Profiling in ATLL Samples
VI. Repression of miR-31 Expression in ATLL
  A. Identification of NF-κB-Inducing Kinase (NIK) as a Target of miR-31 in ATLL
  B. Genetic Deletion and Polycomb-Directed Epigenetic Silencing Cause Loss of miR-31 Expression
  C. The Polycomb Group Regulates NF-κB Activity by Controlling miR-31 Expression
VII. Perspectives
  Acknowledgments
  References

Recent efforts to understand the molecular networks governing normal T cell development and driving the neoplastic transformation of T cells have brought to light the involvement of microRNAs (miRNAs), a class of noncoding RNAs of approximately 22 nucleotides that regulate gene expression at the posttranscriptional level. In the present review, we compare the expression profiles of miRNAs in normal T cell development to that of transformed T cells using as a model adult T cell leukemia/lymphoma, an aggressive malignancy of mature CD4+ T cells that is caused by infection with human T cell leukemia virus type 1. © 2012 Elsevier Inc.

I. INTRODUCTION

miRNAs are a class of small noncoding RNAs that regulate mRNA expression at the posttranscriptional level by hybridizing to complementary sequences on the target transcripts. This interaction generally results in silencing of the mRNA through inhibition of its translation and/or by promoting its degradation. miRNAs have been identified in protozoa, plants, metazoan animals, and viruses. The sequences and genomic locations of the known miRNAs in various organisms are cataloged in the Sanger miRBase at http://www.mirbase.org/ (Griffiths-Jones et al., 2008). The current miRBase (version 18, released in November 2011) contains 1921 human miRNAs whose lengths range from ~16 to 27nt, with 22-nt sequences prevailing. Global miRNA profiling studies (e.g., Basso et al., 2009; Ghisi et al., 2011; Landgraf et al., 2007; Morin et al., 2008) indicate that some miRNAs are specific for a particular cell lineage or differentiation stage, while others are expressed in many cell types and thus probably play broader roles in fine-tuning gene expression.

Most miRNA–mRNA functional interactions involve base pairing of nucleotides ~2–8 of the miRNA (the "seed sequence") with a complementary sequence in the 3' untranslated region (3'UTR) of the mRNA. For many miRNA–mRNA interactions, the critical seed interaction is reinforced by additional base pairing. A search for 6-to-8nt perfect matches between phylogenetically conserved miRNAs and 3'UTRs indicated that almost 60% of all human protein-coding genes may be targeted by miRNAs, an underestimate if one considers that some miRNA–mRNA interactions occur in regions outside the 3'UTR and some involve imperfect seed matches (Friedman et al., 2009).

The fact that a miRNA–mRNA interaction involves such a short stretch of complementary nucleotides renders the miRNA regulatory network extremely complex, as a single miRNA has the potential to regulate multiple genes and a given gene generally contains several target sites for different miRNAs. miRNAs are thus likely to regulate most normal biological processes. Not surprisingly, aberrant miRNA expression or function contributes to the pathogenesis of many diseases, including cancer (Sayed and Abdellatif, 2011). The first direct evidence for the importance of miRNAs in human cancer came from a study of chronic lymphocytic leukemia (CLL), which revealed a tumor suppressor function for miR-15a and miR-16-1 (Calin et al., 2002). Subsequent studies have identified many additional miRNAs with oncogenic or tumor suppressor activities in the context of solid and hematopoietic tumors (Croce, 2009).

All viruses rely on the host gene expression machinery for their replication and may therefore be affected by the host miRNA network at some level. In turn, viruses have evolved mechanisms that exploit the miRNA network to impinge on host cell turnover and immune defenses to promote expansion and persistence of infected cells (Umbach and Cullen, 2009). It is therefore not surprising that some of the miRNAs shown thus far to be exploited by viruses are also involved in neoplastic transformation. One notable example is miR-155, a miRNA that is overexpressed in several solid tumors and hematological malignancies; miR-155 is upregulated by Epstein–Barr virus and expressed as a viral ortholog by human herpesvirus 8 (Lin and Flemington, 2011). In addition, some viruses produce RNAs or proteins that suppress the RNAi pathway and thereby may have general effects on miRNA expression (Strebel *et al.*, 2009). As described in Sections IV–VI, studies of human T cell leukemia virus type 1 (HTLV-1)-infected cell lines and adult T cell leukemia/lymphoma (ATLL) samples have indicated several miRNAs whose effects on specific target genes are likely to reveal novel mechanisms of HTLV-1/host cell interaction and ATLL pathogenesis.

A. miRNA Biogenesis

Mature single-stranded miRNAs are derived from longer RNA precursor molecules through one or more cleavage events (Miyoshi *et al.*, 2010; Fig. 1). The immediate precursors of most mature miRNAs are hairpin structures of ~55–70 nt termed pre-miRNAs (Berezikov *et al.*, 2006). Mature miRNAs are contained within the stem portion of the hairpin, and the suffixes -5p and -3p in the names of some miRNAs indicate the position of the mature miRNA sequence with respect to the loop. Most pre-miRNAs are derived from longer RNA precursors (pri-miRNAs) produced by RNA polymerase II; pri-miRNAs are of varying length, can be noncoding or coding, and can contain more than one pre-miRNA. The pre-miRNA hairpin is released from the pri-miRNA through cleavage by the ribonuclease Drosha and its associated dsRNA-binding protein DGCR8. A 3′ overhang produced by Drosha cleavage is recognized by the nuclear export factor Exportin 5, which transports the pre-miRNA to the cytoplasm (Yi *et al.*, 2003). The pre-miRNA is then recognized by a second ribonuclease (Dicer) and its associated RNA-binding protein TRBP and is cleaved near the base of the loop region, producing a ~22-nt duplex RNA with 3′ overhangs. Through a mechanism that is in part determined by the strength of hybridization of the two ends of the duplex, usually one of the two strands is selected for association with a member of the Argonaute (Ago) family of proteins to form the miRISC (miRNA-induced silencing complex), where it encounters its target mRNA. Although most miRNAs are produced through Drosha- and Dicer-mediated maturation (referred to as the canonical pathway), some pre-miRNAs are released through splicing rather than by Drosha cleavage (the "mirtron" pathway; Berezikov *et al.*, 2007) and some (e.g., the miR-451 pre-miRNA) bypass Dicer cleavage and are processed by Ago2 in the RISC

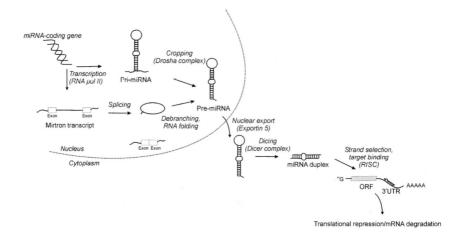

Fig. 1 miRNA biogenesis and function. Shown are the canonical and mirtron pathways of miRNA production from RNA pol II-derived transcripts; other types of RNAs may also give rise to small RNAs with miRNA activity (not shown). The precursor molecule in the canonical pathway (pri-miRNA) is a capped and polyadenylated RNA containing one or more hairpin structures harboring the mature miRNA; cleavage by the Drosha complex releases the 50–70 nt hairpin (pre-miRNA). In the mirtron pathway, the precursor is a splicing lariat that folds into a pre-miRNA upon debranching and in some cases trimming. Upon export into the cytoplasm by Exportin 5, the pre-miRNA undergoes a second processing step by the Dicer complex that releases an RNA duplex, each strand of which may become a functional mature miRNA if selected for loading onto the RISC, where it will encounter a target mRNA. Some miRNAs (e.g., miR-451) are cleaved from a pre-miRNA by Ago2 rather than Dicer (not shown). The mRNA–miRNA interaction generally results in translational repression or degradation of the mRNA target.

(Cheloufi *et al.*, 2010; Cifuentes *et al.*, 2010). Some small RNAs derived from other types of precursors, including small nucleolar RNAs, endogenous short hairpin RNAs, and tRNA precursors may also function as miRNAs (Miyoshi *et al.*, 2010).

B. Consequences of miRNA–mRNA Interactions

Once incorporated into RISC with its attached miRNAs, the target mRNA undergoes a translational block and/or is degraded (Fabian *et al.*, 2010). Studies of miRNA-mediated inhibition of translation indicated that it most likely occurs at the initiation stage through a defect in cap recognition or failure to assemble the 80S ribosome complex, or possibly during the elongation stage, by inducing ribosome detachment or degradation of the nascent peptide. miRNA-mediated decay of target mRNAs commences with removal of the poly(A) tail; this is followed by removal of the 5′ cap and degradation of the mRNA by 5′–3′ exonucleases.

Factors that determine whether a miRNA–mRNA interaction will lead to a translational block or degradation of the target mRNA, as well as the severity of the inhibitory effect, include the number and distribution of miRNA binding sequences and other destabilizing elements in the 3′UTR (Grimson et al., 2007) and the availability of RNA-binding proteins such as HuR (Eulalio et al., 2008). Although the miRNA/mRNA interaction results in inhibition of expression in most cases, recent studies provided evidence for a positive effect of some miRNAs on the translation of their mRNA targets (Fabian et al., 2010). One interesting example is miR-122, whose binding to the 5′UTR of the hepatitis C virus RNA promotes its translation (Jangra et al., 2010).

II. HTLV-1 AND ATLL

HTLV-1 is classified as "carcinogenic to humans" by the International Agency for Research on Cancer (Bouvard et al., 2009). The discovery of HTLV-1 was preceded by the description of ATLL in a population in southwestern Japan (Uchiyama et al., 1977). The connection between HTLV-1 and ATLL was established in the early 1980s, commencing with the discovery of the first human retrovirus in a T cell line (HUT-102) derived from a patient who at the time was diagnosed with cutaneous T-cell lymphoma; the authors named this virus human T cell leukemia virus (HTLV) (Poiesz et al., 1980). This seminal discovery was followed shortly afterward by the identification of antibodies in ATLL patients that recognized antigens in T cell lines established by cocultivation of ATLL cells with cord blood cells (Hinuma et al., 1981; Yoshida et al., 1982) and the isolation of a retrovirus [named adult T cell leukemia virus (ATLV) by the investigators] from one of these cell lines (MT-2) (Yoshida et al., 1982). Further studies demonstrated that HTLV and ATLV were the same retrovirus which was assigned the name HTLV-1 (Watanabe et al., 1984).

Shortly after the discovery of HTLV-1, a second closely related virus named HTLV-2 was identified in a patient with a T cell variant of hairy cell leukemia (Kalyanaraman et al., 1982). The genetic organization and expression strategy of HTLV-2 are similar to that of HTLV-1. However, despite its original isolation from a patient with leukemia and ability to immortalize T cells in culture, HTLV-2 has not been recognized as the causative agent of malignancies in humans (Feuer and Green, 2005). More recent studies revealed genetically related but distinct variant HTLVs named HTLV-3 and HTLV-4 infecting inhabitants of South Cameroon. The prevalence and modes of transmission of these viruses as well as their possible association with human diseases is still under investigation (Mahieux and Gessain, 2009).

A. HTLV-1 Genome Organization and Expression

Sequencing of the HTLV-1 genome (Seiki *et al.*, 1983) revealed the existence of additional viral-encoded nonstructural genes and regulatory elements that were not evident in previously described animal retroviruses. Such increased coding potential was subsequently found in several other retroviruses and led to a classification of "simple" and "complex" retroviruses based on the absence or presence of such additional genes (Cavallari *et al.*, 2011; Cullen, 1991). The extra HTLV-1 ORFs code for the regulatory proteins Tax and Rex and the accessory proteins p21Rex, p30, p13, and p12/p8 (Lairmore *et al.*, 2011) and are produced from alternatively spliced transcripts with an early-late temporal expression pattern (Rende *et al.*, 2011). The recent discovery of minus-strand transcripts coding for a regulatory protein named HBZ (Matsuoka and Green, 2009) adds further complexity to HTLV-1's expression strategy.

Among the extra gene products that define complex retroviruses, HTLV-1 Tax and Rex and their orthologs (e.g., HIV Tat and Rev) stand out as key regulatory proteins that orchestrate viral gene expression at the transcriptional and posttranscriptional levels, respectively. Tax is needed for production of the primary plus-strand transcript, which is driven by the 5'LTR promoter. The promoter contains three elements termed TRE-1 that resemble cyclic-AMP responsive elements (CRE), the sequences recognized by the transcription factor CREB and its associated coactivators. Tax associates with CREB and favors formation of the CREB-coactivator complex on the TRE-1; ensuing chromatin remodeling events induced by the histone acetyltransferase activities of the coactivators together with associations between Tax and proteins bound to the TATA box result in greatly increased transcription (Nyborg *et al.*, 2010). The fate of the primary plus-strand transcript is determined by Rex: by binding to the Rex-responsive element (RXRE) present at the 3' end of the viral RNA, Rex subtracts it from the splicing machinery and favors its nuclear export through interactions with the nuclear export factor CRM1, thus assuring that a proportion of the primary transcripts will remain unspliced or only partially spliced (Hidaka *et al.*, 1988; Inoue *et al.*, 1986).

B. HTLV-1 Transmission and Persistence

HTLV-1 infects an estimated 15–20 million people worldwide, with infection most prevalent (i.e., >5% of the population) in southwestern Japan, the Caribbean basin, and Sub-Saharan Africa (Proietti *et al.*, 2005). Transmission occurs through transfer of blood, semen, and breast milk, with live infected cells rather than free virions most likely representing the main vehicle of transmission. While HTLV-1 can infect many cell types in tissue culture, it shows a strong tropism for CD4+ T cells *in vivo*. Once integrated into the cell genome, viral spread and persistence occurs mainly through mitotic transmission of the integrated viral genome to daughter cells and through cell-to-cell transmission of infectious virions across a specialized molecular structure defined as the "virological synapse" (Nejmeddine *et al.*, 2009).

Infection of PBMC with HTLV-1 yields IL-2-dependent immortalized T cells, some of which progress to a fully transformed phenotype with IL-2-independent growth. The expansion and survival of HTLV-1-infected T cells are driven mainly by Tax. Indeed, the effects of Tax extend well beyond enhancement of CREB to activation of numerous other cellular transcription factors including NF-κB, serum responsive factor, NFAT, and basic helix–loop–helix proteins (Hall and Fujii, 2005). Through the control of these families of transcription factors Tax induces a profound reprogramming of the host cell's transcriptome, resulting in increased cell division, reduced cell death, and genetic instability (Matsuoka and Jeang, 2011). These effects are consistent the ability of Tax to immortalize primary peripheral T cells *in vitro* (Grassmann *et al.*, 1989; Robek and Ratner, 1999) and cause an ATLL-like lymphoproliferative disease in transgenic mice (Hasegawa *et al.*, 2006). Recent studies indicate that HBZ also contributes to HTLV-1-associated oncogenesis; transgenic mice expressing HBZ in CD4+ T cells develop T-cell lymphomas and systemic inflammation which are reminiscent of the clinical pictures observed in HTLV-1-infected individuals (see below) (Satou *et al.*, 2011).

C. ATLL and Other HTLV-1-Associated Diseases

Epidemiological, virological, and molecular evidence support a clear etiological association between HTLV-1 and four diseases: ATLL, HTLV-associated myelopathy/tropical spastic paraparesis (HAM/TSP), HTLV-associated uveitis (HAU), and infective dermatitis (ID) (Proietti *et al.*, 2005). Epidemiological and/or biological evidence also suggests a link between HTLV-1 and several other inflammatory syndromes, including arthropathy, polymyositis, and Sjogren's syndrome (Verdonck *et al.*, 2007).

About 3% of HTLV-1-infected patients develop ATLL, usually several decades after infection. It is more common in males than females, and is linked to transmission in childhood through breast feeding. ATLL is manifested as four clinical subtypes termed as smoldering, chronic, lymphomatous, and acute. Lymphomatous ATLL and acute ATLL are the most aggressive subtypes, with survival times of 9–10 months and 4–6 months, respectively. Patients with chronic ATLL may live for up to 2 years, and patients with smoldering ATLL may live for more than 5 years; both diseases may evolve into the acute form. The most frequent manifestations of acute

ATLL and lymphomatous ATLL are lymphadenopathy, hepatosplenomegaly, and skin lesions. Important complications include hypercalcemia and immune suppression.

ATLL cells have an atypical morphology, with highly lobulated nuclei. These cells, termed "flower cells," contain the integrated provirus, and are a distinguishing feature of the disease. ATLL cells are usually CD3+, CD4+, CD8−, CD25+, CD45RO+ (Kress et al., 2011). This peculiar phenotype, with the concomitant expression of activation (e.g., CD25) and memory cell markers (e.g., CD45RO), is also found in effector regulatory T cells (Treg), a subset of CD4+ CD25+ T cells that play an essential role in maintaining self-tolerance and dampening peripheral immune responses by suppressing the activity and proliferation of other immune system cells (Sakaguchi et al., 2010). Most ATLL cells are also positive for the transcription factor Foxp3, another marker of Tregs (Chen et al., 2006; Karube et al., 2004; Kohno et al., 2005; Roncador et al., 2005) and two additional Treg markers, GITR (Kohno et al., 2005) and CTLA-4 (Matsubara et al., 2005). The possibility that ATLL cells possess some Treg function that contributes to the immune suppression associated with this disease has been debated. Two studies have described an apparent immunosuppressive function of ATLL cells (Kohno et al., 2005; Yano et al., 2007). However, it is difficult to separate ATLL cells from Tregs based on cell surface markers. Toulza et al. showed that functional Foxp3+ Tregs are distinct from CD25+ HTLV-1-infected cells (Toulza et al., 2009) and that HTLV-1 infection results in the expansion of uninfected Tregs through Tax-mediated induction of the chemokine CCL22 (Toulza et al., 2010); these Tregs may decrease the efficiency of immune surveillance in HTLV-1-infected individuals.

HAM/TSP, HAU, and ID reflect the impact of HTLV-1 on the host's immune system and inflammatory responses. The connection between HTLV-1 and the neurological disease HAM/TSP was first made in 1985 (Gessain et al., 1985). Similar to ATLL, HAM/TSP develops in a small percentage of infected individuals, but after a latency period of years rather than decades. It occurs more frequently in females than males, and is thought to arise mainly in persons who have acquired HTLV-1 through sexual transmission. It is characterized by the accumulation of infiltrating mononuclear cells in the grey and white matter of the thoracic spinal cord and progressive demyelination and death of neurons. HAU is characterized by ocular infiltration of HTLV-1-positive cells, opacities in the vitreous body and retinal vasculitis (Mochizuki et al., 1992). ID is a severe recurring eczematous disease with a high prevalence in Jamaican children (LaGrenade et al., 1990).

In the 30 years since its discovery, much has been learned about HTLV-1 replication and pathogenesis. However, many aspects of HTLV-1 pathobiology remain poorly understood. In particular, it is not clear why, despite the proven transforming potential of Tax, the virus causes ATLL in only a small percentage of infected patients after a very long latency period. Most importantly, ATLL and the other diseases associated with HTLV-1 infection are still essentially incurable.

III. miRNAS IN NORMAL CD4+ T CELLS

The generation of a functional T cell population involves a series of molecular events that regulate lineage commitment, cell proliferation, apoptosis, and differentiation and give rise to distinct cellular subsets endowed with specific functional properties (Carpenter and Bosselut, 2010).

This complex process also includes rearrangement of the T cell receptor (TCR) to provide each lymphocyte with a single antigen receptor. T cell development takes place in the thymus, and commences with seeding of progenitor cells migrated from the fetal liver or bone marrow. During the maturation process, thymocytes are distinguished on the basis of CD4 and CD8 expression into double-negative (DN) and double-positive (DP) cells. DN cells are also divided into different stages of development (four stages in mouse and three stages in humans) on the basis of TCR gene rearrangements. Following the DN stages, the contemporary expression of both CD4 and CD8 defines the DP stage of lymphocyte maturation, at which point the thymocytes express a completely rearranged TCR and undergo positive and negative selection. The selection process drives the final step of T lymphocyte maturation and shapes the peripheral T cell repertoire. As the surviving lymphocytes mature, they become CD4 single-positive (CD4 SP) or CD8 single-positive (CD8 SP) cells and finally egress the thymus. The naïve peripheral T cells undergo further differentiation into CD8+ cytotoxic T cells and CD4+ T-helper cells upon engagement of their TCR with a relevant antigen and proliferate to form antigen-specific clones. Following clearance of the antigen, most of the reactive T cells are deleted through activation-induced cell death (AICD); only a small subset of antigen-specific T cells survive this process as "memory" T cells. Peripheral Tregs are derived from thymocytes that have been selected through interaction of their TCR with self-peptides.

A. miRNA Profiles in T Cell Development

The first strong experimental evidence for a role of miRNAs in T cell development and homeostasis came from experiments performed in mice demonstrating that conditional deletion of Dicer at the DN stage resulted in a loss of both CD4+ and CD8+ peripheral T cell populations (Cobb *et al.*, 2006). Deletion of Dicer during the DN/DP transition greatly reduced the number of peripheral CD8+ cells and impaired the ability of peripheral CD4+ T cells to differentiate into mature helper cells (Muljo et al., 2005).

In a recent analysis of the global expression profile of miRNAs in human DP, SP CD4$^+$, and SP CD8$^+$ T cells, the miRNA profile in DP cells was found to be distinct from those of CD4 SP and CD8 SP cells, while the two SP populations showed important similarities; a general upregulation of miRNAs from the DP to the SP stage was noted (Ghisi et al., 2011). Comparison of miRNA profiles in DP, SP, and peripheral SP CD4 and CD8 cells indicated a progressive upregulation of miR-150, miR-146a, and miR-146b and downregulation of miR-128; miR-181 is less abundant in mature peripheral T lymphocytes compared with DP thymocytes (Ghisi et al., 2011).

The study by Ghisi et al. included mass sequencing analyses of small RNA libraries prepared from DP thymocytes and peripheral SP CD4+ cells, with differential expression calculated based on frequencies of sequence reads. The top five miRNAs showing the strongest upregulation from DP thymocytes and peripheral CD4+ T cells were miR-150, miR-29b, miR-27a, miR-29a, and miR-21; the five most strongly downregulated miRNAs were miR-181b, miR-128, miR-181a, miR-93, and miR-20b (Ghisi et al., 2011).

B. miRNAs in Activated T Cells

An early microarray-based comparison of murine CD4+ T cells before and after stimulation with anti-TCR and anti-CD28 indicated profound changes in miRNA expression (Cobb et al., 2006). Comparison of the deep sequencing data obtained for peripheral CD4+ T cells by Ghisi et al. to an additional library of peripheral CD4+ T cells after in vitro activation with phytohemagglutinin and IL-2 revealed 21 differentially regulated miRNAs (>2-fold change). The top five upregulated miRNAs were miR-20b, miR-21, miR-93, miR-155, and miR-25, and the top five downregulated were miR-150, miR-342-3p, miR-101, miR-140, and miR-30b (D'Agostino et al., unpublished).

A recent study by Grigoryev et al. that profiled miRNA expression in human CD2+ T lymphocytes following stimulation with anti-CD3/CD28 beads (Grigoryev et al., 2011) indicated a trend toward upregulation of miRNA expression upon activation, with 44 miRNAs showing a >2-fold increase in expression, versus only 13 showing a >2-fold decrease (including miR-150) (Grigoryev et al., 2011). The top five upregulated miRNAs were miR-221, miR-210, miR-98, miR-29b, and miR-155, and the top five downregulated miRNAs were miR-181a, miR-199a, miR-223, miR-224, and miR-127-3p (Table II in Grigoryev et al., 2011). Integration of miRNA and gene profiling data generated a network with a central hub occupied by the gene *PIK3R1*, which codes for the 85-kDa regulatory subunit of phosphoinositide-3-kinase that is predicted to be targeted by miR-221, miR-155, miR-21, and miR-218 in the study (all upregulated in activated T cells). The authors focused on the functional analysis of miR-155 and miR-221, which showed 71-fold and 7882-fold upregulation, respectively. Knockdown of either of these two miRNAs with locked nucleic acid anti-miRNAs increased proliferation of CD4+ T cells. This effect was accompanied by increased expression of PIK3R1 and IRS2 (insulin receptor substrate-2, an insulin receptor docking protein) induced by anti-miR-155, and the transcription factor FOS, induced by both anti-miRs. The authors proposed that these effects provide a negative feedback loop inhibiting cell proliferation and regulating survival in response to activation (Grigoryev *et al.*, 2011).

C. miRNAs in Tregs

The early studies of mice with deletion of Dicer at the stage of thymocytes also implicated the miRNA pathway in Treg development, with a reduction in their numbers in both the thymus and the periphery (Cobb *et al.*, 2006); deletion of Dicer after Treg lineage commitment (at the time of Foxp3 induction) resulted in a profound impairment of Treg suppressor function and fatal systemic autoimmune disease (Liston *et al.*, 2008; Zhou *et al.*, 2008). A comparison of miRNA profiles in murine Tregs (CD4+, CD25+, GITR+) versus non-Treg CD4+ naïve T cells yielded many miRNAs that were differentially expressed in the Treg population. *In vitro* stimulation of the naïve T cells resulted in a transient shift in their miRNA profile toward that of Tregs. When compared to naïve cells, both Tregs and activated T cells exhibited an upregulation of miR-155, miR-214, miR-23b, miR-22, miR-21, miR-23a, miR-24, miR-27a, miR-103 and downregulation of miR-29c, miR-142-5p, miR-142-3p, let-7 family members, miR-30b, miR-30c, miR-25a, miR-26b, and miR-150 (Cobb *et al.*, 2006).

A comparison of human Tregs (CD4+, CD25+, Foxp3+) versus naïve T cells (CD4+, CD25−) isolated from umbilical cord blood indicated that human Tregs can be distinguished on the basis of their increased levels of miR-21, miR-181c, and miR-374 and reduced levels of miR-31 and miR-125a (Rouas *et al.*, 2009). This study also demonstrated the regulation of Foxp3 by two of the miRNAs in this signature: miR-31 downregulated expression of Foxp3 through direct targeting of its 3′UTR, while miR-21 upregulated its expression through an indirect mechanism (Rouas *et al.*, 2009).

A detailed investigation of the miRNA signatures of 17 lymphocyte subsets purified from human peripheral blood profiled the expression of 664 miRNAs using quantitative RT-PCR (Rossi *et al.*, 2011). Results of this analysis revealed many miRNAs that were not previously reported as differentially expressed in lymphocyte populations and highlighted 29 miRNAs whose expression was subset-specific. Comparison of results obtained for six human subsets with data available for the corresponding murine cell populations revealed important differences in the human and murine miRnomes (Rossi *et al.*, 2011).

Rossi *et al.* placed emphasis on miR-125b, which was enriched in naïve CD4+ T cells compared to the other subsets. They demonstrated that this miRNA controls a network of target genes involved in CD4+ T cell ontogenesis; forced expression of miR-125b resulted in a block in differentiation and favored the naïve phenotype (Rossi *et al.*, 2011).

The report by Rossi *et al.* provided raw RT-PCR datasets which allow further analyses of the different subpopulations. As shown in Fig. 2, a comparison of the most abundant miRNAs expressed in naïve CD4, memory CD4 and Tregs yielded a group of miRNAs that are highly expressed in all three subsets and others that are more enriched in one cell type. Pairwise comparison of Tregs versus naïve CD4 cells confirmed upregulation of miR-21 and downregulation of miR-31 and miR-125a as reported by Rouas *et al.* but also revealed other miRNAs whose up- or downregulation was much more substantial (see Fig. 2).

IV. CELLULAR miRNA EXPRESSION IN HTLV-1-INFECTED CELL LINES

The role of the miRNA network in the HTLV-1 pathobiology has been dissected using freshly isolated ATLL cells, cell lines stabilized from PBMC of HTLV-1-infected patients, and *in vitro*-transformed T cell lines generated by cocultivating normal lymphocytes (PBMC or umbilical cord blood cells) with infected patients' cells. These cell systems have yielded valuable information and also underscore the inherent differences between the stabilized cell lines, most of which are characterized by robust viral expression, and freshly isolated ATLL cells, which often exhibit very low-level expression of viral proteins prior to *in vitro* culture.

Pichler *et al.* (2008) employed quantitative RT-PCR to compare expression of selected miRNAs in panels of T cell lines stabilized from PBMC of patients with ATLL or HAM/TSP (the latter IL-2-dependent), *in vitro*-transformed T cell lines generated by cocultivating umbilical cord blood cells with ATLL cells, as well as a T cell line that expresses Tax in a repressible manner (Tesi cells). Control samples consisted of uninfected PBMC, uninfected CD4+ T cells, CD4+ CD25+ cells, uninfected T-ALL cell lines and the Tesi cells after repression of Tax expression. The investigators focused on

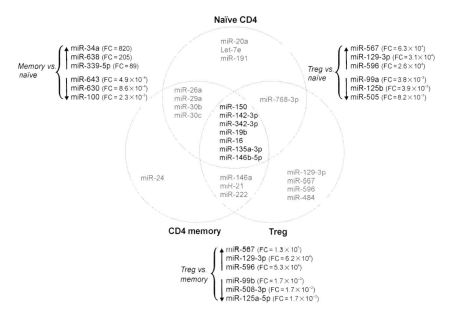

Fig. 2 miRNAs in normal CD4+ T cells. Shown is a Venn diagram that includes the 15 most abundant miRNAs detected in human peripheral naïve CD4 cells, memory CD4 cells, and Tregs in the study by Rossi et al. (2011) along with three miRNAs showing the greatest up- and downregulation in pairwise comparisons of the three subsets. Lists were obtained by elaborating published real-time RT-PCR data (Rossi et al., 2011). The 15 most abundant miRNAs in each cell subset were identified by calculating delta Ct values (=mean Ct value for the miRNA−mean Ct value for the housekeeping RNA RNU48). The miRNAs showing the greatest differential expression were identified based on fold-change (FC) values calculated with the relative quantification method using the second subset indicated in each pair as the calibrator. In agreement with Rouas et al. (2009), miR-21, miR-31, and miR-125a-5p were differentially expressed in Treg versus naïve CD4 cells (FC=7.5, 0.02, and 0.09, respectively); miR-181c and miR-374 showed a less than twofold difference in expression.

miR-21, miR-24, miR-146a, miR-155, miR-191, miR-214, and miR-223, as these miRNAs had been reported to be upregulated in Tregs compared to resting CD4+ T cells in the murine system (Cobb et al., 2006) and had already been implicated in oncogenic transformation in other cell systems. Results revealed significant upregulation of miR-21, miR-24, miR-146a, and miR-155 in the infected cell lines and in Tax-expressing Tesi cells, and downregulation of miR-223 in the infected cell lines; the difference in miR-21, miR-146a, and miR-155 levels was particularly evident ($p<0.0005$) despite the diversity of samples investigated (Pichler et al., 2008). It is noteworthy that all four of these miRNAs have been identified as upregulated in one or more studies of activated uninfected CD4+ T cells (see Section III and Cobb et al., 2006; Grigoryev et al., 2011).

Further studies demonstrated that miR-146a is upregulated by Tax via stimulation of the NF-κB pathway (Pichler *et al.*, 2008). The investigators identified two potential NF-κB binding sites in the miR-146a promoter, one of which was found to be essential for stimulation by Tax (Pichler *et al.*, 2008). Tomita *et al.* confirmed upregulation of miR-146a in HTLV-1-infected cell lines through Tax-dependent NF-κB activation (Tomita *et al.*, 2010). Treatment of infected cell lines with a miR-146a inhibitor resulted in increased expression of TRAF6, a known target of miR-146a that plays a role in the innate immune response (see below). Inhibition of miR-146a interfered with growth of infected cell lines, while a growth-enhancing effect was observed in infected cell lines forced to overexpress miR-146a. These findings led the authors to propose that upregulation of miR-146a expression may be one of the strategies used by HTLV-1 to favor proliferation of infected cells (Tomita *et al.*, 2010).

In the context of normal CD4+ T cells, miR-146a plays an important role in controlling AICD. The switch from TCR-driven clonal expansion to AICD is controlled by the NF-κB pathway, which drives the transition from an AICD-resistant to an AICD-susceptible T cell phenotype (Krammer *et al.*, 2007). Curtale *et al.* showed that stimulation of naïve CD4+ T cells with phorbol-12-myristate 13-acetate (PMA) and ionomycin results in a gradual upregulation of miR-146a expression, which remains high in memory T cells; upregulation of this miRNA was also detected in T-ALL cell line Jurkat stimulated with anti-CD3/CD28 antibodies. Consistent with these findings, miR-146a was found to inhibit AICD by targeting the Fas-associated death domain (Curtale *et al.*, 2010).

The first studies of miR-146a revealed its important role in the innate immune response (Taganov *et al.*, 2006). Its expression is induced by NF-κB after Toll-like and IL-1 receptor (TIR) engagement, and it exerts a negative feedback control on TIR signaling by targeting the adaptor proteins IRAK1 and TRAF6 (Taganov *et al.*, 2006). Another target of miR-146a is the chemokine receptor CXCR4. Controlled downregulation of miR-146a and consequent upregulation of CXCR4 was shown to be required for differentiation of CD34+ progenitors into megakaryocytes (Labbaye *et al.*, 2008).

miR-146a is also upregulated by viral proteins of other oncogenic human viruses, including the EBV LMP-1 protein (Motsch *et al.*, 2007), the HHV8 K13 protein (Punj *et al.*, 2010), and the HPV E5 protein (Greco *et al.*, 2011). In HHV8-infected cells, upregulation of miR-146a by K13 was linked to the downregulation of CXCR4, and was proposed to play a role in HHV8-associated Kaposi's sarcoma through promotion of the premature egress of immature infected endothelial cells from the bone marrow (Punj *et al.*, 2010).

Expression of miR-146a is elevated in various solid cancers (He *et al.*, 2005; Volinia *et al.*, 2006) and in pediatric AML and B-ALL (Zhang *et al.*, 2009). Interestingly, a polymorphism that interferes with processing of the miRNA precursor was found to be a risk factor for development of papillary thyroid carcinoma (Jazdzewski et al., 2008) and several other solid tumors. Altered expression of miR-146a has also been documented in a subtype of myelodysplastic syndrome that is characterized by deletion of the *MIR146A* gene (5q-syndrome) (Taganov et al., 2006) and in autoimmune disorders such as rheumatoid arthritis (elevated miR-146a) (Ceribelli et al., 2011) and systemic lupus erythematosus (reduced miR-146a) (Tang et al., 2009).

Like miR-146a, levels of miR-21 are low in naïve CD4+ T lymphocytes and high in memory T cells (Carissimi et al., 2009). As described in Section III, miR-21 is a positive regulator of Foxp3 expression; therefore upregulation of miR-21 could explain the high levels of Foxp3 transiently detected in TCR-stimulated CD4+ T cells (Rouas et al., 2009; Wang et al., 2007). Upregulated miR-21 in turn targets RASGRP1 (RAS guanyl releasing protein 1), a key activator of Ras in response to TCR signaling (Carissimi et al., 2009). Studies of miR-21 expression in different cell contexts indicate that it is directly upregulated by STAT3 (Loffler et al., 2007), AP-1 (Fujita et al., 2008), and NF-κB (Zhou et al., 2009).

Many human cancers exhibit elevated miR-21 expression, and several of its targets possess tumor suppressor properties (Jazbutyte and Thum, 2010). It is noteworthy that miR-21 upregulation has been described for Sézary syndrome (van der Fits et al., 2011) and mycosis fungoides (van Kester et al., 2011), two cutaneous CD4+ T-cell lymphomas that bear similarities to ATLL. A murine model that exploited the Cre and Tet-off technologies revealed that miR-21 overexpression leads to the development of pre-B malignant lymphoid neoplasms. Interestingly, following inactivation of miR-21, the tumors regressed completely in a few days due to increased apoptosis (Medina et al., 2010). These results demonstrate that continuous expression of miR-21 is necessary to maintain the neoplastic phenotype, making miR-21 the first example of "oncomiR addiction." This finding also has important therapeutic implications, suggesting that targeting miR-21 might be a viable strategy to treat human tumors.

miR-24 is coded in two miRNA clusters that also include miR-23 and miR-27. miR-24 is expressed in many normal tissues and is upregulated in differentiated cells. Forced expression of miR-24 in tumor cell lines and primary cells was shown to cause arrest in G1, while its silencing induced cell proliferation. The ability of miR-24 to block the cell cycle was attributed to its targeting of several genes involved in cell cycle control including *MYC* and *E2F2* (Lal et al., 2009). There is also evidence that miR-24 can exert oncogenic rather that growth-suppressive properties: Qin et al. showed that miR-24 exerts an anti-apoptotic effect in several cell lines and targets the pro-apoptotic factor Fas-associated factor 1 through binding to its coding sequence (Qin et al., 2010); Qian et al. showed that miR-24 inhibits apoptosis of cardiomyocytes by targeting Bim (Qian et al., 2011). Studies in promyelocytic leukemia and erythroleukemia cell lines demonstrated that miR-24 also targets H2AX, resulting in defect in the DNA damage repair response (Fernandez-Capetillo *et al.*, 2004) which could lead to the accumulation of transforming mutations.

High levels of miR-155 are evident in Tregs and activated T cells (see above) as well as in activated B cells, activated macrophages and dendritic cells (reviewed by Faraoni *et al.*, 2009). Mice deficient in miR-155 show impairments in T-, B-, and dendritic cell functions (Rodriguez *et al.*, 2007). CD4+ T cells from miR-155-deficient mice tend to differentiate into Th2 cells and show reduced IL-2 and IFN-γ production in response to antigen stimulation (Thai *et al.*, 2007). miR-155 expression in Tregs depends on the activity of Foxp3; miR-155 in turn blocks expression of SOCS1, a negative regulator of IL-2R signaling, thus maintaining Tregs highly sensitive to IL-2 (Lu *et al.*, 2009). Overexpression of miR-155 in CD4+ T cells renders them resistant to Treg-mediated suppression (Stahl *et al.*, 2009).

Overexpression of miR-155 has been documented in ATLL and several other hematological malignancies (see Table I described below) and solid tumors (e.g., lung, thyroid, pancreas, breast, colon, cervix) (reviewed by Faraoni *et al.*, 2009). In mice, forced expression of miR-155 induces polyclonal pre-B-cell tumors (Costinean *et al.*, 2006). As mentioned in Section I, miR-155 may also be important in the mechanisms of B-cell transformation driven by EBV, which induces miR-155 expression, and HHV-8, which codes for a miR-155 ortholog with target specificity similar to that of the cellular miRNA (Gottwein *et al.*, 2007; Skalsky *et al.*, 2007). The EBV LMP-1 protein upregulates cellular miR-155 through NF-κB (Gatto *et al.*, 2008; Lu *et al.*, 2008; Rahadiani *et al.*, 2008). miR-155 in turn targets IKKε (Lu *et al.*, 2008), a transcriptional target of the NF-κB pathway that is involved in the interferon antiviral response. These properties, together with the observation that knockdown of miR-155 leads to a reduction in EBV copy number, suggest that miR-155 is needed for viral persistence in latently infected cells (Lu *et al.*, 2008).

The oncogenic properties of miR-155 can in part be explained by its ability to block expression of tumor protein 53-induced nuclear protein 1 (TP53INP1), a nuclear protein that promotes cell cycle arrest and apoptosis (Gironella *et al.*, 2007). As described below, TP53INP1 is known to be targeted by two miRNAs that were identified as upregulated in ATLL samples and HTLV-1-infected cell lines through microarray analysis, and was pointed out by Pichler *et al.* as a potential target of both miR-155 and miR-21, miR-24 and miR-146a (Pichler *et al.*, 2008).

miR-223 plays a key role in promyelocytic-to-granulocyte differentiation (Fazi *et al.*, 2005). miR-223-knockout mice exhibit an increase in granulocyte progenitors and atypical neutrophils exhibiting hypersensitivity to activating stimuli, accompanied by a spontaneous lung inflammation (Johnnidis *et al.*, 2008). miR-223 is upregulated in bladder cancer (Gottardo *et al.*, 2007),

Table I Selected miRNAs That Are Differentially Expressed in ATLL and Other Hematological Malignancies Compared to Non-Neoplastic Controls

miRNA	ATLL	Sézary syndrome	Mycosis fungoides	Other hematological tumors
miR-155	↑ (Bellon et al., 2009; Yeung et al., 2008)	—	↑ (van Kester et al., 2011)	↑ ped-ALL (Zhang et al., 2009), ped-AML (Zhang et al., 2009), AML (O'Connell et al., 2008), mantle cell lymphoma (Zhao et al., 2010), CLL (Calin et al., 2004; Fulci et al., 2007; Pallasch et al., 2009), ALCL (ALK−) (Merkel et al., 2010), DLBCL (Eis et al., 2005; Roehle et al., 2008; Burkitt's lymphoma (Metzler et al., 2004),[a] follicular lymphoma (Roehle et al., 2008), NK leukemia/lymphoma (Yamanaka et al., 2009) ↓ AML Garzon et al., 2008), multiple myeloma (Gutierrez et al., 2010)
miR-93	↑ (Yeung et al., 2008)	↓ (Ballabio et al., 2010)	↑ (van Kester et al., 2011)	↑ Multiple myeloma (Pichiorri et al., 2008) ↓ AML (Garzon et al., 2008)
miR-130b	↑ (Yeung et al., 2008)	↑ (Narducci et al., 2011)	—	↑ ped-ALL (Zhang et al., 2009), multiple myeloma (Chi et al., 2011) ↓ AML (Cammarata et al., 2010)
miR-142-5p	↑ (Bellon et al., 2009)	↓ (Ballabio et al., 2010)	—	↑ ped-ALL (Schotte et al., 2009), mantle cell lymphoma (Navarro et al., 2009), multiple myeloma (Chi et al., 2011; Zhou et al., 2010) ↓ AML (Cammarata et al., 2010), ped-AML (Zhang et al., 2009), ped-ALL (Zhang et al., 2009), mantle cell lymphoma (Zhao et al., 2010)
miR-142-3p	↑ (Bellon et al., 2009)	↑ (Narducci et al., 2011) ↓ (Ballabio et al., 2010)	↑ (van Kester et al., 2011)	↑ T-ALL (Lv et al., 2011), ped-ALL (Schotte et al., 2009), mantle cell lymphoma (Navarro et al., 2009), CML (Flamant et al., 2010) ↓ ped-AML (Zhang et al., 2009), mantle cell lymphoma (Zhao et al., 2010)
miR-150	↑ (Bellon et al., 2009; Yeung et al., 2008)	↓ (Ballabio et al., 2010)	—	↑ ped-ALL (Schotte et al., 2009), CLL (Fulci et al., 2007), multiple myeloma (Zhou et al., 2010) ↓ ped-AML (Zhang et al., 2009), DLBCL (Roehle et al., 2008), mantle cell lymphoma (Di Lisio et al., 2010; Zhao et al., 2010) CML (Flamant et al., 2010), NK/T-cell lymphoma (Watanabe et al., 2011)
miR-223	↑ (Bellon et al., 2009)	↓ (Ballabio et al., 2010)	—	↑ Multiple myeloma (Pichiorri et al., 2008) ↓ AML (Cammarata et al., 2010), CLL (Calin et al., 2004), mantle cell lymphoma (Di Lisio et al., 2010), multiple myeloma (Chi et al., 2011)
miR-31	↓ (Yamagishi et al., 2012; Yeung et al., 2008)	↓ (Ballabio et al., 2010; Narducci et al., 2011)	—	↑ Multiple myeloma (Chi et al., 2011) ↓ AML (Cammarata et al., 2010), mantle cell lymphoma (Di Lisio et al., 2010; Navarro et al., 2009)

(*continues*)

Table I (continued)

miRNA	ATLL	Sézary syndrome	Mycosis fungoides	Other hematological tumors
miR-125a	↓ (Yamagishi et al., 2012; Yeung et al., 2008)	↓ (Ballabio et al., 2010)	—	↓ CLL (Pallasch et al., 2009), mantle cell lymphoma (Di Lisio et al., 2010), AML (Garzon et al., 2008), multiple myeloma (Pichiorri et al., 2008)
miR-126	↓ (Yamagishi et al., 2012; Yeung et al., 2008)	—	—	↑ ped-AML (Zhang et al., 2009), ped-ALL (Zhang et al., 2009) ↓ CLL (Pallasch et al., 2009), mantle cell lymphoma (Di Lisio et al., 2010), AML (Cammarata et al., 2010; Garzon et al., 2008)
miR-130a	↓ (Yamagishi et al., 2012; Yeung et al., 2008)	—	—	↑ ped-AML (Zhang et al., 2009), ped-ALL (Zhang et al., 2009), multiple myeloma (Chi et al., 2011; Pichiorri et al., 2008) ↓ AML (Cammarata et al., 2010; Garzon et al., 2008), CLL (Calin et al., 2004; Pallasch et al., 2009)
miR-132	↓ (Bellon et al., 2009)	↓ (Narducci et al., 2011)	—	↓ Mantle cell lymphoma (Di Lisio et al., 2010; Navarro et al., 2009)
miR-146b	↓ (Bellon et al., 2009; Yamagishi et al., 2012)	↓ (Ballabio et al., 2010)	↑ (van Kester et al., 2011)	↑ ped-AML (Zhang et al., 2009) ↓ Mantle cell lymphoma (Di Lisio et al., 2010), AML (Garzon et al., 2008),[b] CLL (Calin et al., 2004)[b]
miR-181a	↓ (Bellon et al., 2009; Yamagishi et al., 2012)	↓ (Ballabio et al., 2010)	↑ (van Kester et al., 2011)	↑ ped-AML (Zhang et al. 2009), ped-ALL (Schotte et al., 2009; Zhang et al., 2009), multiple myeloma (Chi et al., 2011; Pichiorri et al., 2008; Zhou et al., 2010) ↓ CLL (Pallasch et al., 2009), mantle cell lymphoma (Navarro et al., 2009)
miR-335	↓ (Yamagishi et al., 2012; Yeung et al., 2008)	↓ (Ballabio et al., 2010; Narducci et al., 2011)	—	↑ ped-AML (Zhang et al., 2009), multiple myeloma (Chi et al., 2011) ↓ mantle cell lymphoma (Pichiorri et al., 2008), multiple myeloma (Chi et al., 2011; Pichiorri et al., 2008)

↑ Upregulated, ↓ downregulated, — not reported. miRNAs listed for Sézary syndrome, mycosis fungoides, and other hematological tumors showed a fold change of at least 1.5 and a p, q, or FDR value <0.05; fold change not reported in Calin et al. (2004), Yamanaka et al. (2009), and Lv et al. (2011).
Abbreviations: ped, pediatric; ALCL, anaplastic large cell lymphoma; ALL, acute lymphoblastic leukemia; AML, acute myeloid leukemia; CLL, chronic lymphocytic leukemia; DLBCL, diffuse large B-cell lymphoma
[a]Examined miRNA precursor.
[b]Reported on miR-146.

esophageal adenocarcinoma (Mathe *et al.*, 2009), and in recurrent ovarian cancer (Laios *et al.*, 2008), but is downregulated in hepatocellular carcinoma (Wong *et al.*, 2008). miR-223 is downregulated in several hematological tumors, but is upregulated in ATLL (see Table I). miR-223 therefore appears to exert either oncogenic or tumor suppressor properties depending on the cell context.

V. miRNA PROFILING IN ATLL SAMPLES

An overview of selected differentially expressed miRNAs in ATLL samples examined by three studies performed to date and their regulation in other hematological malignancies is presented in Table I. The shared up- or downregulation of many miRNAs indicated in Table I suggests that the study of miRNAs in ATLL pathogenesis might also provide insights into the molecular mechanisms leading to the development of other hematological tumors.

The first study that examined the miRNA profile of ATLL samples employed microarrays to compare the miRnome of PBMC from four acute ATLL patients (>90% CD4+ CD25+ cells in the peripheral blood) to control PBMC from three healthy donors and between seven HTLV-1-transformed cell lines and uninfected cord blood cells (Yeung *et al.*, 2008). The analysis of ATLL samples yielded 22 upregulated and 22 downregulated miRNAs, and the analysis of cell lines yielded 13 upregulated and 30 downregulated miRNAs. Comparison of these results yielded 15 miRNAs that were regulated in the same manner in the ATLL samples and cell lines: miR-18a, miR-9, miR-17-3p, miR-130b, miR-20b, and miR-93 were upregulated and miR-1, miR-130a, miR-199a*, miR-126, miR-144, miR-335, miR-337, miR-338, and miR-432 were downregulated (Fig. 1B in Yeung *et al.*, 2008). The small number of miRNA showing the same trend of expression in the ATLL samples and cell lines underscores the substantial differences between primary samples from ATLL patients and *in vitro*-stabilized cell lines.

To distinguish miRNAs that might be altered in response to a proliferative stimulus, the authors also examined the changes in miRNA expression profiles of uninfected PBMC in response to stimulation with PMA, and found 29 upregulated miRNAs and 12 downregulated miRNAs. Comparison of the three datasets identified miR-93, miR-130b, and miR-18a as upregulated in ATLL cells, HTLV-1-infected cell lines, and PMA-treated cells (Yeung *et al.*, 2008). The authors focused the remainder of their study on miR-93 and miR-130b, by verifying their expression by real-time RT-PCR and by testing their functional properties (Yeung *et al.*, 2008).

miR-93 is a member of the miR-106b-25 cluster, which also includes miR-25 and miR-106b. These miRNAs, which are coded on chromosome 7q22.1 within an intron of the *MCM7* gene, share sequence similarity with those encoded in the oncogenic miR-17-92 cluster (Mendell, 2008). A recent investigation demonstrated that miR-93 and its relatives are highly abundant in hematopoietic progenitor cells and promote their expansion when ectopically expressed (Meenhuis *et al.*, 2011). miR-93 is upregulated in several tumors of epithelial derivation (Ambs *et al.*, 2008; Blenkiron *et al.*, 2007; Kan *et al.*, 2009; Li *et al.*, 2009; Nam *et al.*, 2008; Petrocca *et al.*, 2008b; Yanaihara *et al.*, 2006). Studies performed in gastric, esophageal, and mammary epithelial cell lines (Ivanovska *et al.*, 2008; Kan *et al.*, 2009; Petrocca *et al.*, 2008a,b) showed that, through targeting of E2F1 and p21, the miR-106b-25 and miR-17-92 clusters play an important role in fine-tuning the TGF-β pathway, which is often functionally inactivated in epithelial tumors. miR-93 also targets MICB (major histocompatibility complex class I chain-related B), a ligand for a receptor present on NK, CD8+, and $\gamma\delta$ T cells that is upregulated by stress signals such as viral infection; it has been proposed that overexpression of miR-93 in tumor cells leads to loss of MICB and contributes to immune evasion (Stern-Ginossar *et al.*, 2008). Interestingly, miR-93 along with miR-155 are included in a panel of 49 miRNAs shown to be upregulated in skin biopsies of mycosis fungoides (van Kester *et al.*, 2011).

In addition to ATLL, miR-130b is also upregulated in pediatric ALL and in multiple myeloma (see Table I) and during progression of gliomas to glioblastomas (Malzkorn *et al.*, 2010). Along with miR-93 and other miR-NAs, miR-130b is able to interfere with Ras-induced senescence (Borgdorff *et al.*, 2010) suggesting its role as an "oncomiR."

Yeung *et al.* performed a computational search for targets of miR-93 and miR-130b which yielded 114 genes, including five potential tumor suppressors (Yeung *et al.*, 2008). One of these, *TP53INP1*, stood out since its 3′UTR contains two potential binding sites for each of these miRNAs (Yeung *et al.*, 2008); it is noteworthy that TP53INP1 was also indicated as a potential mRNA target of all four miRNAs identified by Pichler *et al.* as upregulated in the HTLV-1-cell lines (miR-21, miR-24, miR-146a, and miR-155) (Pichler *et al.*, 2008). Results of luciferase-3′UTR reporter assays confirmed that the TP53INP1 3′UTR is indeed a target for miR-93 and miR-130b (Yeung *et al.*, 2008). Accordingly, knockdown of miR-93 and miR-130b with antagomirs in an HTLV-1 cell line resulted in increased expression of TP53INP1 accompanied by an increase in apoptotic cell death. The authors showed that Tax was able to increase expression of a reporter construct driven by the miR-130b promoter (Yeung *et al.*, 2008).

Loss of TP53INP1 expression has been documented during the progression of gastric cancer (Jiang *et al.*, 2006) and early in the development of pancreatic adenocarcinoma, where it is targeted by overexpressed miR-155 (Gironella *et al.*, 2007). Reintroduction of TP53INP1 in a pancreatic adenocarcinoma cell line interfered with its growth in nude mice (Gironella *et al.*, 2007), suggesting its role as a tumor suppressor. TP53INP1 is upregulated by p53 and p73 in response to stress signals and contributes to the cell cycle arrest and apoptosis induced by these tumor suppressors (Tomasini et al., 2003, 2005). ATLL is characterized by inactivation of p53 either as a result of inactivating mutations (Yamada and Kamihira, 2005), or more often through other mechanisms that lead to accumulation of the protein in an inactive state (Tabakin-Fix et al., 2005). Targeting of TP53INP1 by miRNAs provides a further mechanism by which p53/p73 function could be blocked in ATLL cells. As depicted in Fig. 3, TP53INP1 appears to be a common target of upregulated miRNAs in the context of HTLV-1. This potential convergence of several miRNAs on one target suggests that downregulation of TP53INP1 may be fundamental for protecting HTLV-1-infected cells from apoptosis.

In a subsequent study, Bellon et al. used microarrays to compare miRNA expression in PBMC from seven ATLL patients versus control PBMC and CD4+ T cells and then validated selected miRNAs by quantitative RT-PCR on a panel of 15 ATLL patients (14 acute, 1 chronic). Differentially expressed miRNAs identified in this analysis included miR-150, miR-155, miR-223, miR-142-3p, and miR142-5p (upregulated) and miR-181a, miR-132, miR-125a, and miR-146b (downregulated). Analysis of a panel of HTLV-1-infected cell lines revealed an ATLL-like miRNA profile, except that miR-150 and miR-223 were downregulated instead of upregulated. Treatment of one of

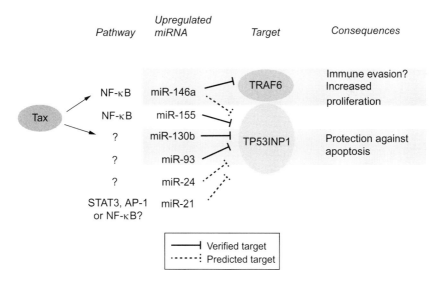

Fig. 3 Targeting of TP53INP1 by HTLV-1-upregulted miRNAs. Shown are miRNAs observed to be upregulated in HTLV-1-infected cell lines and/or ATLL samples which are predicted or known to target the tumor suppressor protein TP53INP1 and would thereby protect cells against apoptosis. The NF-κB pathway is implicated in activation of several of the miRNAs. See text for details. (See Page 1 in Color Section at the back of the book.)

the infected cell lines with an inhibitor of the NF-κB or JNK pathway resulted in a decrease in miR-155 expression (Bellon et al., 2009).

As mentioned in Section III, low expression of miR-125a was identified as a distinguishing feature of human Tregs (Rouas et al., 2009). Downregulation of miR-125a has been reported in various solid tumors such as breast cancer (Guo et al., 2009), medulloblastoma (Ferretti et al., 2009), and gastric carcinoma (Nishida et al., 2011) as well as in several hematological tumors (see Table I), supporting a tumor suppressor function. Experiments carried out in the murine system revealed an important role for miR-125a in hematopoiesis, with its forced expression leading to an expansion of the hematopoietic stem cell pool through a mechanism involving a reduction in apoptosis (Guo et al., 2010).

The *MIR142* locus was originally identified as a rearrangement partner of *MYC* in a prolymphocytic B-cell leukemia (Gauwerky et al., 1989). miR-142 is specifically expressed in hematopoietic tissues and is considered a lymphoid marker (Chen et al., 2004; Merkerova et al., 2008; Ramkissoon et al., 2006). As indicated in Table I, miR-142-5p and miR-142-3p are either up- or downregulated in other hematological malignancies. A study aimed at understanding how murine Tregs maintain elevated levels of cAMP revealed that Foxp3 downregulates expression of miR-142-3p, which would otherwise inhibit the expression of adenylate cyclase-9, the enzyme responsible for cAMP production (Huang et al., 2009). It would therefore be of interest to test for a possible correlation between the levels of Foxp3 and miR-142-3p in ATLL cells.

As mentioned in Section III, miR-181a is temporally regulated during human T cell development, with very low levels detected in mature peripheral T cells. Experiments carried out in the murine system demonstrated that increasing the expression of miR-181a in mature T cells augments both the strength and sensitivity of TCR signaling by targeting tyrosine phosphatases that act as negative feedback modulators in TCR signaling (Li et al., 2007). miR-181a is downregulated in glioblastoma (Ciafre et al., 2005; Shi et al., 2008) and can be either up- or downregulated in hematological tumors (see Table I). In CLL, reduced levels of miR-181a deregulate expression of the oncogenic transcription factor PLAG1 (Pallasch et al., 2009). Conversely, upregulation of miR-181a is observed in multiple myeloma, and inhibition of miR-181a expression in multiple myeloma cell lines leads to a significant suppression of their growth as tumors in nude mice (Pichiorri et al., 2008). In AML, miR-181a expression levels correlate with morphological subclasses (Debernardi et al., 2007).

Both Yeung et al. and Bellon et al. reported an upregulation of miR-150 and miR-155 in ATLL samples (Bellon et al., 2009; Yeung et al., 2008). The upregulation of miR-150 in ATLL is intriguing considering the accumulated knowledge of this miRNA's function. As described in Section III, miR-150 is gradually upregulated during T cell development and downregulated in activated CD4+ T cells. Overexpression of miR-150 was shown to inhibit the growth of B-lymphoma cell lines, indicating its possible function as a tumor suppressor (Chang *et al*., 2008). Recent studies indicated that miR-150 is also downregulated in Sézary syndrome (Ballabio *et al*., 2010), in NK/T-cell lymphomas (Watanabe *et al*., 2011), and in several other hematological tumors (Table I); further, ectopic expression of miR-150 in NK cell lines reduces proliferation and induces apoptosis (Watanabe *et al*., 2011). Ghisi *et al*. likewise observed that forced expression of miR-150 in T-ALL cell lines (which were found to constitutively express very low levels of this miRNA) triggered a reduction in proliferation and promoted apoptotic death, and identified Notch3 as a direct target of miR-150 in T cells (Ghisi *et al*., 2011). The interaction of miR-150 with a component of the Notch pathway is particularly interesting, as this pathway plays a key role in the development of the T cell compartment (Sultana *et al*., 2010). It would be of interest to investigate the influence of miR-150 on the Notch pathway in ATLL cells given the recent finding of a high rate of activating Notch mutations and constitutive activation of the Notch pathway in ATLL patients (Pancewicz *et al*., 2010).

In the most extensive miRNA profiling study of ATLL samples performed to date, Yamagishi *et al*. used microarrays to compare the miRnome in 40 primary ATLL samples (20 acute, 18 chronic, 1 lymphoma, 1 of unknown subtype) versus 22 samples of CD4+ T cells from healthy donors (Yamagishi *et al*., 2012). Results revealed 61 miRNAs with significantly ($p<1\times10^{-5}$) altered levels of expression, 59 of which (96.7%) showed decreased expression. Such a preponderance of downregulated miRNA expression in tumor cells versus their normal counterparts is in agreement with early profiling studies of a broad panel of human tumors (Lu *et al*., 2005).

Only four of the downregulated miRNAs identified by Yamagishi *et al*. were also reported by Yeung *et al*. (2008) (miR-130a, 126, 335, 31) and three were reported by Bellon *et al*. (2009) (miR-181a, 146b, 125a), and neither of the two upregulated miRNAs (miR-451 and miR-144*) had been previously described. The larger number of samples and the newer array system employed by Yamagishi *et al*. may in part explain the limited overlap in results from the previous studies.

VI. REPRESSION OF miR-31 EXPRESSION IN ATLL

Among the downregulated miRNAs identified by Yamagishi *et al*., miR-31 was of particular interest due to its profound repression in all of the ATLL samples (ca. 250-fold downregulation) and its recent identification as a tumor suppressor and/or metastasis-associated miRNA in breast cancer (Valastyan *et al*., 2009).

A. Identification of NF-κB-Inducing Kinase (NIK) as a Target of miR-31 in ATLL

A search for putative target genes of miR-31 using four computational algorithms identified a small number of targets, including *MAP3K14* (NIK). Results of luciferase-3'UTR reporter assays confirmed the interaction of miR-31 with the NIK3'UTR at one of two potential target sequences. Transfection experiments using a NIK cDNA with or without its wild-type 3'UTR sequence confirmed that miR-31 specifically targeted the 3'UTR of NIK (Yamagishi *et al.*, 2012).

NIK plays a central role in noncanonical NF-κB signaling by phosphorylation of IKKα. NIK is overexpressed in ATLL cells, and is considered to be largely responsible for their constitutive NF-κB activation (Saitoh *et al.*, 2008; Watanabe *et al.*, 2005). Experiments carried out in the ATLL-derived cell line TL-Om1 showed that ectopic expression of miR-31 downregulated NIK at the mRNA and protein levels, repressed the levels of phospho-IKKα, and caused a reduction in NF-κB activity, while inhibition of miR-31 expression in HeLa cells triggered an accumulation of NIK mRNA and protein. Forced expression of miR-31 in B cells also attenuated both BAFF- and CD40L-mediated NIK accumulation and subsequent NF-κB signaling with decreased levels of IκBα phosphorylation. On the other hand, TNF-α-triggered canonical NF-κB activation was not affected by miR-31 in Jurkat cells. These findings indicate that miR-31 inhibits the basal and receptor-initiated activities of the noncanonical NF-κB pathway.

TL-Om1 cells ectopically expressing either miR-31 or a NIK-specific shRNA (shNIK) proliferated significantly more slowly than control cells and downregulated genes coding for anti-apoptotic proteins such as Bcl-xl, XIAP, and FLIP, suggesting that miR-31 has a pro-apoptotic function through inhibition of the NF-κB pathway. Ectopic expression of miR-31 in TL-Om1 cells or primary ATLL cells promoted basal and Fas-directed apoptosis; this effect was reversed by introduction of a NIK cDNA lacking the 3'UTR. shRNA-mediated repression of NIK also induced cell death, suggesting that NIK and NF-κB activity are crucial for survival of ATLL tumor cells. These findings indicated that (a) miR-31 acts as a tumor suppressor in T cells and (b) NIK-regulated NF-κB is of crucial importance to the survival of ATLL cells (Yamagishi *et al.*, 2012).

B. Genetic Deletion and Polycomb-Directed Epigenetic Silencing Cause Loss of miR-31 Expression

miR-31 is coded in an intron of the *LOC554202* gene, which maps to chromosome 9p21.3 adjacent to clusters of the *CDKN2* and *IFNA* families, a well-known hotspot of genomic loss in several types of human cancer. Genomewide scans of genetic lesions in 168 ATLL samples demonstrated deletions of 9p21.3 spanning the miR-31 coding region in 12.5% of the cases.

Yamagishi *et al.* next investigated the possible epigenetic silencing of miR-31. Computational analysis revealed a putative TATA box and transcriptional start site 2500 bp upstream of the miR-31 coding region. An assembly of YY1-binding motifs was found upstream of the miR-31 region in human and mouse. YY1 is a pivotal transcription factor and a recruiter of the Polycomb repressive complex (PRC) (Simon and Kingston, 2009). Chromatin immunoprecipitation revealed elevated histone H3K9 and H3K27 methylation in a broad area containing the miR-31 coding region in ATLL cells, indicating that repressive histone methylation, especially that of Polycomb family-dependent H3K27me3, contributes to miR-31 repression. shRNA-mediated knockdown of YY1 increased miR-31 expression and decreased YY1 occupancy in the miR-31 region along with loss of occupancy of EZH2, a key component of PRC2. Thus, YY1 appears to regulate PRC2 localization and initiate miR-31 suppression.

PRC2 components, especially EZH2 and SUZ12, are overexpressed in primary ATLL cells (Sasaki *et al.*, 2011). Yamagishi *et al.* observed an inverse correlation between levels of EZH2 or SUZ12 and miR-31 in ATLL samples. In ATLL cell lines, knockdown of PRC2 using shSUZ12 or shEZH2 increased the levels of miR-31 and was accompanied by histone demethylation at H3K27 in the miR-31 region. EZH2 occupancy and HDAC1 recruitment in the miR-31 genomic region suggest that this multimeric complex leads to completely inaccessible chromatin architecture as a result of histone modifications. Evidence for Polycomb-mediated epigenetic regulation of miR-31 expression was also obtained in experiments carried out in human breast cancer cell lines, suggesting that this phenomenon may be present in tumors other than ATLL (Yamagishi *et al.*, 2012).

C. The Polycomb Group Regulates NF-κB Activity by Controlling miR-31 Expression

The study by Yamagishi *et al.* established a novel connection between Polycomb proteins and regulation of the NF-κB pathway. PRC2 knockdown in ATLL cell lines resulted in decreased levels of NIK, p52, and phospho-IκBα with consequent repression of NF-κB activity, reduced proliferation and increased sensitivity to apoptotic signals, thus recapitulating the effects seen with miR-31 overexpression. Conversely, overexpression of EZH2 induced NF-κB activation, which was partially cancelled by forced expression of miR-31.

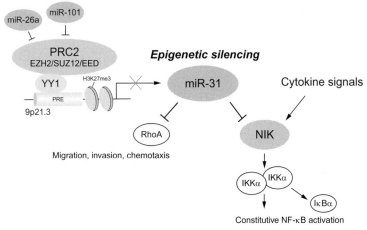

Fig. 4 An oncogenic cascade involving Polycomb repressive factors, miR-31, and NF-κB revealed by analysis of ATLL samples. PRC2 indicates Polycomb Repressive Complex 2; EZH2, SUZ12, EED are components of the PRC2. YY1 is the ubiquitous transcription factor Yin Yang 1 that recruits the PRC2 on the miR31 promoter (PRE). MAP3K14 (NIK) plays a central role in controlling the noncanonical NF-κB pathway by phosphorylating IKKα. (For color version of this figure, the reader is referred to the Web version of this chapter.)

The importance of the Polycomb group in controlling NF-κB signaling and apoptotic death was confirmed in primary ATLL cells, which showed a strong death response upon transduction of miR-31 or knockdown of NIK or EZH2. PRC2 knockdown also affected receptor-initiated accumulation of NIK in B cells. Taken together, these findings identified a novel molecular role for Polycomb-dependent epigenetic gene silencing that involves miR-31 loss, NF-κB activation and other signaling pathways (Yamagishi et al., 2012; Fig. 4).

VII. PERSPECTIVES

Table II reports miRNAs whose differential regulation was reported in at least two studies of ATLL samples or infected cell lines. The limited overlap between these profiling studies indicates that further work is needed to consolidate the miRNA signature of ATLL. Such studies would be aided by careful clinical stratification of the patients and by focusing on purified T cell populations (resting and activated CD4+ T cells and Tregs) as controls. Analysis of small RNAs in HTLV-1-infected cells and ATLL

Table II miRNAs Identified in At Least Two Studies of ATLL Samples or HTLV-1-Infected Cell Lines

miRNA	References
ATLL samples	
↑ miR-150	Yeung *et al.* (2008), Bellon *et al.* (2009)
↑ miR-155	Yeung *et al.* (2008), Bellon *et al.* (2009)
↓ miR-31	Yeung *et al.* (2008), Yamagishi *et al.* (2012)
↓ miR-125a	Bellon *et al.* (2009), Yamagishi *et al.* (2012)
↓ miR-126	Yeung *et al.* (2008), Yamagishi *et al.* (2012)
↓ miR-130a	Yeung *et al.* (2008), Yamagishi *et al.* (2012)
↓ miR-146b	Bellon *et al.* (2009), Yamagishi *et al.* (2012)
↓ miR-181a	Bellon *et al.* (2009), Yamagishi *et al.* (2012)
↓ miR-335	Yeung *et al.* (2008), Yamagishi *et al.* (2012)
Infected cell lines	
↑ miR-155	Bellon *et al.* (2009), Pichler *et al.* (2008)
↓ miR-150	Yeung *et al.* (2008), Bellon *et al.* (2009)
↓ miR-223	Yeung *et al.* (2008), Bellon *et al.* (2009), Pichler *et al.* (2008)

↑ Upregulated, ↓ downregulated.

samples by deep sequencing would also provide valuable information about possible differences in the processing of known miRNAs and reveal the existence of novel small RNA species connected to ATLL and HTLV-1 infection.

In addition to aiding to our understanding of HTLV-1 pathogenesis, unraveling the impact of HTLV-1 infection and transformation on the miRnome of T cells might provide novel predictive markers of disease progression in HTLV-1-infected individuals, as well as reveal key pathways involved in T cell transformation. It is noteworthy that miR-155 and miR-31 are the only "signature" miRNAs listed in Table II with verified targets of demonstrated functional significance in the context of HTLV-1 infection (i.e., TP53INP1 and NIK, respectively); clearly more work needs to be done to identify targets for other differentially regulated miRNAs.

Future studies should also be aimed at dissecting the mechanisms leading to HTLV-1-induced alteration in miRNA expression. Besides Tax, which was already shown to control miR-146a (Pichler *et al.*, 2008) and miR-130b through the NF-κB pathway (Yeung *et al.*, 2008), the accessory proteins p12, p13, p30, and HBZ would be interesting candidates, given their complex effects on signal transduction pathways, transcriptional control, and cell turnover (Bai *et al.*, 2010; Matsuoka, 2010; Silic-Benussi *et al.*, 2010; Van Prooyen *et al.*, 2010).

ACKNOWLEDGMENTS

We thank Luigi Chieco-Bianchi and Katia Ruggero for valuable discussions and Riccardo Rossi for sharing unpublished information. This work was supported by Grants-in-Aid for Scientific Research from the Ministry of Education, Culture, Sports, Science, and Technology of Japan (TW, no. 23390250), Grants-in-Aid from the Ministry of Health, Labour, and Welfare (TW, H21-G-002 and H22-AIDS-I-002), the Associazione Italiana per la Ricerca sul Cancro (PZ and VC), the European Union (VC, "The role of chronic infections in the development of cancer"; contract no. 2005-018704), the Ministero per l'Università e la Ricerca Scientifica, e Tecnologica Progetti di Ricerca di Interesse Nazionale (DMD and VC), the Fondazione Cariverona (VC), and the University of Padova (DMD).

REFERENCES

Ambs, S., Prueitt, R. L., Yi, M., Hudson, R. S., Howe, T. M., Petrocca, F., Wallace, T. A., Liu, C. G., Volinia, S., Calin, G. A., Yfantis, H. G., Stephens, R. M., et al. (2008). Genomic profiling of microRNA and messenger RNA reveals deregulated microRNA expression in prostate cancer. *Cancer Res.* **68**, 6162–6170.

Bai, X. T., Baydoun, H. H., and Nicot, C. (2010). HTLV-I p30: a versatile protein modulating virus replication and pathogenesis. *Mol. Aspects Med.* **31**, 344–349.

Ballabio, E., Mitchell, T., van Kester, M. S., Taylor, S., Dunlop, H. M., Chi, J., Tosi, I., Vermeer, M. H., Tramonti, D., Saunders, N. J., Boultwood, J., Wainscoat, J. S., et al. (2010). MicroRNA expression in Sezary syndrome: identification, function, and diagnostic potential. *Blood* **116**, 1105–1113.

Basso, K., Sumazin, P., Morozov, P., Schneider, C., Maute, R. L., Kitagawa, Y., Mandelbaum, J., Haddad, J., Jr., Chen, C. Z., Califano, A., and Dalla-Favera, R. (2009). Identification of the human mature B cell miRNome. *Immunity* **30**, 744–752.

Bellon, M., Lepelletier, Y., Hermine, O., and Nicot, C. (2009). Deregulation of microRNA involved in hematopoiesis and the immune response in HTLV-I adult T-cell leukemia. *Blood* **113**, 4914–4917.

Berezikov, E., van Tetering, G., Verheul, M., van de Belt, J., van Laake, L., Vos, J., Verloop, R., van de Wetering, M., Guryev, V., Takada, S., van Zonneveld, A. J., Mano, H., et al. (2006). Many novel mammalian microRNA candidates identified by extensive cloning and RAKE analysis. *Genome Res.* **16**, 1289–1298.

Berezikov, E., Chung, W. J., Willis, J., Cuppen, E., and Lai, E. C. (2007). Mammalian mirtron genes. *Mol. Cell* **28**, 328–336.

Blenkiron, C., Goldstein, L. D., Thorne, N. P., Spiteri, I., Chin, S. F., Dunning, M. J., Barbosa-Morais, N. L., Teschendorff, A. E., Green, A. R., Ellis, I. O., Tavare, S., Caldas, C., et al. (2007). MicroRNA expression profiling of human breast cancer identifies new markers of tumor subtype. *Genome Biol.* **8**, R214.

Borgdorff, V., Lleonart, M. E., Bishop, C. L., Fessart, D., Bergin, A. H., Overhoff, M. G., and Beach, D. H. (2010). Multiple microRNAs rescue from Ras-induced senescence by inhibiting p21(Waf1/Cip1). *Oncogene* **29**, 2262–2271.

Bouvard, V., Baan, R., Straif, K., Grosse, Y., Secretan, B., El Ghissassi, F., Benbrahim-Tallaa, L., Guha, N., Freeman, C., Galichet, L., and Cogliano, V. (2009). A review of human carcinogens—part B: biological agents. *Lancet Oncol.* **10**, 321–322.

Calin, G. A., Dumitru, C. D., Shimizu, M., Bichi, R., Zupo, S., Noch, E., Aldler, H., Rattan, S., Keating, M., Rai, K., Rassenti, L., Kipps, T., *et al.* (2002). Frequent deletions and downregulation of micro-RNA genes miR15 and miR16 at 13q14 in chronic lymphocytic leukemia. *Proc. Natl. Acad. Sci. USA* **99**, 15524–15529.

Calin, G. A., Sevignani, C., Dumitru, C. D., Hyslop, T., Noch, E., Yendamuri, S., Shimizu, M., Rattan, S., Bullrich, F., Negrini, M., and Croce, C. M. (2004). Human microRNA genes are frequently located at fragile sites and genomic regions involved in cancers. *Proc. Natl. Acad. Sci. USA* **101**, 2999–3004.

Cammarata, G., Augugliaro, L., Salemi, D., Agueli, C., La Rosa, M., Dagnino, L., Civiletto, G., Messana, F., Marfia, A., Bica, M. G., Cascio, L., Floridia, P. M., *et al.* (2010). Differential expression of specific microRNA and their targets in acute myeloid leukemia. *Am. J. Hematol.* **85**, 331–339.

Carissimi, C., Fulci, V., and Macino, G. (2009). MicroRNAs: novel regulators of immunity. *Autoimmun. Rev.* **8**, 520–524.

Carpenter, A. C., and Bosselut, R. (2010). Decision checkpoints in the thymus. *Nat. Immunol.* **11**, 666–673.

Cavallari, I., Rende, F., Donna, M., D'Agostino, D. M., and Ciminale, V. (2011). Converging strategies in expression of human complex retroviruses. *Viruses* **3**, 1395–1414.

Ceribelli, A., Yao, B., Dominguez-Gutierrez, P. R., Nahid, M. A., Satoh, M., and Chan, E. K. (2011). MicroRNAs in systemic rheumatic diseases. *Arthritis Res. Ther.* **13**, 229.

Chang, T. C., Yu, D., Lee, Y. S., Wentzel, E. A., Arking, D. E., West, K. M., Dang, C. V., Thomas-Tikhonenko, A., and Mendell, J. T. (2008). Widespread microRNA repression by Myc contributes to tumorigenesis. *Nat. Genet.* **40**, 43–50.

Cheloufi, S., Dos Santos, C. O., Chong, M. M., and Hannon, G. J. (2010). A dicer-independent miRNA biogenesis pathway that requires Ago catalysis. *Nature* **465**, 584–589.

Chen, C. Z., Li, L., Lodish, H. F., and Bartel, D. P. (2004). MicroRNAs modulate hematopoietic lineage differentiation. *Science* **303**, 83–86.

Chen, S., Ishii, N., Ine, S., Ikeda, S., Fujimura, T., Ndhlovu, L. C., Soroosh, P., Tada, K., Harigae, H., Kameoka, J., Kasai, N., Sasaki, T., *et al.* (2006). Regulatory T cell-like activity of Foxp3+ adult T cell leukemia cells. *Int. Immunol.* **18**, 269–277.

Chi, J., Ballabio, E., Chen, X. H., Kusec, R., Taylor, S., Hay, D., Tramonti, D., Saunders, N. J., Littlewood, T., Pezzella, F., Boultwood, J., Wainscoat, J. S., *et al.* (2011). MicroRNA expression in multiple myeloma is associated with genetic subtype, isotype and survival. *Biol. Direct* **6**, 23.

Ciafre, S. A., Galardi, S., Mangiola, A., Ferracin, M., Liu, C. G., Sabatino, G., Negrini, M., Maira, G., Croce, C. M., and Farace, M. G. (2005). Extensive modulation of a set of microRNAs in primary glioblastoma. *Biochem. Biophys. Res. Commun.* **334**, 1351–1358.

Cifuentes, D., Xue, H., Taylor, D. W., Patnode, H., Mishima, Y., Cheloufi, S., Ma, E., Mane, S., Hannon, G. J., Lawson, N. D., Wolfe, S. A., and Giraldez, A. J. (2010). A novel miRNA processing pathway independent of Dicer requires Argonaute2 catalytic activity. *Science* **328**, 1694–1698.

Cobb, B. S., Hertweck, A., Smith, J., O'Connor, E., Graf, D., Cook, T., Smale, S. T., Sakaguchi, S., Livesey, F. J., Fisher, A. G., and Merkenschlager, M. (2006). A role for Dicer in immune regulation. *J. Exp. Med.* **203**, 2519–2527.

Costinean, S., Zanesi, N., Pekarsky, Y., Tili, E., Volinia, S., Heerema, N., and Croce, C. M. (2006). Pre-B cell proliferation and lymphoblastic leukemia/high-grade lymphoma in E(mu)-miR155 transgenic mice. *Proc. Natl. Acad. Sci. USA* **103**, 7024–7029.

Croce, C. M. (2009). Causes and consequences of microRNA dysregulation in cancer. *Nat. Rev. Genet.* **10**, 704–714.

Cullen, B. R. (1991). Human immunodeficiency virus as a prototypic complex retrovirus. *J. Virol.* **65**, 1053–1056.

Curtale, G., Citarella, F., Carissimi, C., Goldoni, M., Carucci, N., Fulci, V., Franceschini, D., Meloni, F., Barnaba, V., and Macino, G. (2010). An emerging player in the adaptive immune response: microRNA-146a is a modulator of IL-2 expression and activation-induced cell death in T lymphocytes. *Blood* **115**, 265–273.

Debernardi, S., Skoulakis, S., Molloy, G., Chaplin, T., Dixon-McIver, A., and Young, B. D. (2007). MicroRNA miR-181a correlates with morphological sub-class of acute myeloid leukaemia and the expression of its target genes in global genome-wide analysis. *Leukemia* **21**, 912–916.

Di Lisio, L., Gomez-Lopez, G., Sanchez-Beato, M., Gomez-Abad, C., Rodriguez, M. E., Villuendas, R., Ferreira, B. I., Carro, A., Rico, D., Mollejo, M., Martinez, M. A., Menarguez, J., *et al.* (2010). Mantle cell lymphoma: transcriptional regulation by microRNAs. *Leukemia* **24**, 1335–1342.

Eis, P. S., Tam, W., Sun, L., Chadburn, A., Li, Z., Gomez, M. F., Lund, E., and Dahlberg, J. E. (2005). Accumulation of miR-155 and BIC RNA in human B cell lymphomas. *Proc. Natl. Acad. Sci. USA* **102**, 3627–3632.

Eulalio, A., Huntzinger, E., and Izaurralde, E. (2008). Getting to the root of miRNA-mediated gene silencing. *Cell* **132**, 9–14.

Fabian, M. R., Sonenberg, N., and Filipowicz, W. (2010). Regulation of mRNA translation and stability by microRNAs. *Annu. Rev. Biochem.* **79**, 351–379.

Faraoni, I., Antonetti, F. R., Cardone, J., and Bonmassar, E. (2009). miR-155 gene: a typical multifunctional microRNA. *Biochim. Biophys. Acta* **1792**, 497–505.

Fazi, F., Rosa, A., Fatica, A., Gelmetti, V., De Marchis, M. L., Nervi, C., and Bozzoni, I. (2005). A minicircuitry comprised of microRNA-223 and transcription factors NFI-A and C/EBPalpha regulates human granulopoiesis. *Cell* **123**, 819–831.

Fernandez-Capetillo, O., Lee, A., Nussenzweig, M., and Nussenzweig, A. (2004). H2AX: the histone guardian of the genome. *DNA Repair (Amst.)* **3**, 959–967.

Ferretti, E., De Smaele, E., Po, A., Di Marcotullio, L., Tosi, E., Espinola, M. S., Di Rocco, C., Riccardi, R., Giangaspero, F., Farcomeni, A., Nofroni, I., Laneve, P., *et al.* (2009). MicroRNA profiling in human medulloblastoma. *Int. J. Cancer* **124**, 568–577.

Feuer, G., and Green, P. L. (2005). Comparative biology of human T-cell lymphotropic virus type 1 (HTLV-1) and HTLV-2. *Oncogene* **24**, 5996–6004.

Flamant, S., Ritchie, W., Guilhot, J., Holst, J., Bonnet, M. L., Chomel, J. C., Guilhot, F., Turhan, A. G., and Rasko, J. E. (2010). Micro-RNA response to imatinib mesylate in patients with chronic myeloid leukemia. *Haematologica* **95**, 1325–1333.

Friedman, R. C., Farh, K. K., Burge, C. B., and Bartel, D. P. (2009). Most mammalian mRNAs are conserved targets of microRNAs. *Genome Res.* **19**, 92–105.

Fujita, S., Ito, T., Mizutani, T., Minoguchi, S., Yamamichi, N., Sakurai, K., and Iba, H. (2008). miR-21 gene expression triggered by AP-1 is sustained through a double-negative feedback mechanism. *J. Mol. Biol.* **378**, 492–504.

Fulci, V., Chiaretti, S., Goldoni, M., Azzalin, G., Carucci, N., Tavolaro, S., Castellano, L., Magrelli, A., Citarella, F., Messina, M., Maggio, R., Peragine, N., *et al.* (2007). Quantitative technologies establish a novel microRNA profile of chronic lymphocytic leukemia. *Blood* **109**, 4944–4951.

Garzon, R., Volinia, S., Liu, C. G., Fernandez-Cymering, C., Palumbo, T., Pichiorri, F., Fabbri, M., Coombes, K., Alder, H., Nakamura, T., Flomenberg, N., Marcucci, G., *et al.* (2008). MicroRNA signatures associated with cytogenetics and prognosis in acute myeloid leukemia. *Blood* **111**, 3183–3189.

Gatto, G., Rossi, A., Rossi, D., Kroening, S., Bonatti, S., and Mallardo, M. (2008). Epstein-Barr virus latent membrane protein 1 trans-activates miR-155 transcription through the NF-kappaB pathway. *Nucleic Acids Res.* **36**, 6608–6619.

Gauwerky, C. E., Huebner, K., Isobe, M., Nowell, P. C., and Croce, C. M. (1989). Activation of MYC in a masked t(8;17) translocation results in an aggressive B-cell leukemia. *Proc. Natl. Acad. Sci. USA* **86**, 8867–8871.

Gessain, A., Barin, F., Vernant, J. C., Gout, O., Maurs, L., Calender, A., and de The, G. (1985). Antibodies to human T-lymphotropic virus type-I in patients with tropical spastic paraparesis. *Lancet* **2**, 407–410.

Ghisi, M., Corradin, A., Basso, K., Frasson, C., Serafin, V., Mukherjee, S., Mussolin, L., Ruggero, K., Bonanno, L., Guffanti, A., De Bellis, G., Gerosa, G., *et al.* (2011). Modulation of microRNA expression in human T-cell development: targeting of NOTCH3 by miR-150. *Blood* **117**, 7053–7062.

Gironella, M., Seux, M., Xie, M. J., Cano, C., Tomasini, R., Gommeaux, J., Garcia, S., Nowak, J., Yeung, M. L., Jeang, K. T., Chaix, A., Fazli, L., *et al.* (2007). Tumor protein 53-induced nuclear protein 1 expression is repressed by miR-155, and its restoration inhibits pancreatic tumor development. *Proc. Natl. Acad. Sci. USA* **104**, 16170–16175.

Gottardo, F., Liu, C. G., Ferracin, M., Calin, G. A., Fassan, M., Bassi, P., Sevignani, C., Byrne, D., Negrini, M., Pagano, F., Gomella, L. G., Croce, C. M., *et al.* (2007). MicroRNA profiling in kidney and bladder cancers. *Urol. Oncol.* **25**, 387–392.

Gottwein, E., Mukherjee, N., Sachse, C., Frenzel, C., Majoros, W. H., Chi, J. T., Braich, R., Manoharan, M., Soutschek, J., Ohler, U., and Cullen, B. R. (2007). A viral microRNA functions as an orthologue of cellular miR-155. *Nature* **450**, 1096–1099.

Grassmann, R., Dengler, C., Muller-Fleckenstein, I., Fleckenstein, B., McGuire, K., Dokhelar, M. C., Sodroski, J. G., and Haseltine, W. A. (1989). Transformation to continuous growth of primary human T lymphocytes by human T-cell leukemia virus type I X-region genes transduced by a Herpesvirus saimiri vector. *Proc. Natl. Acad. Sci. USA* **86**, 3351–3355.

Greco, D., Kivi, N., Qian, K., Leivonen, S. K., Auvinen, P., and Auvinen, E. (2011). Human papillomavirus 16 E5 modulates the expression of host microRNAs. *PLoS One* **6**, e21646.

Griffiths-Jones, S., Saini, H. K., van Dongen, S., and Enright, A. J. (2008). miRBase: tools for microRNA genomics. *Nucleic Acids Res.* **36**, D154–D158.

Grigoryev, Y. A., Kurian, S. M., Hart, T., Nakorchevsky, A. A., Chen, C., Campbell, D., Head, S. R., Yates, J. R., 3rd, and Salomon, D. R. (2011). MicroRNA regulation of molecular networks mapped by global microRNA, mRNA, and protein expression in activated T lymphocytes. *J. Immunol.* **187**, 2233–2243.

Grimson, A., Farh, K. K., Johnston, W. K., Garrett-Engele, P., Lim, L. P., and Bartel, D. P. (2007). MicroRNA targeting specificity in mammals: determinants beyond seed pairing. *Mol. Cell* **27**, 91–105.

Guo, X., Wu, Y., and Hartley, R. S. (2009). MicroRNA-125a represses cell growth by targeting HuR in breast cancer. *RNA Biol.* **6**, 575–583.

Guo, S., Lu, J., Schlanger, R., Zhang, H., Wang, J. Y., Fox, M. C., Purton, L. E., Fleming, H. H., Cobb, B., Merkenschlager, M., Golub, T. R., and Scadden, D. T. (2010). MicroRNA miR-125a controls hematopoietic stem cell number. *Proc. Natl. Acad. Sci. USA* **107**, 14229–14234.

Gutierrez, N. C., Sarasquete, M. E., Misiewicz-Krzeminska, I., Delgado, M., De Las Rivas, J., Ticona, F. V., Ferminan, E., Martin-Jimenez, P., Chillon, C., Risueno, A., Hernandez, J. M., Garcia-Sanz, R., *et al.* (2010). Deregulation of microRNA expression in the different genetic subtypes of multiple myeloma and correlation with gene expression profiling. *Leukemia* **24**, 629–637.

Hall, W. W., and Fujii, M. (2005). Deregulation of cell-signaling pathways in HTLV-1 infection. *Oncogene* **24**, 5965–5975.

Hasegawa, H., Sawa, H., Lewis, M. J., Orba, Y., Sheehy, N., Yamamoto, Y., Ichinohe, T., Tsunetsugu-Yokota, Y., Katano, H., Takahashi, H., Matsuda, J., Sata, T., *et al.* (2006). Thymus-derived leukemia-lymphoma in mice transgenic for the Tax gene of human T-lymphotropic virus type I. *Nat. Med.* **12**, 466–472.

He, H., Jazdzewski, K., Li, W., Liyanarachchi, S., Nagy, R., Volinia, S., Calin, G. A., Liu, C. G., Franssila, K., Suster, S., Kloos, R. T., Croce, C. M., *et al.* (2005). The role of microRNA genes in papillary thyroid carcinoma. *Proc. Natl. Acad. Sci. USA* **102**, 19075–19080.

Hidaka, M., Inoue, J., Yoshida, M., and Seiki, M. (1988). Post-transcriptional regulator (rex) of HTLV-1 initiates expression of viral structural proteins but suppresses expression of regulatory proteins. *EMBO J.* 7, 519–523.

Hinuma, Y., Nagata, K., Hanaoka, M., Nakai, M., Matsumoto, T., Kinoshita, K. I., Shirakawa, S., and Miyoshi, I. (1981). Adult T-cell leukemia: antigen in an ATL cell line and detection of antibodies to the antigen in human sera. *Proc. Natl. Acad. Sci. USA* 78, 6476–6480.

Huang, B., Zhao, J., Lei, Z., Shen, S., Li, D., Shen, G. X., Zhang, G. M., and Feng, Z. H. (2009). miR-142-3p restricts cAMP production in CD4+CD25- T cells and CD4+CD25+ TREG cells by targeting AC9 mRNA. *EMBO Rep.* 10, 180–185.

Inoue, J., Seiki, M., and Yoshida, M. (1986). The second pX product p27 chi-III of HTLV-1 is required for gag gene expression. *FEBS Lett.* 209, 187–190.

Ivanovska, I., Ball, A. S., Diaz, R. L., Magnus, J. F., Kibukawa, M., Schelter, J. M., Kobayashi, S. V., Lim, L., Burchard, J., Jackson, A. L., Linsley, P. S., and Cleary, M. A. (2008). MicroRNAs in the miR-106b family regulate p21/CDKN1A and promote cell cycle progression. *Mol. Cell. Biol.* 28, 2167–2174.

Jangra, R. K., Yi, M., and Lemon, S. M. (2010). Regulation of hepatitis C virus translation and infectious virus production by the microRNA miR-122. *J. Virol.* 84, 6615–6625.

Jazbutyte, V., and Thum, T. (2010). MicroRNA-21: from cancer to cardiovascular disease. *Curr. Drug Targets* 11, 926–935.

Jazdzewski, K., Murray, E. L., Franssila, K., Jarzab, B., Schoenberg, D. R., and de la Chapelle, A. (2008). Common SNP in pre-miR-146a decreases mature miR expression and predisposes to papillary thyroid carcinoma. *Proc. Natl. Acad. Sci. USA* 105, 7269–7274.

Jiang, P. H., Motoo, Y., Garcia, S., Iovanna, J. L., Pebusque, M. J., and Sawabu, N. (2006). Down-expression of tumor protein p53-induced nuclear protein 1 in human gastric cancer. *World J. Gastroenterol.* 12, 691–696.

Johnnidis, J. B., Harris, M. H., Wheeler, R. T., Stehling-Sun, S., Lam, M. H., Kirak, O., Brummelkamp, T. R., Fleming, M. D., and Camargo, F. D. (2008). Regulation of progenitor cell proliferation and granulocyte function by microRNA-223. *Nature* 451, 1125–1129.

Kalyanaraman, V. S., Sarngadharan, M. G., Robert-Guroff, M., Miyoshi, I., Golde, D., and Gallo, R. C. (1982). A new subtype of human T-cell leukemia virus (HTLV-II) associated with a T-cell variant of hairy cell leukemia. *Science* 218, 571–573.

Kan, T., Sato, F., Ito, T., Matsumura, N., David, S., Cheng, Y., Agarwal, R., Paun, B. C., Jin, Z., Olaru, A. V., Selaru, F. M., Hamilton, J. P., et al. (2009). The miR-106b-25 polycistron, activated by genomic amplification, functions as an oncogene by suppressing p21 and Bim. *Gastroenterology* 136, 1689–1700.

Karube, K., Ohshima, K., Tsuchiya, T., Yamaguchi, T., Kawano, R., Suzumiya, J., Utsunomiya, A., Harada, M., and Kikuchi, M. (2004). Expression of FoxP3, a key molecule in CD4CD25 regulatory T cells, in adult T-cell leukaemia/lymphoma cells. *Br. J. Haematol.* 126, 81–84.

Kohno, T., Yamada, Y., Akamatsu, N., Kamihira, S., Imaizumi, Y., Tomonaga, M., and Matsuyama, T. (2005). Possible origin of adult T-cell leukemia/lymphoma cells from human T lymphotropic virus type-1-infected regulatory T cells. *Cancer Sci.* 96, 527–533.

Krammer, P. H., Arnold, R., and Lavrik, I. N. (2007). Life and death in peripheral T cells. *Nat. Rev. Immunol.* 7, 532–542.

Kress, A. K., Grassmann, R., and Fleckenstein, B. (2011). Cell surface markers in HTLV-1 pathogenesis. *Viruses* 3, 1439–1459.

Labbaye, C., Spinello, I., Quaranta, M. T., Pelosi, E., Pasquini, L., Petrucci, E., Biffoni, M., Nuzzolo, E. R., Billi, M., Foa, R., Brunetti, E., Grignani, F., et al. (2008). A three-step pathway comprising PLZF/miR-146a/CXCR4 controls megakaryopoiesis. *Nat. Cell Biol.* 10, 788–801.

LaGrenade, L., Hanchard, B., Fletcher, V., Cranston, B., and Blattner, W. (1990). Infective dermatitis of Jamaican children: a marker for HTLV-I infection. *Lancet* **336**, 1345–1347.

Laios, A., O'Toole, S., Flavin, R., Martin, C., Kelly, L., Ring, M., Finn, S. P., Barrett, C., Loda, M., Gleeson, N., D'Arcy, T., McGuinness, E., et al. (2008). Potential role of miR-9 and miR-223 in recurrent ovarian cancer. *Mol. Cancer* **7**, 35.

Lairmore, M. D., Anupam, R., Bowden, N., Haines, R., Haynes, R. A. H., Ratner, L., and Green, P. (2011). Molecular determinants of human T-lymphotropic virus type 1 transmission and spread. *Viruses* **3**, 1131–1165.

Lal, A., Navarro, F., Maher, C. A., Maliszewski, L. E., Yan, N., O'Day, E., Chowdhury, D., Dykxhoorn, D. M., Tsai, P., Hofmann, O., Becker, K. G., Gorospe, M., et al. (2009). miR-24 inhibits cell proliferation by targeting E2F2, MYC, and other cell-cycle genes via binding to "seedless" 3'UTR microRNA recognition elements. *Mol. Cell* **35**, 610–625.

Landgraf, P., Rusu, M., Sheridan, R., Sewer, A., Iovino, N., Aravin, A., Pfeffer, S., Rice, A., Kamphorst, A. O., Landthaler, M., Lin, C., Socci, N. D., et al. (2007). A mammalian microRNA expression atlas based on small RNA library sequencing. *Cell* **129**, 1401–1414.

Li, Q. J., Chau, J., Ebert, P. J., Sylvester, G., Min, H., Liu, G., Braich, R., Manoharan, M., Soutschek, J., Skare, P., Klein, L. O., Davis, M. M., et al. (2007). miR-181a is an intrinsic modulator of T cell sensitivity and selection. *Cell* **129**, 147–161.

Li, Y., Tan, W., Neo, T. W., Aung, M. O., Wasser, S., Lim, S. G., and Tan, T. M. (2009). Role of the miR-106b-25 microRNA cluster in hepatocellular carcinoma. *Cancer Sci.* **100**, 1234–1242.

Lin, Z., and Flemington, E. K. (2011). miRNAs in the pathogenesis of oncogenic human viruses. *Cancer Lett.* **305**, 186–199.

Liston, A., Lu, L. F., O'Carroll, D., Tarakhovsky, A., and Rudensky, A. Y. (2008). Dicer-dependent microRNA pathway safeguards regulatory T cell function. *J. Exp. Med.* **205**, 1993–2004.

Loffler, D., Brocke-Heidrich, K., Pfeifer, G., Stocsits, C., Hackermuller, J., Kretzschmar, A. K., Burger, R., Gramatzki, M., Blumert, C., Bauer, K., Cvijic, H., Ullmann, A. K., et al. (2007). Interleukin-6 dependent survival of multiple myeloma cells involves the Stat3-mediated induction of microRNA-21 through a highly conserved enhancer. *Blood* **110**, 1330–1333.

Lu, J., Getz, G., Miska, E. A., Alvarez-Saavedra, E., Lamb, J., Peck, D., Sweet-Cordero, A., Ebert, B. L., Mak, R. H., Ferrando, A. A., Downing, J. R., Jacks, T., et al. (2005). MicroRNA expression profiles classify human cancers. *Nature* **435**, 834–838.

Lu, F., Weidmer, A., Liu, C. G., Volinia, S., Croce, C. M., and Lieberman, P. M. (2008). Epstein-Barr virus-induced miR-155 attenuates NF-kappaB signaling and stabilizes latent virus persistence. *J. Virol.* **82**, 10436–10443.

Lu, L. F., Thai, T. H., Calado, D. P., Chaudhry, A., Kubo, M., Tanaka, K., Loeb, G. B., Lee, H., Yoshimura, A., Rajewsky, K., and Rudensky, A. Y. (2009). Foxp3-dependent microRNA155 confers competitive fitness to regulatory T cells by targeting SOCS1 protein. *Immunity* **30**, 80–91.

Lv, M., Zhang, X., Jia, H., Li, D., Zhang, B., Zhang, H., Hong, M., Jiang, T., Jiang, Q., Lu, J., Huang, X., and Huang, B. (2011). An oncogenic role of miR-142-3p in human T-cell acute lymphoblastic leukemia (T-ALL) by targeting glucocorticoid receptor-alpha and cAMP/PKA pathways. *Leukemia* .

Mahieux, R., and Gessain, A. (2009). The human HTLV-3 and HTLV-4 retroviruses: new members of the HTLV family. *Pathol. Biol. (Paris)* **57**, 161–166.

Malzkorn, B., Wolter, M., Liesenberg, F., Grzendowski, M., Stuhler, K., Meyer, H. E., and Reifenberger, G. (2009). Identification and functional characterization of microRNAs involved in the malignant progression of gliomas. *Brain Pathol.* **20**, 539–550.

Mathe, E. A., Nguyen, G. H., Bowman, E. D., Zhao, Y., Budhu, A., Schetter, A. J., Braun, R., Reimers, M., Kumamoto, K., Hughes, D., Altorki, N. K., Casson, A. G., et al. (2009). MicroRNA expression in squamous cell carcinoma and adenocarcinoma of the esophagus: associations with survival. *Clin. Cancer Res.* **15**, 6192–6200.

Matsubara, Y., Hori, T., Morita, R., Sakaguchi, S., and Uchiyama, T. (2005). Phenotypic and functional relationship between adult T-cell leukemia cells and regulatory T cells. *Leukemia* **19**, 482–483.

Matsuoka, M. (2010). HTLV-1 bZIP factor gene: its roles in HTLV-1 pathogenesis. *Mol. Aspects Med.* **31**, 359–366.

Matsuoka, M., and Green, P. L. (2009). The HBZ gene, a key player in HTLV-1 pathogenesis. *Retrovirology* **6**, 71.

Matsuoka, M., and Jeang, K. T. (2011). Human T-cell leukemia virus type 1 (HTLV-1) and leukemic transformation: viral infectivity, Tax, HBZ and therapy. *Oncogene* **30**, 1379–1389.

Medina, P. P., Nolde, M., and Slack, F. J. (2010). OncomiR addiction in an in vivo model of microRNA-21-induced pre-B-cell lymphoma. *Nature* **467**, 86–90.

Meenhuis, A., van Veelen, P. A., de Looper, H., van Boxtel, N., van den Berge, I. J., Sun, S. M., Taskesen, E., Stern, P., de Ru, A. H., van Adrichem, A. J., Demmers, J., Jongen-Lavrencic, M., *et al.* (2011). MiR-17/20/93/106 promote hematopoietic cell expansion by targeting sequestosome 1-regulated pathways in mice. *Blood* **118**, 916–925.

Mendell, J. T. (2008). miRiad roles for the miR-17-92 cluster in development and disease. *Cell* **133**, 217–222.

Merkel, O., Hamacher, F., Laimer, D., Sifft, E., Trajanoski, Z., Scheideler, M., Egger, G., Hassler, M. R., Thallinger, C., Schmatz, A., Turner, S. D., Greil, R., *et al.* (2010). Identification of differential and functionally active miRNAs in both anaplastic lymphoma kinase (ALK)+ and ALK− anaplastic large-cell lymphoma. *Proc. Natl. Acad. Sci. USA* **107**, 16228–16233.

Merkerova, M., Belickova, M., and Bruchova, H. (2008). Differential expression of microRNAs in hematopoietic cell lineages. *Eur. J. Haematol.* **81**, 304–310.

Metzler, M., Wilda, M., Busch, K., Viehmann, S., and Borkhardt, A. (2004). High expression of precursor microRNA-155/BIC RNA in children with Burkitt lymphoma. *Genes Chromosomes Cancer* **39**, 167–169.

Miyoshi, K., Miyoshi, T., and Siomi, H. (2010). Many ways to generate microRNA-like small RNAs: non-canonical pathways for microRNA production. *Mol. Genet. Genomics* **284**, 95–103.

Mochizuki, M., Watanabe, T., Yamaguchi, K., Takatsuki, K., Yoshimura, K., Shirao, M., Nakashima, S., Mori, S., Araki, S., and Miyata, N. (1992). HTLV-I uveitis: a distinct clinical entity caused by HTLV-I. *Jpn. J. Cancer Res.* **83**, 236–239.

Morin, R. D., O'Connor, M. D., Griffith, M., Kuchenbauer, F., Delaney, A., Prabhu, A. L., Zhao, Y., McDonald, H., Zeng, T., Hirst, M., Eaves, C. J., and Marra, M. A. (2008). Application of massively parallel sequencing to microRNA profiling and discovery in human embryonic stem cells. *Genome Res.* **18**, 610–621.

Motsch, N., Pfuhl, T., Mrazek, J., Barth, S., and Grasser, F. A. (2007). Epstein-Barr virus-encoded latent membrane protein 1 (LMP1) induces the expression of the cellular microRNA miR-146a. *RNA Biol.* **4**, 131–137.

Muljo, S. A., Ansel, K. M., Kanellopoulou, C., Livingston, D. M., Rao, A., and Rajewsky, K. (2005). Aberrant T cell differentiation in the absence of Dicer. *J. Exp. Med.* **202**, 261–269.

Nam, E. J., Yoon, H., Kim, S. W., Kim, H., Kim, Y. T., Kim, J. H., Kim, J. W., and Kim, S. (2008). MicroRNA expression profiles in serous ovarian carcinoma. *Clin. Cancer Res.* **14**, 2690–2695.

Narducci, M. G., Arcelli, D., Picchio, M. C., Lazzeri, C., Pagani, E., Sampogna, F., Scala, E., Fadda, P., Cristofoletti, C., Facchiano, A., Frontani, M., Monopoli, A., *et al.* (2011). MicroRNA profiling reveals that miR-21, miR486 and miR-214 are upregulated and involved in cell survival in Sezary syndrome. *Cell Death Dis.* **2**, e151.

Navarro, A., Bea, S., Fernandez, V., Prieto, M., Salaverria, I., Jares, P., Hartmann, E., Mozos, A., Lopez-Guillermo, A., Villamor, N., Colomer, D., Puig, X., *et al.* (2009). MicroRNA expression, chromosomal alterations, and immunoglobulin variable heavy chain hypermutations in Mantle cell lymphomas. *Cancer Res.* **69**, 7071–7078.

Nejmeddine, M., Negi, V. S., Mukherjee, S., Tanaka, Y., Orth, K., Taylor, G. P., and Bangham, C. R. (2009). HTLV-1-Tax and ICAM-1 act on T-cell signal pathways to polarize the microtubule-organizing center at the virological synapse. *Blood* **114**, 1016–1025.

Nishida, N., Mimori, K., Fabbri, M., Yokobori, T., Sudo, T., Tanaka, F., Shibata, K., Ishii, H., Doki, Y., and Mori, M. (2011). MicroRNA-125a-5p is an independent prognostic factor in gastric cancer and inhibits the proliferation of human gastric cancer cells in combination with trastuzumab. *Clin. Cancer Res.* **17**, 2725–2733.

Nyborg, J. K., Egan, D., and Sharma, N. (2010). The HTLV-1 Tax protein: revealing mechanisms of transcriptional activation through histone acetylation and nucleosome disassembly. *Biochim. Biophys. Acta* **1799**, 266–274.

O'Connell, R. M., Rao, D. S., Chaudhuri, A. A., Boldin, M. P., Taganov, K. D., Nicoll, J., Paquette, R. L., and Baltimore, D. (2008). Sustained expression of microRNA-155 in hematopoietic stem cells causes a myeloproliferative disorder. *J. Exp. Med.* **205**, 585–594.

Pallasch, C. P., Patz, M., Park, Y. J., Hagist, S., Eggle, D., Claus, R., Debey-Pascher, S., Schulz, A., Frenzel, L. P., Claasen, J., Kutsch, N., Krause, G., *et al.* (2009). miRNA deregulation by epigenetic silencing disrupts suppression of the oncogene PLAG1 in chronic lymphocytic leukemia. *Blood* **114**, 3255–3264.

Pancewicz, J., Taylor, J. M., Datta, A., Baydoun, H. H., Waldmann, T. A., Hermine, O., and Nicot, C. (2010). Notch signaling contributes to proliferation and tumor formation of human T-cell leukemia virus type 1-associated adult T-cell leukemia. *Proc. Natl. Acad. Sci. USA* **107**, 16619–16624.

Petrocca, F., Vecchione, A., and Croce, C. M. (2008a). Emerging role of miR-106b-25/miR-17-92 clusters in the control of transforming growth factor beta signaling. *Cancer Res.* **68**, 8191–8194.

Petrocca, F., Visone, R., Onelli, M. R., Shah, M. H., Nicoloso, M. S., de Martino, I., Iliopoulos, D., Pilozzi, E., Liu, C. G., Negrini, M., Cavazzini, L., Volinia, S., *et al.* (2008b). E2F1-regulated microRNAs impair TGFbeta-dependent cell-cycle arrest and apoptosis in gastric cancer. *Cancer Cell* **13**, 272–286.

Pichiorri, F., Suh, S. S., Ladetto, M., Kuehl, M., Palumbo, T., Drandi, D., Taccioli, C., Zanesi, N., Alder, H., Hagan, J. P., Munker, R., Volinia, S., *et al.* (2008). MicroRNAs regulate critical genes associated with multiple myeloma pathogenesis. *Proc. Natl. Acad. Sci. USA* **105**, 12885–12890.

Pichler, K., Schneider, G., and Grassmann, R. (2008). MicroRNA miR-146a and further oncogenesis-related cellular microRNAs are dysregulated in HTLV-1-transformed T lymphocytes. *Retrovirology* **5**, 100.

Poiesz, B. J., Ruscetti, F. W., Gazdar, A. F., Bunn, P. A., Minna, J. D., and Gallo, R. C. (1980). Detection and isolation of type C retrovirus particles from fresh and cultured lymphocytes of a patient with cutaneous T-cell lymphoma. *Proc. Natl. Acad. Sci. USA* **77**, 7415–7419.

Proietti, F. A., Carneiro-Proietti, A. B., Catalan-Soares, B. C., and Murphy, E. L. (2005). Global epidemiology of HTLV-I infection and associated diseases. *Oncogene* **24**, 6058–6068.

Punj, V., Matta, H., Schamus, S., Tamewitz, A., Anyang, B., and Chaudhary, P. M. (2010). Kaposi's sarcoma-associated herpesvirus-encoded viral FLICE inhibitory protein (vFLIP) K13 suppresses CXCR4 expression by upregulating miR-146a. *Oncogene* **29**, 1835–1844.

Qian, L., Van Laake, L. W., Huang, Y., Liu, S., Wendland, M. F., and Srivastava, D. (2011). miR-24 inhibits apoptosis and represses Bim in mouse cardiomyocytes. *J. Exp. Med.* **208**, 549–560.

Qin, W., Shi, Y., Zhao, B., Yao, C., Jin, L., Ma, J., and Jin, Y. (2010). miR-24 regulates apoptosis by targeting the open reading frame (ORF) region of FAF1 in cancer cells. *PLoS One* **5**, e9429.

Rahadiani, N., Takakuwa, T., Tresnasari, K., Morii, E., and Aozasa, K. (2008). Latent membrane protein-1 of Epstein-Barr virus induces the expression of B-cell integration cluster, a precursor form of microRNA-155, in B lymphoma cell lines. *Biochem. Biophys. Res. Commun.* **377**, 579–583.

Ramkissoon, S. H., Mainwaring, L. A., Ogasawara, Y., Keyvanfar, K., McCoy, J. P., Jr., Sloand, E. M., Kajigaya, S., and Young, N. S. (2006). Hematopoietic-specific microRNA expression in human cells. *Leuk. Res.* **30,** 643–647.

Rende, F., Cavallari, I., Corradin, A., Silic-Benussi, M., Toulza, F., Toffolo, G. M., Tanaka, Y., Jacobson, S., Taylor, G. P., D'Agostino, D. M., Bangham, C. R., and Ciminale, V. (2011). Kinetics and intracellular compartmentalization of HTLV-1 gene expression: nuclear retention of HBZ mRNA. *Blood* **117,** 4855–4859.

Robek, M. D., and Ratner, L. (1999). Immortalization of CD4(+) and CD8(+) T lymphocytes by human T-cell leukemia virus type 1 Tax mutants expressed in a functional molecular clone. *J. Virol.* **73,** 4856–4865.

Rodriguez, A., Vigorito, E., Clare, S., Warren, M. V., Couttet, P., Soond, D. R., van Dongen, S., Grocock, R. J., Das, P. P., Miska, E. A., Vetrie, D., Okkenhaug, K., *et al.* (2007). Requirement of bic/microRNA-155 for normal immune function. *Science* **316,** 608–611.

Roehle, A., Hoefig, K. P., Repsilber, D., Thorns, C., Ziepert, M., Wesche, K. O., Thiere, M., Loeffler, M., Klapper, W., Pfreundschuh, M., Matolcsy, A., Bernd, H. W., *et al.* (2008). MicroRNA signatures characterize diffuse large B-cell lymphomas and follicular lymphomas. *Br. J. Haematol.* **142,** 732–744.

Roncador, G., Garcia, J. F., Garcia, J. F., Maestre, L., Lucas, E., Menarguez, J., Ohshima, K., Nakamura, S., Banham, A. H., and Piris, M. A. (2005). FOXP3, a selective marker for a subset of adult T-cell leukaemia/lymphoma. *Leukemia* **19,** 2247–2253.

Rossi, R. L., Rossetti, G., Wenandy, L., Curti, S., Ripamonti, A., Bonnal, R. J., Birolo, R. S., Moro, M., Crosti, M. C., Gruarin, P., Maglie, S., Marabita, F., *et al.* (2011). Distinct microRNA signatures in human lymphocyte subsets and enforcement of the naive state in CD4+ T cells by the microRNA miR-125b. *Nat. Immunol.* **12,** 796–803.

Rouas, R., Fayyad-Kazan, H., El Zein, N., Lewalle, P., Rothe, F., Simion, A., Akl, H., Mourtada, M., El Rifai, M., Burny, A., Romero, P., Martiat, P., *et al.* (2009). Human natural Treg microRNA signature: role of microRNA-31 and microRNA-21 in FOXP3 expression. *Eur. J. Immunol.* **39,** 1608–1618.

Saitoh, Y., Yamamoto, N., Dewan, M. Z., Sugimoto, H., Martinez Bruyn, V. J., Iwasaki, Y., Matsubara, K., Qi, X., Saitoh, T., Imoto, I., Inazawa, J., Utsunomiya, A., *et al.* (2008). Overexpressed NF-kappaB-inducing kinase contributes to the tumorigenesis of adult T-cell leukemia and Hodgkin Reed-Sternberg cells. *Blood* **111,** 5118–5129.

Sakaguchi, S., Miyara, M., Costantino, C. M., and Hafler, D. A. (2010). FOXP3+ regulatory T cells in the human immune system. *Nat. Rev. Immunol.* **10,** 490–500.

Sasaki, D., Imaizumi, Y., Hasegawa, H., Osaka, A., Tsukasaki, K., Choi, Y. L., Mano, H., Marquez, V. E., Hayashi, T., Yanagihara, K., Moriwaki, Y., Miyazaki, Y., *et al.* (2011). Overexpression of Enhancer of zeste homolog 2 with trimethylation of lysine 27 on histone H3 in adult T-cell leukemia/lymphoma as a target for epigenetic therapy. *Haematologica* **96,** 712–719.

Satou, Y., Yasunaga, J., Zhao, T., Yoshida, M., Miyazato, P., Takai, K., Shimizu, K., Ohshima, K., Green, P. L., Ohkura, N., Yamaguchi, T., Ono, M., *et al.* (2011). HTLV-1 bZIP factor induces T-cell lymphoma and systemic inflammation in vivo. *PLoS Pathog.* **7,** e1001274.

Sayed, D., and Abdellatif, M. (2011). MicroRNAs in development and disease. *Physiol. Rev.* **91,** 827–887.

Schotte, D., Chau, J. C., Sylvester, G., Liu, G., Chen, C., van der Velden, V. H., Broekhuis, M. J., Peters, T. C., Pieters, R., and den Boer, M. L. (2009). Identification of new microRNA genes and aberrant microRNA profiles in childhood acute lymphoblastic leukemia. *Leukemia* **23,** 313–322.

Seiki, M., Hattori, S., Hirayama, Y., and Yoshida, M. (1983). Human adult T-cell leukemia virus: complete nucleotide sequence of the provirus genome integrated in leukemia cell DNA. *Proc. Natl. Acad. Sci. USA* **80,** 3618–3622.

Shi, L., Cheng, Z., Zhang, J., Li, R., Zhao, P., Fu, Z., and You, Y. (2008). hsa-mir-181a and hsa-mir-181b function as tumor suppressors in human glioma cells. *Brain Res.* **1236**, 185–193.

Silic-Benussi, M., Biasiotto, R., Andresen, V., Franchini, G., D'Agostino, D. M., and Ciminale, V. (2010). HTLV-1 p13, a small protein with a busy agenda. *Mol. Aspects Med.* **31**, 350–358.

Simon, J. A., and Kingston, R. E. (2009). Mechanisms of polycomb gene silencing: knowns and unknowns. *Nat. Rev. Mol. Cell Biol.* **10**, 697–708.

Skalsky, R. L., Samols, M. A., Plaisance, K. B., Boss, I. W., Riva, A., Lopez, M. C., Baker, H. V., and Renne, R. (2007). Kaposi's sarcoma-associated herpesvirus encodes an ortholog of miR-155. *J. Virol.* **81**, 12836–12845.

Stahl, H. F., Fauti, T., Ullrich, N., Bopp, T., Kubach, J., Rust, W., Labhart, P., Alexiadis, V., Becker, C., Hafner, M., Weith, A., Lenter, M. C., *et al.* (2009). miR-155 inhibition sensitizes CD4+ Th cells for TREG mediated suppression. *PLoS One* **4**, e7158.

Stern-Ginossar, N., Gur, C., Biton, M., Horwitz, E., Elboim, M., Stanietsky, N., Mandelboim, M., and Mandelboim, O. (2008). Human microRNAs regulate stress-induced immune responses mediated by the receptor NKG2D. *Nat. Immunol.* **9**, 1065–1073.

Strebel, K., Luban, J., and Jeang, K. T. (2009). Human cellular restriction factors that target HIV-1 replication. *BMC Med.* **7**, 48.

Sultana, D. A., Bell, J. J., Zlotoff, D. A., De Obaldia, M. E., and Bhandoola, A. (2010). Eliciting the T cell fate with Notch. *Semin. Immunol.* **22**, 254–260.

Tabakin-Fix, Y., Azran, I., Schavinky-Khrapunsky, Y., Levy, O., and Aboud, M. (2005). Functional inactivation of p53 by human T-cell leukemia virus type 1 Tax protein: mechanisms and clinical implications. *Carcinogenesis* **27**, 673–681.

Taganov, K. D., Boldin, M. P., Chang, K. J., and Baltimore, D. (2006). NF-kappaB-dependent induction of microRNA miR-146, an inhibitor targeted to signaling proteins of innate immune responses. *Proc. Natl. Acad. Sci. USA* **103**, 12481–12486.

Tang, Y., Luo, X., Cui, H., Ni, X., Yuan, M., Guo, Y., Huang, X., Zhou, H., de Vries, N., Tak, P. P., Chen, S., and Shen, N. (2009). MicroRNA-146A contributes to abnormal activation of the type I interferon pathway in human lupus by targeting the key signaling proteins. *Arthritis Rheum.* **60**, 1065–1075.

Thai, T. H., Calado, D. P., Casola, S., Ansel, K. M., Xiao, C., Xue, Y., Murphy, A., Frendewey, D., Valenzuela, D., Kutok, J. L., Schmidt-Supprian, M., Rajewsky, N., *et al.* (2007). Regulation of the germinal center response by microRNA-155. *Science* **316**, 604–608.

Tomasini, R., Samir, A. A., Carrier, A., Isnardon, D., Cecchinelli, B., Soddu, S., Malissen, B., Dagorn, J. C., Iovanna, J. L., and Dusetti, N. J. (2003). TP53INP1s and homeodomain-interacting protein kinase-2 (HIPK2) are partners in regulating p53 activity. *J. Biol. Chem.* **278**, 37722–37729.

Tomasini, R., Seux, M., Nowak, J., Bontemps, C., Carrier, A., Dagorn, J. C., Pebusque, M. J., Iovanna, J. L., and Dusetti, N. J. (2005). TP53INP1 is a novel p73 target gene that induces cell cycle arrest and cell death by modulating p73 transcriptional activity. *Oncogene* **24**, 8093–8104.

Tomita, M., Tanaka, Y., and Mori, N. (2009). MicroRNA miR-146a is induced by HTLV-1 Tax and increases the growth of HTLV-1-infected T-cells. *Int. J. Cancer.*

Toulza, F., Nosaka, K., Takiguchi, M., Pagliuca, T., Mitsuya, H., Tanaka, Y., Taylor, G. P., and Bangham, C. R. (2009). FoxP3+ regulatory T cells are distinct from leukemia cells in HTLV-1-associated adult T-cell leukemia. *Int. J. Cancer* **125**, 2375–2382.

Toulza, F., Nosaka, K., Tanaka, Y., Schioppa, T., Balkwill, F., Taylor, G. P., and Bangham, C. R. (2010). Human T-lymphotropic virus type 1-induced CC chemokine ligand 22 maintains a high frequency of functional FoxP3+ regulatory T cells. *J. Immunol.* **185**, 183–189.

Uchiyama, T., Yodoi, J., Sagawa, K., Takatsuki, K., and Uchino, H. (1977). Adult T-cell leukemia: clinical and hematologic features of 16 cases. *Blood* **50**, 481–492.

Umbach, J. L., and Cullen, B. R. (2009). The role of RNAi and microRNAs in animal virus replication and antiviral immunity. *Genes Dev.* **23**, 1151–1164.

Valastyan, S., Reinhardt, F., Benaich, N., Calogrias, D., Szasz, A. M., Wang, Z. C., Brock, J. E., Richardson, A. L., and Weinberg, R. A. (2009). A pleiotropically acting microRNA, miR-31, inhibits breast cancer metastasis. *Cell* **137**, 1032–1046.

van der Fits, L., van Kester, M. S., Qin, Y., Out-Luiting, J. J., Smit, F., Zoutman, W. H., Willemze, R., Tensen, C. P., and Vermeer, M. H. (2011). MicroRNA-21 expression in CD4 + T cells is regulated by STAT3 and is pathologically involved in Sezary syndrome. *J. Invest. Dermatol.* **131**, 762–768.

van Kester, M. S., Ballabio, E., Benner, M. F., Chen, X. H., Saunders, N. J., van der Fits, L., van Doorn, R., Vermeer, M. H., Willemze, R., Tensen, C. P., and Lawrie, C. H. (2011). miRNA expression profiling of mycosis fungoides. *Mol. Oncol.* **5**, 273–280.

Van Prooyen, N., Andresen, V., Gold, H., Bialuk, I., Pise-Masison, C., and Franchini, G. (2010). Hijacking the T-cell communication network by the human T-cell leukemia/lymphoma virus type 1 (HTLV-1) p12 and p8 proteins. *Mol. Aspects Med.* **31**, 333–343.

Verdonck, K., Gonzalez, E., Van Dooren, S., Vandamme, A. M., Vanham, G., and Gotuzzo, E. (2007). Human T-lymphotropic virus 1: recent knowledge about an ancient infection. *Lancet Infect. Dis.* **7**, 266–281.

Volinia, S., Calin, G. A., Liu, C. G., Ambs, S., Cimmino, A., Petrocca, F., Visone, R., Iorio, M., Roldo, C., Ferracin, M., Prueitt, R. L., Yanaihara, N., et al. (2006). A microRNA expression signature of human solid tumors defines cancer gene targets. *Proc. Natl. Acad. Sci. USA* **103**, 2257–2261.

Wang, J., Ioan-Facsinay, A., van der Voort, E. I., Huizinga, T. W., and Toes, R. E. (2007). Transient expression of FOXP3 in human activated nonregulatory CD4+ T cells. *Eur. J. Immunol.* **37**, 129–138.

Watanabe, T., Seiki, M., and Yoshida, M. (1984). HTLV type I (U. S. isolate) and ATLV (Japanese isolate) are the same species of human retrovirus. *Virology* **133**, 238–241.

Watanabe, M., Ohsugi, T., Shoda, M., Ishida, T., Aizawa, S., Maruyama-Nagai, M., Utsunomiya, A., Koga, S., Yamada, Y., Kamihira, S., Okayama, A., Kikuchi, H., et al. (2005). Dual targeting of transformed and untransformed HTLV-1-infected T cells by DHMEQ, a potent and selective inhibitor of NF-kappaB, as a strategy for chemoprevention and therapy of adult T-cell leukemia. *Blood* **106**, 2462–2471.

Watanabe, A., Tagawa, H., Yamashita, J., Teshima, K., Nara, M., Iwamoto, K., Kume, M., Kameoka, Y., Takahashi, N., Nakagawa, T., Shimizu, N., and Sawada, K. (2011). The role of microRNA-150 as a tumor suppressor in malignant lymphoma. *Leukemia* **25**, 1324–1334.

Wong, Q. W., Lung, R. W., Law, P. T., Lai, P. B., Chan, K. Y., To, K. F., and Wong, N. (2008). MicroRNA-223 is commonly repressed in hepatocellular carcinoma and potentiates expression of Stathmin1. *Gastroenterology* **135**, 257–269.

Yamada, Y., and Kamihira, S. (2005). Inactivation of tumor suppressor genes and the progression of adult T-cell leukemia-lymphoma. *Leuk. Lymphoma* **46**, 1553–1559.

Yamagishi, M., Nakano, K., Miyake, A., Yamochi, T., Kagami, Y., Tsutsumi, A., Matsuda, Y., Matsubara, A., Muto, S., Utsunomiya, A., Yamaguchi, K., Uchimaru, K., et al. (2012). Polycomb-mediated loss of miR-31 activates NIK-dependent NF-κB pathway in adult T-cell leukemia and other cancers. *Cancer Cell* **21**, 121–135.

Yamanaka, Y., Tagawa, H., Takahashi, N., Watanabe, A., Guo, Y. M., Iwamoto, K., Yamashita, J., Saitoh, H., Kameoka, Y., Shimizu, N., Ichinohasama, R., and Sawada, K. (2009). Aberrant overexpression of microRNAs activate AKT signaling via down-regulation of tumor suppressors in natural killer-cell lymphoma/leukemia. *Blood* **114**, 3265–3275.

Yanaihara, N., Caplen, N., Bowman, E., Seike, M., Kumamoto, K., Yi, M., Stephens, R. M., Okamoto, A., Yokota, J., Tanaka, T., Calin, G. A., Liu, C. G., *et al.* (2006). Unique microRNA molecular profiles in lung cancer diagnosis and prognosis. *Cancer Cell* **9**, 189–198.

Yano, H., Ishida, T., Inagaki, A., Ishii, T., Kusumoto, S., Komatsu, H., Iida, S., Utsunomiya, A., and Ueda, R. (2007). Regulatory T-cell function of adult T-cell leukemia/lymphoma cells. *Int. J. Cancer* **120**, 2052–2057.

Yeung, M. L., Yasunaga, J., Bennasser, Y., Dusetti, N., Harris, D., Ahmad, N., Matsuoka, M., and Jeang, K. T. (2008). Roles for microRNAs, miR-93 and miR-130b, and tumor protein 53-induced nuclear protein 1 tumor suppressor in cell growth dysregulation by human T-cell lymphotrophic virus 1. *Cancer Res.* **68**, 8976–8985.

Yi, R., Qin, Y., Macara, I. G., and Cullen, B. R. (2003). Exportin-5 mediates the nuclear export of pre-microRNAs and short hairpin RNAs. *Genes Dev.* **17**, 3011–3016.

Yoshida, M., Miyoshi, I., and Hinuma, Y. (1982). Isolation and characterization of retrovirus from cell lines of human adult T-cell leukemia and its implication in the disease. *Proc. Natl. Acad. Sci. USA* **79**, 2031–2035.

Zhang, H., Luo, X. Q., Zhang, P., Huang, L. B., Zheng, Y. S., Wu, J., Zhou, H., Qu, L. H., Xu, L., and Chen, Y. Q. (2009). MicroRNA patterns associated with clinical prognostic parameters and CNS relapse prediction in pediatric acute leukemia. *PLoS One* **4**, e7826.

Zhao, J. J., Lin, J., Lwin, T., Yang, H., Guo, J., Kong, W., Dessureault, S., Moscinski, L. C., Rezania, D., Dalton, W. S., Sotomayor, E., Tao, J., *et al.* (2010). MicroRNA expression profile and identification of miR-29 as a prognostic marker and pathogenetic factor by targeting CDK6 in mantle cell lymphoma. *Blood* **115**, 2630–2639.

Zhou, X., Jeker, L. T., Fife, B. T., Zhu, S., Anderson, M. S., McManus, M. T., and Bluestone, J. A. (2008). Selective miRNA disruption in T reg cells leads to uncontrolled autoimmunity. *J. Exp. Med.* **205**, 1983–1991.

Zhou, R., Hu, G., Liu, J., Gong, A. Y., Drescher, K. M., and Chen, X. M. (2009). NF-kappaB p65-dependent transactivation of miRNA genes following Cryptosporidium parvum infection stimulates epithelial cell immune responses. *PLoS Pathog.* **5**, e1000681.

Zhou, Y., Chen, L., Barlogie, B., Stephens, O., Wu, X., Williams, D. R., Cartron, M. A., van Rhee, F., Nair, B., Waheed, S., Pineda-Roman, M., Alsayed, Y., *et al.* (2010). High-risk myeloma is associated with global elevation of miRNAs and overexpression of EIF2C2/AGO2. *Proc. Natl. Acad. Sci. USA* **107**, 7904–7909.

The Multifaceted Oncoprotein Tax: Subcellular Localization, Posttranslational Modifications, and NF-κB Activation

Youmna Kfoury,* Rihab Nasr,[†] Chloé Journo,[‡,§] Renaud Mahieux,[‡,§,1] Claudine Pique,[¶,1] and Ali Bazarbachi*

*Department of Internal Medicine, Faculty of Medicine, American University of Beirut, Beirut, Lebanon
[†]Department of Anatomy, Cell Biology and Physiological Sciences, Faculty of Medicine, American University of Beirut, Beirut, Lebanon
[‡]Retroviral Oncogenesis Laboratory, INSERM-U758 Human Virology, Lyon Cedex 07, France
[§]Ecole Normale Supérieure de Lyon, Lyon Cedex 07, France
[¶]INSERM-U1016, CNRS UMR8104, Université Paris Descartes, Paris, France

I. The Oncogenic Retrovirus HTLV-I
II. The Viral Oncoprotein Tax: Structural and Functional Domains
III. Tax: A Potent Transactivator and a Deregulator of the Cellular Machinery
IV. The NF-κB Pathway: Generalities
V. Tax: A Powerful Activator of the NF-κB Pathway
VI. Tax Posttranslational Modifications and NF-κB Activation
   A. Tax Phosphorylation
   B. Tax Acetylation
   C. Tax Ubiquitination
   D. Tax SUMOylation
VII. Tax Posttranslational Modifications and Intracellular Localization
VIII. Conclusion
   Acknowledgment
   References

The human T-cell lymphotropic virus type-I (HTLV-I) is the etiologic agent of adult T-cell leukemia/lymphoma (ATL) and of tropical spastic paraparesis/HTLV-I-associated myelopathy. Constitutive NF-κB activation by the viral oncoprotein Tax plays a crucial role in the induction and maintenance of cellular proliferation, transformation, and inhibition of apoptosis. In an attempt to provide a general view of the molecular mechanisms of constitutive Tax-induced NF-κB activation, we summarize in this review the recent body of literature that supports a major role for Tax posttranslational modifications, chiefly ubiquitination, and SUMOylation, in the NF-κB activity of Tax.

---

[1] Equal contribution.

These modifications indeed participate in the control of Tax subcellular localization and modulate its protein–protein interaction potential. Tax posttranslational modifications, which highlight the ability of HTLV-I to optimize its limited viral genome size, might represent an attractive target for the design of new therapies for ATL. © 2012 Elsevier Inc.

I. THE ONCOGENIC RETROVIRUS HTLV-I

Discovered in the early 1980s by two-independent groups in the United States and Japan, HTLV-I was the first human retrovirus to be associated with a disease and remains the only one directly associated with a malignancy (Poiesz et al., 1980; Yoshida et al., 1982). Along with other HTLVs that were discovered later (HTLV-II, -III, and -IV; Calattini et al., 2005; Kalyanaraman et al., 1982; Mahieux and Gessain, 2009; Wolfe et al., 2005), the virus is a member of the *deltaretrovirus* family.

It is estimated that around 15–20 million people are infected with the virus worldwide. These populations are distributed among regions of high endemicity, that is, southwestern Japan, intertropical Africa, the Caribbean, and South America. Other foci of the infection such as Melanesia, Romania, and areas of the Middle East, mainly in the area of Mashhad in Iran have also been described (Abbaszadegan et al., 2003; Meytes et al., 1990; Proietti et al., 2005). The virus has three major routes of transmission: (1) mother to child (mainly through prolonged breast-feeding), (2) sexual contact (chiefly from male to female), and (3) parenteral transmission (Proietti et al., 2005).

While HTLV-I can infect several different cell types of various origins *in vitro* (Feuer et al., 1996; Lo et al., 1992), it was long thought to preferentially infect $CD4^+$ T lymphocytes *in vivo*. However, a series of data clearly indicate that $CD8^+$ T cells and B cells might constitute an additional reservoir for the virus *in vivo* (Franchini et al., 1985; Hanon et al., 2000; Koyanagi et al., 1993; Makino et al., 1999; Nagai et al., 2001; Richardson et al., 1990). Cell-to-cell contact is a must for viral transmission to uninfected lymphocytes although recent findings indicate that free viral particles can infect dendritic cells (DCs); (Jones et al., 2008). HTLV-I entry involves three cell-surface proteins that appear to function in concert: heparan sulfate proteoglycans, Neuropiline-1, and GLUcose transporter-1 (Ghez et al., 2006, 2010; Jones et al., 2005; Lambert et al., 2009; Manel et al., 2003). DC-Specific adhesion molecule-3 ICAM-3 Grabbing Nonintegrin (DC-SIGN) has also been recently shown to mediate HTLV-I entry in DC (Jain et al., 2009).

Like all retroviruses, HTLV-I provirus is flanked by a long terminal repeat sequence at both the 5′ and 3′ ends. Both sequences contain promoters that drive the transcription of all the viral genes. HTLV-I genome contains three ORFs (*gag*, *pol*, and *env*) encoding structural and enzymatic proteins that are essential for viral replication. The antisense strand of the HTLV-I provirus encodes for the regulatory protein HBZ which has been shown to play a key role in HTLV-I pathogenesis (Gaudray *et al.*, 2002; Matsuoka and Green, 2009). The virus also possesses an additional pX region that contains at least four ORFs encoding the p12 (ORFI), p13 and p30 (ORFII) auxiliary proteins in addition to the regulatory proteins Rex/p27 (ORFIII), and Tax (ORFIV) (reviewed in Baydoun *et al.*, 2008; Journo *et al.*, 2009a; Nicot *et al.*, 2005). Indeed, the viral protein Tax is a powerful oncoprotein that plays a crucial role in the pathogenesis of HTLV-I-associated diseases: Tax is sufficient to transform murine fibroblasts (Tanaka *et al.* 1990), to immortalize primary human lymphocytes (Akagi and Shimotohno, 1993), and to induce severe pathologies in transgenic mice such as an adult T-cell leukemia/lymphoma (ATL)-like disease, mesenchymal tumors, and neurofibromas (Coscoy *et al.*, 1998; Grossman *et al.*, 1995; Hasegawa *et al.*, 2006). The high oncogenic potential of Tax can be attributed to its ability to perturb a large variety of signaling pathways involved in proliferation, DNA repair and apoptosis, and also to its ability to activate several transcription factors such as NF-κB (Journo *et al.*, 2009a). Despite that Tax proteins have been identified in HTLV-II, -III and -IV (Calattini *et al.*, 2006; Chevalier *et al.*, 2006; Higuchi *et al.*, 2009; Switzer *et al.*, 2009), none of them has the oncogenic potential as Tax of HTLV-I mainly due to differences in structural and functional domains (Higuchi, 2009).

Only 2–5% of all infected individuals end up developing an HTLV-I-associated disease, mainly tropical spastic paraparesis/HTLV-I-associated myelopathy (TSP/HAM) or ATL. First described in 1969 before the discovery of HTLV-I (Mani *et al.*, 1969), TSP/HAM is characterized by inflammatory manifestations and incomplete paralysis of the inferior limbs (Gessain *et al.*, 1986). ATL, first described in 1977 (Takatsuki *et al.*, 1977), is a malignant proliferation of cells displaying a mature activated $CD3^+$, $CD4^+$, $CD8^-$, $CD7^-$, and $CD25^+$ T-cell phenotype (Dahmoush *et al.*, 2002; Nagatani *et al.*, 1990). It develops after a long period of latency and is characterized by the presence of atypical lymphocytes called flower cells due to their multilobulated nucleus. ATL presents four subtypes: acute, lymphoma, chronic, and smoldering (Shimoyama, 1991; Takatsuki *et al.*, 1985).

In fact, several cellular and viral factors determine the outcome of HTLV-I infection and contribute to ATL leukemogenesis. Among the cellular factors are somatic mutations in the cellular DNA (Hatta *et al.*, 1995; Morosetti *et al.*, 1995; Nosaka *et al.*, 2000; Sakashita *et al.*, 1992; Takeuchi *et al.*, 2003; Tamiya *et al.*, 1998), epigenetic changes such as hypermethylation and hypomethylation of the promoter region of tumor suppressor genes (Nosaka *et al.*, 2000), and a genetic predisposition for the disease (Miyamoto *et al.*, 1985; Tsukasaki *et al.*, 2001; Yashiki *et al.*, 2001). The site of viral integration (Doi *et al.*, 2005; Ozawa *et al.*, 2004) and the viral replication through the clonal expansion of $CD4^+/CD8^+$ cells (Zane et al., 2009) constitute the viral factors in addition to viral proteins, notably HBZ (Matsuoka and Green, 2009) and Tax which has emerged as a key role player in the pathogenesis of both HTLV-I diseases: TSP/HAM and ATL.

In this review, we emphasize on the regulatory protein Tax and the recent discoveries in the molecular mechanisms through which this protein induces a constitutive activation of the NF-κB pathway, which is a corner stone in ATL leukemogenesis.

II. THE VIRAL ONCOPROTEIN TAX: STRUCTURAL AND FUNCTIONAL DOMAINS

As a protein of 353 amino acids, Tax is rich in structural and functional domains that allow the protein to exert its transactivation and transformation properties (Fig. 1A). Tax was shown to be located in both the cytoplasm and the nucleus and to be able to shuttle between these two compartments.

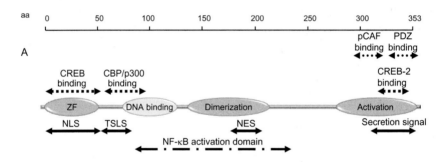

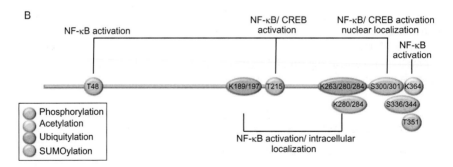

Fig. 1 (A) Schematic representation of the structural and functional domains of Tax. (B) Schematic representation of Tax posttranslational modifications and their role in transcriptional activities and Tax subcellular localization. (See Page 2 in Color Section at the back of the book.)

This relies on the presence of both a leucine-rich nuclear export signal (NES) located between amino acids 188 and 200 (Alefantis *et al.*, 2003; Burton *et al.*, 2000; Chevalier *et al.*, 2005) and a nuclear localization signal (NLS) spanning the first 48 amino acids of the protein (Gitlin *et al.*, 1991; Semmes and Jeang, 1992). Recently, a new signal targeting Tax to Tax speckled structures in the nucleus has been mapped between amino acids 50 and 75 (Fryrear *et al.*, 2009). In addition to its dual subcellular localization, cell-free Tax has been detected in the cerebrospinal fluid of HAM/TSP patients (Cartier and Ramirez, 2005). Recently, two potent secretion signals (YTNI and DHE) have been identified in the C-terminal part of Tax (Alefantis *et al.*, 2005; Jain *et al.*, 2007).

In addition to these shuttling domains, several regions have been identified in Tax. They play an important role in the formation of homodimers, an essential event for the protein to attain optimal transcriptional activity (Gitlin *et al.*, 1991; Jin and Jeang, 1997a; Tie *et al.*, 1996). Most of the regions were identified in the central region of Tax such as residues 153, 174, 212, 261 (Gitlin *et al.*, 1991), the region between threonine 123 and alanine 204 (Tie *et al.*, 1996) in addition to three additional domains named Tax DD1 (127–146), Tax DD2 (181–194), and Tax DD3 (213–228) (Basbous *et al.*, 2003). A zinc finger (ZF) between residues 22 and 53 in the N-terminal region of Tax is also implicated in the homodimerization of Tax (Jin and Jeang, 1997b).

The four last residues of Tax (aa 350–353) correspond to a PDZ-domain-binding motif (PBM) and the ability of Tax to bind a variety of PDZ-domain proteins was indeed demonstrated (Rousset *et al.*, 1998). Tax PBM was shown to be necessary for HTLV-I-mediated primary T-cell immortalization or IL-2-independent growth of T cells (Tsubata *et al.*, 2005; Xie *et al.*, 2006). Strikingly, the Tax protein of HTLV-II (Tax2) lacks a PBM and this has been correlated with reduced transforming activity relative to Tax1 (Higuchi *et al.*, 2007; Hirata *et al.*, 2004). The PBM also regulates Tax maturation as shown by recent data showing that the PDZ-domain protein PDLIM-2 induces Tax ubiquitination and degradation in the nuclear matrix (Fu *et al.*, 2010).

Finally, the central region of Tax was also shown to play an important role in the activation of the NF-κB pathway (Tsuchiya *et al.*, 1994).

III. TAX: A POTENT TRANSACTIVATOR AND A DEREGULATOR OF THE CELLULAR MACHINERY

As a potent transactivator, Tax drives the transcription of viral genes through two out of the three Tax responsive elements (TxRE-1 and TxRE-2) that are located in the 5′LTR (Brady *et al.*, 1987; Rosen *et al.*, 1985; Shimotohno *et al.*, 1986). This effect is exerted through cellular proteins that function as transcriptional effectors such as CREB, ATF, CBP/p300, and p/CAF (Baranger *et al.*, 1995; Beimling and Moelling, 1989; Giam and Xu, 1989; Goren *et al.*, 1995; Harrod *et al.*, 2000; Jiang *et al.*, 1999; Park *et al.*, 1988).

In addition to the transcriptional activation of viral genes, the activation of several cellular transcription factors involved in proliferation and cell cycle control constitutes the hallmark of Tax expression in HTLV-I-infected cells. Among these are: SRF (serum response factor) (Fujii *et al.*, 1991; Matsumoto *et al.*, 1997), NF-AT (Good *et al.*, 1997; Rivera *et al.*, 1998; Song *et al.*, 2005), AP-1 (Fujii *et al.*, 2000), and NF-κB (reviewed in Kfoury *et al.*, 2005; Peloponese and Jeang, 2006; Sun and Yamaoka, 2005).

Tax deregulates the cellular machinery at multiple levels. Despite the tight control of the cell cycle by the cyclin/CDK complexes and CDK inhibitors, Tax manages to push the cell toward an accelerated cell cycle by activating the transcription of *cyclins* and *cdk* genes, by directly binding to them or by stabilizing their complexes (Akagi *et al.*, 1996; Haller *et al.*, 2002; Huang *et al.*, 2001; Iwanaga *et al.*, 2001; Lemoine and Marriott, 2001; Santiago *et al.*, 1999). In addition, Tax alters cell cycle checkpoint such as the G2/M checkpoint (Haoudi *et al.*, 2003; Liang *et al.*, 2002) by physically interacting with the Chk1 and Chk2 proteins (Haoudi and Semmes, 2003; Park *et al.*, 2004) and the M checkpoint by the interaction with MAD1 (Jin *et al.*, 1998), and APC^{cdc20p}, leading to aneuploidy (Afonso *et al.*, 2007; Liu *et al.*, 2003).

As for many other viruses and as a strategy to evade the immune response, Tax activates two cell survival pathways in HTLV-I-infected cells: NF-κB and AKT. AKT activation by Tax was recently shown to increase the expression of Bcl3, leading to higher proliferation of HTLV-I-infected T cells (Saito *et al.*, 2010). Through the activation of the NF-κB pathway, Tax induces the expression of antiapoptotic proteins such as Bcl-xL that is equally induced by the CREB pathway (Mori *et al.*, 2001; Nicot and Harrod, 2000; Tsukahara *et al.*, 1999). Through the same pathway, Tax also induces the expression of the antiapoptotic protein family IAP (Waldele *et al.*, 2006) and of CXCR-7, a chemokine receptor involved in proliferation and cell survival (Jin *et al.*, 2009). Tax also inhibits the innate immune signaling through the NF-κB-dependent induction of expression of SOCS1, which is an inhibitor of interferon signaling (Charoenthongtrakul *et al.*, 2011). Beyond the NF-κB and the AKT pathways, Tax induces the expression of the antiapoptotic proteins FAP-1, XIAP (Kawakami *et al.*, 1999), and c-FLIP (Krueger *et al.*, 2006). In addition, Tax represses the function of major tumor suppressor genes such as p53, mainly through NF-κB activation (Ariumi *et al.*, 2000; Dreyfus *et al.*, 2005; Jeong *et al.*, 2004, 2005; Pise-Masison *et al.*, 1998), and of Rb (Hangaishi *et al.*, 1996; Kehn *et al.*, 2005).

Several reports also suggest that Tax induces apoptosis through Fas and TRAIL in addition to the inhibition of the DNA repair mechanisms

(Kao et al., 2000; Nicot and Harrod, 2000; Rivera et al., 1998; Rivera-Walsh et al., 2001). This proapoptotic effect of Tax could be an initial event, helping malignant cells to acquire resistance. A report by Kuo et al. presented evidence suggesting that Tax-induced I-kappa B kinase (IKK) and NF-κB hyperactivation is the cause of Tax-induced cellular senescence. The authors show that via a RelA-dependent mechanism, Tax triggers a checkpoint mediated by p21 and p27, which independently from pRb and p53, induces cellular senescence. However, this effect is alleviated through the expression of the viral antisense protein HBZ (Kuo and Giam, 2006; Zhi et al., 2011).

Despite the lack of evidence that Tax directly induces DNA damage (Majone and Jeang, 2000), ATL cells, like many cancer cells, are unstable genetically and are characterized by several deletions, translocations, duplications, and aneuploidy (Chieco-Bianchi et al., 1988; Itoyama et al., 1990; Whang-Peng et al., 1985). Actually, Tax inhibits the DNA damage response such as nucleotide excision repair, base excision repair, and mismatch repair, through the inhibition of CHK1, CHK2, and DNA β-polymerase (Haoudi et al., 2003; Jeang et al., 1990; Kao et al., 2001; Park et al., 2004). In addition, Tax inhibits the activity and decreases the expression of the KU 80 protein (Ducu et al., 2011; Majone and Jeang, 2000; Majone et al., 2005). The effect of Tax on hTERT activity is controversial. First, Gabet et al. described a transcriptional repression of hTERT by Tax (Gabet et al., 2003). One year later, another team demonstrated that Tax-expressing primary lymphocytes were able to activate telomerase expression and maintain telomere length in an NF-κB-dependent manner (Sinha-Datta et al., 2004). The discrepancy of the data was then attributed to different experimental settings where in the former study the experiments were done with PHA-activated cells, whereas in the latter, the experiments were performed with primary-inactivated lymphocytes. Tax also induces centrosome overduplication and multipolar mitosis by targeting Tax1BP2 (Ching et al., 2006). A recent indirect way of eliciting DNA damage by Tax is the production of reactive oxygen species (ROS). Importantly, a recent report demonstrated that Tax expression in primary human cells induces the production of ROS which elicits DNA damage and the expression of senescence markers (Kinjo et al., 2010).

IV. THE NF-κB PATHWAY: GENERALITIES

First described in 1986, B lymphocytes as a nuclear factor that can bind the enhancer of the immunoglobulin Kappa light chain gene (Sen and Baltimore, 1986), NF-κB is a family of transcription factors that play a crucial role in proliferation, apoptosis, oncogenesis, and immune response. Five members have been described until now, p65 (RelA), c-Rel, RelB, p50/p105, and p52/p100 (Ghosh and Hayden, 2008). All five members share a common Rel homology domain, which is a conserved domain of 300 amino acids that contains a DNA-binding domain, a dimerization domain, a region of interaction with inhibitory proteins IκB, and a NLS (Baeuerle and Henkel, 1994; Baldwin, 1996). These proteins are capable of homo- or heterodimerization using all possible combinations, except for RelB which dimerizes only with p50 or p52 (Ryseck et al., 1992).

Despite the fact that all NF-κB family members are ubiquitously expressed, they are all sequestered in the cytoplasm in an inactive form bound to a member of the IκB family of proteins. Until now, eight members have been identified: IκB-α, IκB-β, IκB-δ, IκB-=ε, IκB-γ, Bcl3, and the NF-κB precursors p100 and p105 (Inoue et al., 1992; Whiteside and Israel, 1997; Zabel and Baeuerle, 1990). They all share a five to seven repeated ankyrin motifs permitting their interaction with the NF-κB family members. This interaction masks the NLS of the NF-κB proteins leading to their sequestration in the cytoplasm (Henkel et al., 1992). In response to activating signal, the IκB proteins are phosphorylated by the IKK complex, which is a high molecular weight complex composed of one regulatory subunit IKK-γ (NEMO) in addition to two catalytic subunits IKK-α and IKK-β (Israel, 2010). Upon activation, the IKK complex is able to induce the phosphorylation of the IκB proteins leading to their ubiquitination and degradation by the proteosome.

NF-κB is activated by a wide variety of signals through two distinct pathways: the canonical and the noncanonical pathway. The canonical pathway is activated by pathogens, cytokines, and antigen receptors and involves the degradation of one of the three canonical IκB molecules: IκB-α, IκB-β, and IκB-ε and the nuclear translocation of the heterodimers that essentially contain RelA (Silverman and Maniatis, 2001). All this is executed through and the catalytic subunit IKK-β. On the other hand, the noncanonical pathway regulates lymphoid organogenesis, B cell survival and maturation, DC activation, and bone metabolism (Sun, 2011) and is executed through the catalytic subunit IKK-α. Through this pathway, the precursor p100 is phosphorylated and cleaved by the proteasome into p52 (Beinke and Ley, 2004). As a consequence, heterodimers containing p52 and most commonly RelB activate the transcription of a limited set of genes.

The canonical NF-κB activating pathway requires IKK-γ (NEMO) whose gene is localized on the X chromosome. NEMO is composed of 419 amino acids and has a molecular weight of 50 kDa. It has two coiled-coil domains (CC1 and CC2), a NEMO ubiquitin-binding domain (NUB), a LZ, and a ZF (Fig. 2). The CC1 domain contains the site for the interaction with the effector subunits of the complex, IKK-α and IKK-β. The CC2 and the LZ domains of NEMO represent the oligomerization domains of the molecule.

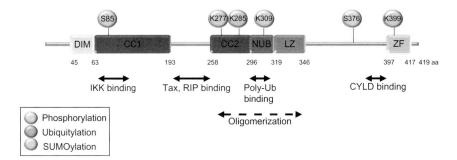

Fig. 2 Schematic representation of the structural domains of NEMO. DIM: dimerization domain, CC1: coiled coil 1, CC2: coiled coil 2, NUB: NEMO ubiquitin binding, LZ: leucine zipper, and ZF: zinc finger. Also indicated are the amino acid residues that are posttranslationally modified. (See Page 2 in Color Section at the back of the book.)

Importantly, the NUB domain participates in the interaction of NEMO with ubiquitinated proteins such as RIP in the TNF-R1 pathway (Wu *et al.*, 2006).

NEMO acts as a platform for the recruitment of activators and inhibitors of the IKK complex, but its exact role in the activation and inhibition of the IKKs is still ambiguous. Recently, several studies pointed that the posttranslational modifications of NEMO play a crucial role in the activation of the NF-κB pathway. For example, NEMO has been shown to be SUMOylated following genotoxic stress (Huang *et al.*, 2003) but not after activation through TNF-α or IL-1β.

V. TAX: A POWERFUL ACTIVATOR OF THE NF-κB PATHWAY

ATL is characterized by the increase in the expression of lymphokines and lymphokine receptors that are products of genes controlled by NF-κB (Arima *et al.*, 1996; Maruyama *et al.*, 1987; Siekevitz *et al.*, 1987). Under normal conditions, NF-κB activation is transitory and occurs most of the time through the canonical pathway. In contrast, in HTLV-I-transformed cells, in Tax-expressing cells, and in ATL cells isolated from patients, both NF-κB pathways are constitutively activated (Arima *et al.*, 1991; Ballard *et al.*, 1988; Xiao *et al.*, 2001). Indeed, Tax acts at multiple levels to induce and maintain NF-κB activity (Fig. 3). In addition, the status of Tax posttranslational modifications and Tax subcellular localization play a crucial role in Tax-induced NF-κB activation.

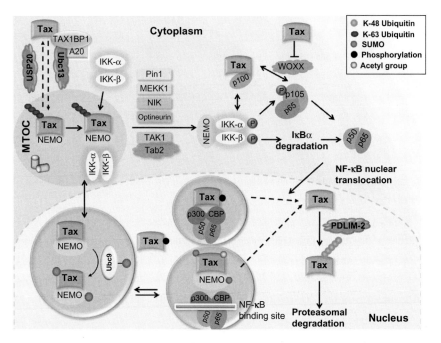

Fig. 3 Molecular mechanisms of Tax-induced activation of the NF-κB pathway. Tax acts at multiple levels to initiate and maintain NF-κB activation; Cytoplasmic steps: Tax dimers activate IKK by interacting with the noncatalytic subunit NEMO, facilitating the recruitment of Tax to the catalytic subunits IKK-α and IKK-β. Tax interaction with and activation of IKK depends on Tax conjugation to Ub-K63 chains, which are formed by Ubc13 and removed by USP20. Ub-K63 Tax ubiquitination also allows the targeting of Tax and IKK component to a MTOC-associated cytoplasmic compartment believed to represent the Tax signaling platform. Tax can also activate IKK by interacting with upstream mediators such as MEKK1, NIK, optineurin, TAK1, and PIN1. This leads to activation of IKK resulting in IκBs phosphorylation, ubiquitination, and proteasomal degradation. IκB-degradation exposes the NLS of NF-κB dimers, inducing their rapid translocation to the nucleus. In addition, Tax disrupts the Ubc13/A20 complex through the interaction with Tax1BP1 thus making Ubc13 available for its ubiquitination. At the level of NF-κB inhibitors: Tax binds to the ankyrin domain of IκB-α, which prevents its interaction with NF-κB factors, recruits p100 to the IKK complex, and enhances the proteasome-dependent processing of p105 into p50. Tax also represses the expression of WWOX which is an inhibitor of the canonical NF-κB pathway. Nuclear steps: Tax physically interacts with the p65/RelA subunit of NF-κB, and recruits the transcriptional coactivators CBP/p300 to p65/RelA. This favors NF-κB binding to consensus sequences presents in the promoters of several cellular genes and activates their transcription. These events occur in particular structures called Tax speckles/nuclear bodies whose formation is facilitated by Tax SUMOylation. The fraction of Tax present in the nuclear matrix is degraded through the formation of K48-Ub chains mediated by PDLIM-2. The same Tax molecule is able to move between nuclear bodies and between nuclear body and the MTOC-associated Tax compartment (arrows). (See Page 3 in Color Section at the back of the book.)

The phosphorylation, ubiquitination and the subsequent degradation of IκB-α are essential in the activation of the NF-κB pathway by Tax. The fact that no kinase activity has been assigned to Tax so far suggests that Tax induces persistent phosphorylation and activity of the IKK complex to sustain activation of the NF-κB pathway. In fact, shortly after the discovery of the constituents of the IKK complex, several reports described the role of Tax in the activation of the IKK subunits α and β in cells infected with HTLV-I or transfected with Tax (Geleziunas et al., 1998; Uhlik et al., 1998; Yamaoka et al., 1998). Actually, Tax binds to the IKK complex (Sun and Ballard, 1999). This mainly relies on the direct interaction between Tax and the regulatory subunit of the IKK complex NEMO, which is mediated by the leucine zipper domains present in the C- and N-terminal parts of NEMO and the leucine repeat region (LRR) in Tax (Chu et al., 1999; Harhaj and Sun, 1999; Jin et al., 1999; Xiao et al., 2000). This interaction allows the subsequent recruitment of the catalytic subunits of the IKK complex. Interestingly, and in contrast to the physiological situation which requires only IKK-α, Tax requires both IKK-α and NEMO to activate NF-κB via the noncanonical pathway, forming a complex that also contains p100 (Hayden and Ghosh, 2004; Pomerantz and Baltimore, 2002; Xiao et al., 2001). In fact, Tax cooperates with IKK by physically recruiting IKK-α to p100, triggering phosphorylation-dependent ubiquitination and processing of p100 to p52 (Xiao et al., 2001).

Hence, by targeting different IKK complexes, Tax activates both the canonical and the noncanonical pathways and seems to modulate the relationships between the two pathways. A recent report by Fu et al. demonstrates that Tax activation of the noncanonical pathway represses the expression of the WWOX tumor suppressor gene, which specifically inhibits Tax activation of the canonical NF-κB pathway (Fu et al., 2011). At the level of IKK activation, Tax was shown to affect the serine/threonine protein phosphatase 2A (PP2A), the cellular inhibitor of IKK. Tax mutants incapable of binding PP2A, whether they are capable or not of binding to IKK, have no effect on the NF-κB pathway (Fu et al., 2003). Tax can also indirectly activate the IKK complex through the activation of the upstream kinases such as the MAP3K family members MEKK1 and NIK. In fact, Tax directly binds to MEKK1, which is able to phosphorylate IKK-β (Yin et al., 1998). On the other hand, there is contradictory evidence concerning the role of NIK in the activation of the NF-κB pathway by Tax. Whereas it has been shown that dominant negative mutants of NIK inhibit the activation of the NF-κB pathway suggesting a role of Tax upstream NIK (Geleziunas et al., 1998; Uhlik et al., 1998), others reported that the activation of the noncanonical pathway by Tax does not require NIK (Xiao et al., 2001).

It is clear that the activation of the NF-κB complex by Tax is not a simple process and may involve many effectors. It has been recently shown that the activation and the interaction with TAK1 (TGF-beta kinase 1) by Tax, through the adaptor molecule Tab2, is essential to maintain a constitutive activation of the IKK complex (Wu and Sun, 2007; Yu et al., 2008). Moreover, it has been demonstrated that Tab2 colocalizes and forms complexes with Tax and RelA in cytoplasmic punctuate structures that contain NEMO, calreticulin, which is a multiprocess calcium buffering chaperone, and TAX1BP1 (Avesani et al., 2010). These findings came after the demonstration that Tax interacts with TAX1BP1 to disrupt its interactions with A20 and Itch5, and thus inactivating the A20/TAX1BP1/Ubc13 ubiquitin-editing complex leading to the stabilization of the E2 ubiquitin-conjugating enzyme Ubc13 involved in Tax ubiquitination (see below) (Shembade et al., 2008, 2010). In the same line of evidence, NEMO-related protein (NRP), also known as optineurin, has been shown to interact with the ubiquitinated forms of Tax through its ubiquitin-binding domain (UBD), leading to the stabilization of these forms and to the potentiation of NF-κB activation by Tax in collaboration with TAX1BP1 (Journo et al., 2009b). Potentiation of NF-κB activation by Tax has also been attributed to PIN1 (peptidylproline cis–trans isomerase), which is present at high levels in HTLV-I-infected cells and in Tax-transfected cells. This activation is believed to be mediated through the interaction with Tax, via its activation domain located between amino acids 99 and 198 (Peloponese et al., 2009).

The fact that Tax can directly bind to ankyrin motifs and that IκB proteins contain such motifs is highly suggestive that Tax can directly bind to IκB proteins through these motifs. In fact, it has been shown that Tax can interact with IκB-α (Hirai et al., 1994; Petropoulos and Hiscott, 1998) and p105 (Hirai et al., 1992). This interaction activates NF-κB by disrupting NF-κB/IκB complexes or by recruiting IκB members directly to proteasome (proteasome targeting model). One of Tax's targets is the precursor p105 (Hirai et al., 1992). Tax targets p105 to the proteasome, to accelerate its cleavage to the active form p50. Indeed, the weak interaction between the HC9 subunit of the proteasome and the precursor p105 is enforced by the formation of a tertiary complex with Tax (Rousset et al., 1996). The precursor p100 is another target of Tax, as well as IκB-β (Lanoix et al., 1994; Suzuki et al., 1995). In addition to phosphorylation-dependent ubiquitination, Tax also induces IκB-α degradation by directly favoring its interaction with the proteasome (Hirai et al., 1994). Moreover, Tax interaction with these inhibitory proteins can influence Tax localization within the cell, as Tax is essentially cytoplasmic when bound to p100, while it is nuclear when bound to p52 (Pepin et al., 1994).

At the level of the NF-κB proteins, Tax can interact with p50, p52, p65, and c-Rel (Lanoix et al., 1994; Suzuki et al., 1994). Tax can bind to their homology domain, and favor dimer formation, resulting in an increase in their DNA-binding and transcriptional activity. Moreover, Tax, RelA, p50, RNA polymerase II, and CBP/p300 colocalize in transcriptionally active small discrete nuclear foci (Bex and Gaynor, 1998; Bex *et al.*, 1997). Indeed, Tax recruits the transcriptional coactivators CBP/p300 to p65/RelA, thereby significantly increasing NF-κB transcriptional activity (Bex *et al.*, 1998).

VI. TAX POSTTRANSLATIONAL MODIFICATIONS AND NF-κB ACTIVATION

Genomic DNA encodes for 20 amino acids only but more than 140 derived amino acids exist due to posttranslational modification of proteins (Uy and Wold, 1977). These posttranslational modifications are in general reversible modifications that permit a rapid response to different signals. Earlier on, modifications with small molecules such as phosphorylation, acetylation, and methylation were described. Now the field has widely expanded, and modifications with either lipid or protein are described to affect nearly every protein in the cell. The chemical and structural changes that affect the targeted protein influence its subcellular localization and its interaction with other cellular partners. The pleotropic effect of Tax and its wide impact on nearly every aspect in cellular physiology, strongly suggests that this protein may exploit the functional potential of posttranslational modifications (Fig. 1B).

A. Tax Phosphorylation

The first described posttranslational modification of Tax is phosphorylation. In 1988, phosphorylated forms of Tax were described in the nuclear and cytoplasmic fractions of the cell (Nyunoya *et al.*, 1988). Another study showed that phorbol esters, okaidic acid, and serum induce Tax phosphorylation (Fontes *et al.*, 1993). According to both reports, the target residue is a serine. Tax contains 26 serines that were all mutated into alanines to assess their role in the activation of the CREB and NF-κB pathways. First, the mutation of serines 113, 116, and 258 was shown to exclusively repress NF-κB activity with no effect on the CREB pathway and according to this study, no single mutation changed the nuclear localization of Tax (Semmes and Jeang, 1992). Later on, the phosphorylation of residue 113 was excluded and serines 300 and 301 in the C-terminal part of Tax were identified as targets of phosphorylation (Krause Boehm *et al.*, 1999). Unlike the earlier study, phosphorylation on at least one serine residue was found to be necessary for Tax localization in nuclear bodies in addition to the activation of transcription through both CREB and NF-κB pathways (Bex *et al.*, 1999; Krause Boehm *et al.*, 1999). It was later possible to refine the phosphorylation map of Tax. Four new phosphorylation sites were described. Three of them are threonine (aa 48, 184, and 215) and one is a serine (aa 336) (Durkin et al., 2006). Through PCR-based mutagenesis, the role of these residues in the activation of the CREB and NF-κB pathways was investigated (Durkin et al., 2006). Mutation on residues 184 and 336 had no effect on both pathways. On the other hand, mutation of threonine 215 was associated with a loss of activity of both pathways, while the mutation of threonine 48 affected the NF-κB pathway exclusively. In the same study, the authors quantified the amount of phosphorylated Tax for each residue and found that the majority of Tax was phosphorylated on residues 48 and 336 with a minor fraction of Tax phosphorylated on residues 184 and 215. Phosphorylation on residues 300 and 301 seemed to be a rare event. At the end, the authors proposed a model where the phosphorylation of Tax on serines 300/301 and the dephosphorylation on threonines 48/215 confers its transcriptional activity to Tax. The stoichiometry between the forms phosphorylated on either serines 300/301 or threonines 48/215 indicates that the active forms of Tax are less abundant than the inactive forms, which has a major effect on the modulation of Tax activity (Durkin et al., 2006). Recently, the pleiotropic human serine/threonine kinase CK2 has been assigned the role of Tax phosphorylation. These data showed that Tax is phosphorylated *in vitro* at three residues: serines 336, 344, and threonine 351 within its C-terminal PBM which plays a role, though indirectly, in the activation of noncanonical NF-κB pathway (Bidoia et al., 2010; Higuchi et al., 2007).

B. Tax Acetylation

Another form of Tax posttranslational modification is acetylation. Acetylation was first described to affect histones and to be is essential for their transcriptional regulation (Allfrey et al., 1964; Vidali et al., 1968). Tax acetylation was demonstrated to affect lysine 346 (Lodewick et al., 2009). Despite the technical difficulties that hindered the detection of endogenously acetylated Tax in HTLV-I-transformed lymphocytes, acetylated Tax species were detected under highly denaturing conditions in both Tax-transfected 293 T cells and T lymphocytes (Lodewick et al., 2009). Phosphorylation of serines 300/301 was suggested to be a prerequisite for Tax acetylation. The authors argued that Tax phosphorylation on serines 300/301 is essential for its nuclear translocation and thus for its interaction with p300 which was found to be involved in Tax acetylation (Lodewick et al., 2009). Tax acetylation was also found to correlate positively with its SUMOylation status mostly because of the important role of Tax SUMOylation in the formation of nuclear bodies where both Tax and p300 colocalize (Lodewick et al., 2009) (see below). A functional consequence of Tax acetylation was found to be required for NF-κYB activation (Lodewick et al., 2009).

C. Tax Ubiquitination

Ubiquitination is a reversible form of posttranslational modification that controls several aspects of cellular physiology such as protein stability and proteasomal degradation, subcellular localization of proteins, differentiation, endocytosis, chromatin remodeling, cellular division, and importantly signal transduction (Kerscher et al., 2006; Mukhopadhyay and Riezman, 2007). Through a series of enzymatic reactions, a ubiquitin moiety is covalently attached to the target protein. Proteins can be modified by the attachment of one ubiquitin molecule on one lysine (monoubiquitination), or on multiple lysines (multi-monoubiquitination). Alternatively, a unique lysine can be conjugated to a chain of ubiquitin molecules (polyubiquitination), in which each monomer is linked to each other through one of several lysines in positions 6, 11, 26, 27, 33, 48, and 63 of the ubiquitin itself (Kirisako et al., 2006; Pickart and Fushman, 2004; Xu et al., 2009). Branched, mixed, and linear Ub chains where ubiquitin moieties are attached to each other in a chain head to tail and not through lysine residues exist but physiological preference is still unclear (Ben-Saadon et al., 2006; Kirisako et al., 2006). The different forms of ubiquitin chains have been associated with different cellular processes not necessarily linked to protein degradation. Lysine-48 chains or (K-48) are the most abundant and target conjugated proteins for proteasomal degradation, whereas lysine 63 (K-63) chains are involved in DNA damage repair, signal transduction, intracellular trafficking, and ribosomal biogenesis (Hochstrasser, 2004; Kerscher et al., 2006; Pickart and Fushman, 2004; Welchman et al., 2005). Lysine 6, 29, and 33 chains play a role in protein stability, lysosomal degradation, and regulation of the enzymatic activity of kinases (Chastagner et al., 2006; Nishikawa et al., 2004). Interestingly, the eukaryotic genome encodes for a broad range of protein containing UBDs. UBDs differentially bind to the different ubiquitin chains through non-covalent bonds (Harper and Schulman, 2006; Hicke et al., 2005; Hurley et al., 2006).

Tax interaction with the proteasome was described long time before the description of Tax ubiquitination. This interaction is thought to play a role in Tax-induced processing of p105, in the degradation of the NF-κB inhibitory protein IκB-α, and in the degradation of Tax itself which facilitates the presentation of Tax by the MHC-I complex (Beraud and Greene, 1996). Studies show that Tax not only interacts with the proteasome but also increases its proteolytic activity (Hemelaar et al., 2001). In fact, Tax ubiquitination has been demonstrated by two-independent teams, one of which showed that Tax is monoubiquitinated in Tax-transfected cells and in HTLV-I-transformed lymphocytes. In one study, monoubiquitinated species were only detected after the treatment of the cells with isopeptidase inhibitors or after the boiling of the lysates in SDS denaturing buffer. According to the authors, this ubiquitination inhibits the activation of the NF-κB pathway and does not lead to the proteasomal degradation of Tax (Peloponese et al., 2004). On the other hand, another team described mono- and polyubiquitination of both transfected and endogenous Tax in HTLV-I-transformed lymphocytes (Chiari et al., 2004). Through site directed mutagenesis, the authors showed that the C-terminal lysines (263, 280, and 284) are the major target for ubiquitination and that Tax ubiquitination is necessary for the interaction with the proteasome (Chiari et al., 2004). This was later confirmed by our findings that Tax is K-48 ubiquitinated, resulting in its proteasomal degradation (Kfoury et al., 2008). In addition, PDLIM-2 was identified as the E3 ubiquitin ligase responsible for the K-48 ubiquitination of Tax in the nuclear matrix. In fact, the same team showed that Tax interacts directly with PDLIM-2 through its α helix motif at amino acids 236–254. The authors also demonstrated that PDLIM-2 suppressed the NF-κB transcriptional activity of Tax and that HTLV-I-transformed T lymphocytes had low levels of PDLIM-2 (Fu et al., 2010; Yan et al., 2009).

We recently showed that Tax is also modified by K-63 ubiquitin chains. This results in the recruitment of the IKK complex to the centrosome, which is the platform for Tax/IKK interaction (Kfoury et al., 2008). In fact, the cytoplasmic Tax ubiquitination on its C-terminal lysines is critical for Tax binding to the IKK complex and for the subsequent activation of the NF-κB pathway (Lamsoul et al., 2005; Nasr et al., 2006). In agreement with these findings, Tax/NEMO interaction is abrogated in cells where the expression of the E2 ubiquitin-conjugating enzyme Ubc13 was silenced using small interfering RNAs, which is highly suggestive that NEMO binds to Tax through the recognition of Tax's polyubiquitin Ub-K63 chains through its ubiquitin-binding motifs (Shembade et al., 2007). We have recently shown that Tax induces NEMO SUMOylation and that this phenomenon is positively regulated by Tax ubiquitination (Kfoury et al., 2011). Moreover Tax nuclear bodies contain NEMO, small ubiquitin-like modifier (SUMO), and Ubc9, which is the E2 conjugating enzyme in the SUMOylation machinery, which is highly suggestive of an enzymatic role for these bodies (Kfoury et al., 2011). As discussed previously, Tax expression in primary human cells induces the production of ROS, which elicits DNA damage and the expression of senescence markers (Kinjo et al., 2010). Whether NEMO SUMOylation by Tax is a consequence of ROS production and the DNA damaged induced, or it is a totally independent event, remains unknown.

Until recently, PDLIM-2 and Ubc13 remained the only enzymes identified to be involved in Tax ubiquitination. A recent report by Fryrear et al.

demonstrates that Tax is a substrate for the really interesting new gene (RING) finger protein (RNF4), which is known to target SUMO proteins for ubiquitin modification. This modification resulted in the cytoplasmic relocalization of Tax into the cytoplasm accompanied by an increase in Tax-induced NF-κB activity (Fryrear et al., 2011). The overexpression of TRAF2, TRAF5, and TRAF6, all of which are E3 ubiquitin ligases, strongly enhances Tax ubiquitination, but it is still unclear whether these enzymes directly execute endogenous Tax ubiquitination (Yu et al., 2008). Despite the recent discoveries in Tax ubiquitination, it is still unclear whether there exists a ubiquitin-editing enzyme that regulates Tax ubiquitination. It was suggested that the interaction between Tax and Tax1BP1, which is an adaptor protein in the ubiquitin-editing complex A20, plays a role in its deubiquitination (Shembade et al., 2007). However, this argument was then turned down by the same team when they showed that Tax interacts with Tax1BP1 and disrupts the Tax1BP1, A20, and Itch5 complex, thus maintaining a constitutive NF-κB activity (Shembade et al., 2008). Tax interaction with the E2 conjugating enzyme Ubc13, which is involved in NF-κB activation, turned out to protect this enzyme from TNF-α-mediated ubiquitination and subsequent degradation by the proteasome (Shembade et al., 2010). Tax also prevented TRAF6 from interacting with A20 following IL-1 stimulation (Shembade et al., 2010). The authors conclude that Tax preserves the E2:E3 complexes that are essential for NF-κB activation and prevent the degradation of Ubc13 which is required for Tax polyubiquitination (Shembade and Harhaj, 2010). Only recently, Yasunaga et al. described the ubiquitin specific peptidase USP20 as a ubiquitin-editing enzyme that deubiquitylates Tax and abrogates Tax-induced NF-κB activation. Interestingly, the authors show lower levels of this enzyme in HTLV-I-infected ATL cell lines and they propose it as a target for potential ATL therapy (Yasunaga et al., 2011). There still remain many questions that need to be investigated such as the existence of a UBD in Tax, or whether Tax-induced NF-κB activation requires other nonconventional forms of polyubiquitin chains such as linear chains ubiquitination.

D. Tax SUMOylation

Another form of posttranslational modification that seems to be highly synchronized with the status of Tax ubiquitination is SUMOylation. This process is the reversible attachment of a SUMO moiety to the target protein, on a lysine residue. Similar to ubiquitination, SUMOylation is executed through a series of enzymatic reactions that will covalently attach a SUMO molecule to the target protein. In higher eukaryotes, four SUMO molecules have been identified so far. SUMO2 and SUMO3 share 95% sequence identity with each other and are considered as functionally equivalent, and around 50% sequence identity with SUMO1. Unlike SUMO1, SUMO2, and SUMO3, very little is known about SUMO4 whose function is still not well identified *in vivo* (Ulrich, 2009). As for ubiquitin, SUMO2/3 molecules can form SUMO chains, while SUMO1 does not since it does not possess a consensus sequence for SUMOylation (Bayer *et al.*, 1998; Bohren *et al.*, 2004; Tatham *et al.*, 2001). Unlike ubiquitination which targets most of its substrates to proteasomal degradation, SUMOylation plays a major role in subcellular localization of its substrates, their interaction with other proteins, their modification by other forms of posttranslational modifications such as ubiquitination and acetylation, as well as signal transduction, gene expression, and genomic stability (Zhao, 2007).

The first reports about Tax SUMOylation came out in parallel with Tax ubiquitination. We and others showed that Tax is a target for SUMOylation on lysines 280 and 284, which are also implicated in Tax ubiquitination. Unlike ubiquitinated Tax, SUMOylated Tax exists exclusively in the nucleus (Lamsoul *et al.*, 2005; Nasr *et al.*, 2006). In fact, a Tax mutant where lysines 280 and 284 were mutated into arginines failed to form nuclear bodies, which led to the conclusion that Tax SUMOylation is necessary for the formation of nuclear bodies and for the complete transcriptional activation of NF-κB (Kfoury *et al.*, 2008; Lamsoul *et al.*, 2005; Nasr *et al.*, 2006). Apparently, the model is not as simple as it first appeared. Recently, we showed that the fusion of a ubiquitin moiety to Tax lysine mutants, that are neither ubiquitinated nor SUMOylated, can restore the wild phenotype of Tax nuclear bodies that contain NEMO (Kfoury *et al.*, 2011). The fusion of a SUMO1 moiety partially restored the formation of very small nuclear bodies (Kfoury *et al.*, 2011). In fact, the more data accumulate, the more the complexity of the image is revealed. Whether there is a chronological order for the ubiquitination and SUMOylation of Tax, where one modification is a prerequisite for the other to occur, or whether these modifications occur on the same residue/molecule is still unclear. In addition, it is still ambiguous whether the ubiquitination or SUMOylation of the different lysine residues in Tax have distinct functions. The most obvious example on the importance of Tax posttranslational modifications is Tax-induced degradation by the combination of arsenic trioxide and interferon-alpha, which led to the cure of murine ATL derived from Tax transgenic mice (El Hajj *et al.*, 2010; Nasr *et al.*, 2003). Our results strongly suggest an essential role of ubiquitination and probably SUMOylation in Tax degradation.

VII. TAX POSTTRANSLATIONAL MODIFICATIONS AND INTRACELLULAR LOCALIZATION

The nucleus is a highly organized organelle. Molecules implicated in similar pathways tend to be compartmentalized in structures that are not bound by a membrane like in the cytoplasm. Several domains have been described in the nucleus such as speckles, cajal bodies, PML bodies, nuclear bodies, and the nucleolus (Spector, 1993). These domains are associated with different functions, where some are involved in transcriptional repression and others are involved in transcriptional activation.

Early studies described Tax as a protein localized in specific nuclear compartments, and colocalizing with what was called intrachromatin granules containing sc-35, which is a protein involved in splicing (Semmes and Jeang, 1996). These structures were called spliceosomal speckles. In addition, it was demonstrated that Tax recruits proteins involved in the detection and the repair of DNA damage into these structures (Durkin *et al.*, 2008; Gupta *et al.*, 2007; Haoudi *et al.*, 2003). As discussed earlier in this review, a NLS has been identified in the N-terminal region of Tax that spans the first 48 amino acids of Tax (Gitlin *et al.*, 1991; Semmes and Jeang, 1992; Smith and Greene, 1992). Moreover, a sequence responsible for the targeting of Tax to the Tax speckles has been identified recently (Fryrear *et al.*, 2009). Despite its presence in the N-terminal region of Tax between amino acids 50 and 75, this sequence is adjacent to but distinct from the NLS.

Several reports support a major role for Tax posttranslational modifications in its localization to the nucleus and the formation of nuclear bodies. Most importantly is SUMOylation, since SUMOylated Tax is present exclusively in the nucleus (Lamsoul *et al.*, 2005; Nasr *et al.*, 2006). On the other hand, ubiquitinated and phosphorylated Tax have been reported in both the nuclear and the cytoplasmic fractions (Kfoury *et al.*, 2008, 2011; Nasr *et al.*, 2006; Yan *et al.*, 2009). Tax acetylation occurs most probably in the nucleus, and this is supported by the fact that both p300 that is present in Tax nuclear bodies and Tax SUMOylation play a crucial role in its acetylation (Lodewick *et al.*, 2009). However, whether these posttranslational modifications are important for Tax nuclear localization, or whether Tax nuclear localization is necessary for these modifications to take place, is still not completely understood. The first scenario is supported by the fact that Tax mutants that are not ubiquitinated (K4-8R), or not SUMOylated (K7-8R), do not localize to the nucleus and do not form nuclear bodies (Kfoury *et al.*, 2011; Lamsoul *et al.*, 2005; Nasr *et al.*, 2006). Also, the Tax mutant F2, where serine 300 is mutated into leucine and serine 301 is mutated into alanine and which is not efficiently phosphorylated, displays a diffuse nuclear distribution but does not localize to nuclear bodies (Bex *et al.*, 1999). It is highly probable that there is an interdependence among these modifications, where one modification is a prerequisite for the other to occur, and where one modification is crucial for Tax localization to the nucleus which is necessary for the other modification to take place and hence the targeting of Tax into nuclear bodies. We recently demonstrated that the fusion of a ubiquitin moiety to the ubiquitin and SUMO deficient Tax mutants (K4-8R and K6-8R) restores the formation of Tax nuclear bodies and the recruitment of NEMO to these bodies, whereas the fusion of a SUMO moiety partially restores the phenotype which might be an indication that SUMOylation is a prerequisite for ubiquitination (Kfoury et al., 2011). In the nucleus and mainly in nuclear bodies, Tax has been shown to colocalize with several proteins such as RNA polymerase II and several transcription factors such as the NF-κB proteins (Bex et al., 1997; Semmes and Jeang, 1996). In fact, Tax binds the NF-κB subunits RelA and p50 in the nucleus leading to the stabilization of the transcriptional complexes and to the enhancement of the transcriptional activity of Tax.

In addition to its NLS, Tax possesses a NES (Burton et al., 2000), which has been shown to be dependent on CRM-1. Surprisingly, this NES is not functional in the context of full-length Tax, which suggests that it is not constitutively exposed to the nuclear export machinery due to a tertiary structure or an interaction with another protein (Alefantis et al., 2003). In fact, the authors showed that the signal is masked since a Tax mutant in which the C-terminal part was truncated directly after the NES, had a dominant cytoplasmic localization. Recently, Gatza et al. showed that the UV treatment of cells induces Tax monoubiquitination on lysines K-280 and K-284, which controls its nucleo-cytoplasmic trafficking. The authors also showed that the fusion of a ubiquitin moiety renders the protein sensitive to the CRM-1 inhibitor LMB. They conclude that UV irradiation-induced ubiquitination on residues K-280 and K-284, facilitates the dissociation of Tax from sc-35 containing foci, and induces its cytoplasmic export through the CRM-1 pathway (Gatza et al., 2007).

We and others have shown that in addition to its nuclear localization, Tax is present in two extranuclear compartments: (i) around the microtubule organizing center (MTOC), colocalizing with the centrosome and in close association with the Golgi and (ii) in the region of contact between two cells at the virological synapse (Kfoury et al., 2008; Nejmeddine et al., 2005). In fact, Tax synergizes with ICAM-I to induce the polarization of the microtubules observed in the virological synapse (Nejmeddine et al., 2005). This is through the Ras-MEK-ERK pathway and through Tax-mediated stimulation of T-cell activation pathways in synergy with ICAM-I. These signaling pathways are different from the ones used to trigger MTOC polarization in the so-called immunologic synapse (Nejmeddine et al., 2005, 2009). Also, Lamsoul et al. showed that the overexpression of ubiquitin favors the localization of Tax to the cytoplasm in inclusion bodies (Lamsoul et al., 2005). On the other hand, we showed that the fusion of a ubiquitin moiety to Tax mutants K4-8R and K6-8R, which are not ubiquitinated, restores the wild phenotype of Tax localization to nuclear bodies (Kfoury et al., 2011).

It is important to stress the fact that Tax localization to the cytoplasm is as important as Tax nuclear localization regarding the activation of the NF-κB pathway. In the cytoplasm, Tax has the ability to directly bind the NF-κB inhibitors IκB-α and p105 (Hirai et al., 1992, 1994; Petropoulos and Hiscott, 1998), which leads to the dissociation of the NF-κB dimers from their inhibitors, or the direct recruitment of IκB to the proteasome and its subsequent degradation. In addition, the fact that IκB-α is constitutively phosphorylated and undergoes degradation by the proteasome in HTLV-I-infected T lymphocytes (Lacoste et al., 1995; Sun et al., 1994), strongly suggests the involvement of the IKK complex in Tax-induced NF-κB activation. We have recently demonstrated that K-63 ubiquitinated Tax is essential for the recruitment of NEMO to the centrosome, which is the platform for the subsequent recruitment and interaction with the catalytic subunits IKK-α and IKK-β. We have also shown that Tax SUMOylation controls the recruitment of NEMO to the centrosome even in the absence of Tax in this compartment (Kfoury et al., 2008, 2011). This is highly suggestive of a specific interaction between NEMO and SUMOylated Tax, although a SUMO interacting motif (SIM) has not been yet identified in NEMO.

Through the adaptor molecule NEMO, Tax has also been shown to recruit the IKK subunits to lipid rafts, and the destruction of these rafts inhibits the activation of the IKK complex (Huang et al., 2009). A recent report demonstrates that Tax activates NEMO through mechanisms that are different from signaling pathways involved in cytokines-induced NF-κB signaling, which supports the possibility of therapeutic targeting of Tax/NEMO interaction without deleterious effects on normal NF-κB activating pathways (Shimizu et al., 2011).

In light of all these facts, we were highly tempted to answer a long remaining unanswered question is whether this is the same molecule of Tax that shuttles between the different subcellular compartments, mainly the nuclear bodies and the centrosome? To answer this question, we recently used new technologies with fluorescent proteins and confocal microscopy. We fused Tax to an irreversibly photoconvertible protein Dendra-2, which under normal conditions is a green fluorophore, but once converted with a 405 laser becomes a red fluorophore. Our live cell imaging experiments clearly showed that the same molecule of Tax shuttles between the centrosome and SUMO1/Ubc9-containing nuclear bodies, as well as among different nuclear bodies (Kfoury et al., 2011). These findings clearly support the role of multistep mechanisms that lead to the complete fulfillment of Tax transcriptional activity. They also provide evidence for the existence of different Tax nuclear bodies that might be fulfilling different activities ranging from transcriptional activation to enzymatic conjugation of posttranslational modifications.

VIII. CONCLUSION

Being the product of a relatively small genome, Tax possesses a protein structure rich in structural domains which facilitate its interaction with a broad range of cellular partners in addition to functional domains that allows its shuttling between the nuclear and the cytoplasmic compartments (Journo et al., 2009a). In addition, Tax is rich in amino acids that are the target of different forms of posttranslational modifications. As a consequence, Tax can exert pleotropic effects on the cellular machinery such as the activation of the NF-κB pathway, which is the most widely described pathway in HTLV-I pathogenesis, thus favoring the proliferation and transformation of the infected cells.

Tax activation of the NF-κB pathway seems to involve complex cellular and biochemical mechanisms (Fig. 3), and what we know seems to be the tip of the iceberg. A better understanding of the chronological order of the events that lead to the constitutive activation of NF-κB by Tax, in addition to the enzymes involved in the different posttranslational modifications affecting Tax, is crucial for the design of novel targeted therapies against ATL and other HTLV-I-associated diseases.

ACKNOWLEDGMENT

This work was supported by the American University of Beirut Medical Practice Plan and University Research Board, the Lebanese National Council for Scientific Research, and the Lady TATA Memorial Trust. R.M. and C.P. are supported by Grants from the Ligue contre le Cancer (Comité de Paris), Institut National du cancer (INCA), Cancéropôle Lyon Auvergne Rhône Alpes (CLARA), and Fondation de France.

REFERENCES

Abbaszadegan, M. R., Gholamin, M., Tabatabaee, A., Farid, R., Houshmand, M., and Abbaszadegan, M. (2003). Prevalence of human T-lymphotropic virus type 1 among blood donors from Mashhad, Iran. *J. Clin. Microbiol.* **41**, 2593–2595.

Afonso, P. V., Zamborlini, A., Saib, A., and Mahieux, R. (2007). Centrosome and retroviruses: the dangerous liaisons. *Retrovirology* **4**, 27.

Akagi, T., and Shimotohno, K. (1993). Proliferative response of Tax1-transduced primary human T cells to anti-CD3 antibody stimulation by an interleukin-2-independent pathway. *J. Virol.* **67**, 1211–1217.

Akagi, T., Ono, H., and Shimotohno, K. (1996). Expression of cell-cycle regulatory genes in HTLV-I infected T-cell lines: possible involvement of Tax1 in the altered expression of cyclin D2, p18Ink4 and p21Waf1/Cip1/Sdi1. *Oncogene* **12**, 1645–1652.

Alefantis, T., Barmak, K., Harhaj, E. W., Grant, C., and Wigdahl, B. (2003). Characterization of a nuclear export signal within the human T cell leukemia virus type I transactivator protein Tax. *J. Biol. Chem.* **278**, 21814–21822.

Alefantis, T., Mostoller, K., Jain, P., Harhaj, E., Grant, C., and Wigdahl, B. (2005). Secretion of the human T cell leukemia virus type I transactivator protein tax. *J. Biol. Chem.* **280**, 17353–17362.

Allfrey, V. G., Faulkner, R., and Mirsky, A. E. (1964). Acetylation and methylation of histones and their possible role in the regulation of RNA synthesis. *Proc. Natl. Acad. Sci. USA* **51**, 786–794.

Arima, N., Molitor, J. A., Smith, M. R., Kim, J. H., Daitoku, Y., and Greene, W. C. (1991). Human T-cell leukemia virus type I Tax induces expression of the Rel-related family of kappa B enhancer-binding proteins: evidence for a pretranslational component of regulation. *J. Virol.* **65**, 6892–6899.

Arima, N., Hidaka, S., Fujiwara, H., Matsushita, K., Ohtsubo, H., Arimura, K., Kukita, T., Fukumori, J., and Tanaka, H. (1996). Relation of autonomous and interleukin-2-responsive growth of leukemic cells to survival in adult T-cell leukemia. *Blood* **87**, 2900–2904.

Ariumi, Y., Kaida, A., Lin, J. Y., Hirota, M., Masui, O., Yamaoka, S., Taya, Y., and Shimotohno, K. (2000). HTLV-1 tax oncoprotein represses the p53-mediated transactivation function through coactivator CBP sequestration. *Oncogene* **19**, 1491–1499.

Avesani, F., Romanelli, M. G., Turci, M., Di Gennaro, G., Sampaio, C., Bidoia, C., Bertazzoni, U., and Bex, F. (2010). Association of HTLV Tax proteins with TAK1-binding protein 2 and RelA in calreticulin-containing cytoplasmic structures participates in Tax-mediated NF-kappaB activation. *Virology* **408**, 39–48.

Baeuerle, P. A., and Henkel, T. (1994). Function and activation of NF-kappa B in the immune system. *Annu. Rev. Immunol.* **12**, 141–179.

Baldwin, A. S., Jr. (1996). The NF-kappa B and I kappa B proteins: new discoveries and insights. *Annu. Rev. Immunol.* **14**, 649–683.

Ballard, D. W., Bohnlein, E., Lowenthal, J. W., Wano, Y., Franza, B. R., and Greene, W. C. (1988). HTLV-I tax induces cellular proteins that activate the kappa B element in the IL-2 receptor alpha gene. *Science* **241**, 1652–1655.

Baranger, A. M., Palmer, C. R., Hamm, M. K., Giebler, H. A., Brauweiler, A., Nyborg, J. K., and Schepartz, A. (1995). Mechanism of DNA-binding enhancement by the human T-cell leukaemia virus transactivator Tax. *Nature* **376**, 606–608.

Basbous, J., Bazarbachi, A., Granier, C., Devaux, C., and Mesnard, J. M. (2003). The central region of human T-cell leukemia virus type 1 Tax protein contains distinct domains involved in subunit dimerization. *J. Virol.* **77**, 13028–13035.

Baydoun, H. H., Bellon, M., and Nicot, C. (2008). HTLV-1 Yin and Yang: Rex and p30 master regulators of viral mRNA trafficking. *AIDS Rev.* **10**, 195–204.

Bayer, P., Arndt, A., Metzger, S., Mahajan, R., Melchior, F., Jaenicke, R., and Becker, J. (1998). Structure determination of the small ubiquitin-related modifier SUMO-1. *J. Mol. Biol.* **280**, 275–286.

Beimling, P., and Moelling, K. (1989). Isolation and characterization of the tax protein of HTLV-I. *Oncogene* **4**, 511–516.

Beinke, S., and Ley, S. C. (2004). Functions of NF-kappaB1 and NF-kappaB2 in immune cell biology. *Biochem. J.* **382**, 393–409.

Ben-Saadon, R., Zaaroor, D., Ziv, T., and Ciechanover, A. (2006). The polycomb protein Ring1B generates self atypical mixed ubiquitin chains required for its in vitro histone H2A ligase activity. *Mol. Cell* **24**, 701–711.

Beraud, C., and Greene, W. C. (1996). Interaction of HTLV-I Tax with the human proteasome: implications for NF-kappa B induction. *J. Acquir. Immune Defic. Syndr. Hum. Retrovirol.* **13** (Suppl. 1), S76–S84.

Bex, F., and Gaynor, R. B. (1998). Regulation of gene expression by HTLV-I Tax protein. *Methods* **16**, 83–94.

Bex, F., McDowall, A., Burny, A., and Gaynor, R. (1997). The human T-cell leukemia virus type 1 transactivator protein Tax colocalizes in unique nuclear structures with NF-kappaB proteins. *J. Virol.* **71**, 3484–3497.

Bex, F., Yin, M. J., Burny, A., and Gaynor, R. B. (1998). Differential transcriptional activation by human T-cell leukemia virus type 1 Tax mutants is mediated by distinct interactions with CREB binding protein and p300. *Mol. Cell. Biol.* **18**, 2392–2405.

Bex, F., Murphy, K., Wattiez, R., Burny, A., and Gaynor, R. B. (1999). Phosphorylation of the human T-cell leukemia virus type 1 transactivator tax on adjacent serine residues is critical for tax activation. *J. Virol.* **73**, 738–745.

Bidoia, C., Mazzorana, M., Pagano, M. A., Arrigoni, G., Meggio, F., Pinna, L. A., and Bertazzoni, U. (2010). The pleiotropic protein kinase CK2 phosphorylates HTLV-1 Tax protein in vitro, targeting its PDZ-binding motif. *Virus Genes* **41**, 149–157.

Bohren, K. M., Nadkarni, V., Song, J. H., Gabbay, K. H., and Owerbach, D. (2004). A M55V polymorphism in a novel SUMO gene (SUMO-4) differentially activates heat shock transcription factors and is associated with susceptibility to type I diabetes mellitus. *J. Biol. Chem.* **279**, 27233–27238.

Brady, J., Jeang, K. T., Duvall, J., and Khoury, G. (1987). Identification of p40x-responsive regulatory sequences within the human T-cell leukemia virus type I long terminal repeat. *J. Virol.* **61**, 2175–2181.

Burton, M., Upadhyaya, C. D., Maier, B., Hope, T. J., and Semmes, O. J. (2000). Human T-cell leukemia virus type 1 Tax shuttles between functionally discrete subcellular targets. *J. Virol.* **74**, 2351–2364.

Calattini, S., Chevalier, S. A., Duprez, R., Bassot, S., Froment, A., Mahieux, R., and Gessain, A. (2005). Discovery of a new human T-cell lymphotropic virus (HTLV-3) in Central Africa. *Retrovirology* **2**, 30.

Calattini, S., Chevalier, S. A., Duprez, R., Afonso, P., Froment, A., Gessain, A., and Mahieux, R. (2006). Human T-cell lymphotropic virus type 3: complete nucleotide sequence and characterization of the human tax3 protein. *J. Virol.* **19**, 9876–9888.

Cartier, L., and Ramirez, E. (2005). Presence of HTLV-I Tax protein in cerebrospinal fluid from HAM/TSP patients. *Arch. Virol.* **150**, 743–753.

Charoenthongtrakul, S., Zhou, Q., Shembade, N., Harhaj, N. S., and Harhaj, E. W. (2011). Human T cell leukemia virus type 1 Tax inhibits innate antiviral signaling via NF-{kappa}B-dependent induction of SOCS1. *J. Virol.* **85**, 6955–6962.

Chastagner, P., Israel, A., and Brou, C. (2006). Itch/AIP4 mediates Deltex degradation through the formation of K29-linked polyubiquitin chains. *EMBO Rep.* **7**, 1147–1153.

Chevalier, S. A., Meertens, L., Calattini, S., Gessain, A., Kiemer, L., and Mahieux, R. (2005). Presence of a functional but dispensable nuclear export signal in the HTLV-2 Tax protein. *Retrovirology* **2**, 70.

Chevalier, S. A., Meertens, L., Pise-Masison, C., Calattini, S., Park, H., Alhaj, A. A., Zhou, M., Gessain, A., Kashanchi, F., Brady, J. N., and Mahieux, R. (2006). The tax protein from the primate T-cell lymphotropic virus type 3 is expressed in vivo and is functionally related to HTLV-1 Tax rather than HTLV-2 Tax. *Oncogene* **25**(32), 4470–4482.

Chiari, E., Lamsoul, I., Lodewick, J., Chopin, C., Bex, F., and Pique, C. (2004). Stable ubiquitination of human T-cell leukemia virus type 1 tax is required for proteasome binding. *J. Virol.* **78**, 11823–11832.

Chieco-Bianchi, L., Saggioro, D., Del Mistro, A., Montaldo, A., Majone, F., and Levis, A. G. (1988). Chromosome damage induced in cord blood T-lymphocytes infected in vitro by HTLV-I. *Leukemia* **2**, 223S–232S.

Ching, Y. P., Chan, S. F., Jeang, K. T., and Jin, D. Y. (2006). The retroviral oncoprotein Tax targets the coiled-coil centrosomal protein TAX1BP2 to induce centrosome overduplication. *Nat. Cell Biol.* **8**, 717–724.

Chu, Z. L., Shin, Y. A., Yang, J. M., DiDonato, J. A., and Ballard, D. W. (1999). IKKgamma mediates the interaction of cellular IkappaB kinases with the tax transforming protein of human T cell leukemia virus type 1. *J. Biol. Chem.* **274**, 15297–15300.

Coscoy, L., Gonzalez-Dunia, D., Tangy, F., Syan, S., Brahic, M., and Ozden, S. (1998). Molecular mechanism of tumorigenesis in mice transgenic for the human T cell leukemia virus Tax gene. *Virology* **248**, 332–341.

Dahmoush, L., Hijazi, Y., Barnes, E., Stetler Stevenson, M., and Abati, A. (2002). Adult T-cell leukemia/lymphoma: a cytopathologic, immunocytochemical, and flow cytometric study. *Cancer* **96**, 110–116.

Doi, K., Wu, X., Taniguchi, Y., Yasunaga, J., Satou, Y., Okayama, A., Nosaka, K., and Matsuoka, M. (2005). Preferential selection of human T-cell leukemia virus type I provirus integration sites in leukemic versus carrier states. *Blood* **106**, 1048–1053.

Dreyfus, D. H., Nagasawa, M., Gelfand, E. W., and Ghoda, L. Y. (2005). Modulation of p53 activity by IkappaBalpha: evidence suggesting a common phylogeny between NF-kappaB and p53 transcription factors. *BMC Immunol.* **6**, 12.

Ducu, R. I., Dayaram, T., and Marriott, S. J. (2011). The HTLV-1 Tax oncoprotein represses Ku80 gene expression. *Virology* **416**, 1–8.

Durkin, S. S., Ward, M. D., Fryrear, K. A., and Semmes, O. J. (2006). Site-specific phosphorylation differentiates active from inactive forms of the human T-cell leukemia virus type 1 Tax oncoprotein. *J. Biol. Chem.* **281**, 31705–31712.

Durkin, S. S., Guo, X., Fryrear, K. A., Mihaylova, V. T., Gupta, S. K., Belgnaoui, S. M., Haoudi, A., Kupfer, G. M., and Semmes, O. J. (2008). HTLV-1 Tax oncoprotein subverts the cellular DNA damage response via binding to DNA-dependent protein kinase. *J. Biol. Chem.* **283**, 36311–36320.

El Hajj, H., El-Sabban, M., Hasegawa, H., Zaatari, G., Ablain, J., Saab, S. T., Janin, A., Mahfouz, R., Nasr, R., Kfoury, Y., et al. (2010). Therapy-induced selective loss of leukemia-initiating activity in murine adult T cell leukemia. *J. Exp. Med.* **207**, 2785–2792.

Feuer, G., Fraser, J. K., Zack, J. A., Lee, F., Feuer, R., and Chen, I. S. (1996). Human T-cell leukemia virus infection of human hematopoietic progenitor cells: maintenance of virus infection during differentiation in vitro and in vivo. *J. Virol.* **70**, 4038–4044.

Fontes, J. D., Strawhecker, J. M., Bills, N. D., Lewis, R. E., and Hinrichs, S. H. (1993). Phorbol esters modulate the phosphorylation of human T-cell leukemia virus type I Tax. *J. Virol.* **67**, 4436–4441.

Franchini, G., Mann, D. L., Popovic, M., Zicht, R. R., Gallo, R. C., and Wong-Staal, F. (1985). HTLV-I infection of T and B cells of a patient with adult T-cell leukemia-lymphoma (ATLL) and transmission of HTLV-I from B cells to normal T cells. *Leuk. Res.* **9**, 1305–1314.

Fryrear, K. A., Durkin, S. S., Gupta, S. K., Tiedebohl, J. B., and Semmes, O. J. (2009). Dimerization and a novel Tax speckled structure localization signal are required for Tax nuclear localization. *J. Virol.* **83**, 5339–5352.

Fryrear, K. A., Kerscher, O., and Semmes, O. J. (2011). The SUMO-targeted ubiquitin ligase RNF4 regulates localization and function of the HTLV-1 oncoprotein Tax. *Retrovirology* **8** (Suppl. 1), A126.

Fu, D. X., Kuo, Y. L., Liu, B. Y., Jeang, K. T., and Giam, C. Z. (2003). Human T-lymphotropic virus type I tax activates I-kappa B kinase by inhibiting I-kappa B kinase-associated serine/threonine protein phosphatase 2A. *J. Biol. Chem.* **278**, 1487–1493.

Fu, J., Yan, P., Li, S., Qu, Z., and Xiao, G. (2010). Molecular determinants of PDLIM2 in suppressing HTLV-I Tax-mediated tumorigenesis. *Oncogene* **29**, 6499–6507.

Fu, J., Qu, Z., Yan, P., Ishikawa, C., Aqeilan, R. I., Rabson, A. B., and Xiao, G. (2011). The tumor suppressor gene WWOX links the canonical and noncanonical NF-kappaB pathways in HTLV-I Tax-mediated tumorigenesis. *Blood* **117**, 1652–1661.

Fujii, M., Niki, T., Mori, T., Matsuda, T., Matsui, M., Nomura, N., and Seiki, M. (1991). HTLV-1 Tax induces expression of various immediate early serum responsive genes. *Oncogene* **6**, 1023–1029.

Fujii, M., Iwai, K., Oie, M., Fukushi, M., Yamamoto, N., Kannagi, M., and Mori, N. (2000). Activation of oncogenic transcription factor AP-1 in T cells infected with human T cell leukemia virus type 1. *AIDS Res. Hum. Retroviruses* **16**, 1603–1606.

Gabet, A. S., Mortreux, F., Charneau, P., Riou, P., Duc-Dodon, M., Wu, Y., Jeang, K. T., and Wattel, E. (2003). Inactivation of hTERT transcription by Tax. *Oncogene* **22**, 3734–3741.

Gatza, M. L., Dayaram, T., and Marriott, S. J. (2007). Ubiquitination of HTLV-I Tax in response to DNA damage regulates nuclear complex formation and nuclear export. *Retrovirology* **4**, 95.

Gaudray, G., Gachon, F., Basbous, J., Biard-Piechaczyk, M., Devaux, C., and Mesnard, J. M. (2002). The complementary strand of the human T-cell leukemia virus type 1 RNA genome encodes a bZIP transcription factor that down-regulates viral transcription. *J. Virol.* **76**, 12813–12822.

Geleziunas, R., Ferrell, S., Lin, X., Mu, Y., Cunningham, E. T., Jr., Grant, M., Connelly, M. A., Hambor, J. E., Marcu, K. B., and Greene, W. C. (1998). Human T-cell leukemia virus type 1 Tax induction of NF-kappaB involves activation of the IkappaB kinase alpha (IKKalpha) and IKKbeta cellular kinases. *Mol. Cell. Biol.* **18**, 5157–5165.

Gessain, A., Abel, L., De-The, G., Vernant, J. C., Raverdy, P., and Guillard, A. (1986). Lack of antibody to HTLV-I and HIV in patients with multiple sclerosis from France and French West Indies. *Br. Med. J. (Clin. Res. Ed.)* **293**, 424–425.

Ghez, D., Lepelletier, Y., Lambert, S., Fourneau, J. M., Blot, V., Janvier, S., Arnulf, B., van Endert, P. M., Heveker, N., Pique, C., and Hermine, O. (2006). Neuropilin-1 is involved in human T-cell lymphotropic virus type 1 entry. *J. Virol.* **80**, 6844–6854.

Ghez, D., Lepelletier, Y., Jones, K. S., Pique, C., and Hermine, O. (2010). Current concepts regarding the HTLV-1 receptor complex. *Retrovirology* **7**, 99.

Ghosh, S., and Hayden, M. S. (2008). New regulators of NF-kappaB in inflammation. *Nat. Rev. Immunol.* **8**, 837–848.

Giam, C. Z., and Xu, Y. L. (1989). HTLV-I tax gene product activates transcription via pre-existing cellular factors and cAMP responsive element. *J. Biol. Chem.* **264**, 15236–15241.

Gitlin, S. D., Lindholm, P. F., Marriott, S. J., and Brady, J. N. (1991). Transdominant human T-cell lymphotropic virus type I TAX1 mutant that fails to localize to the nucleus. *J. Virol.* **65**, 2612–2621.

Good, L., Maggirwar, S. B., Harhaj, E. W., and Sun, S. C. (1997). Constitutive dephosphorylation and activation of a member of the nuclear factor of activated T cells, NF-AT1, in Tax-expressing and type I human T-cell leukemia virus-infected human T cells. *J. Biol. Chem.* **272**, 1425–1428.

Goren, I., Semmes, O. J., Jeang, K. T., and Moelling, K. (1995). The amino terminus of Tax is required for interaction with the cyclic AMP response element binding protein. *J. Virol.* **69**, 5806–5811.

Grossman, W. J., Kimata, J. T., Wong, F. H., Zutter, M., Ley, T. J., and Ratner, L. (1995). Development of leukemia in mice transgenic for the tax gene of human T-cell leukemia virus type I. *Proc. Natl. Acad. Sci. USA* **92**, 1057–1061.

Gupta, S. K., Guo, X., Durkin, S. S., Fryrear, K. F., Ward, M. D., and Semmes, O. J. (2007). Human T-cell leukemia virus type 1 Tax oncoprotein prevents DNA damage-induced chromatin egress of hyperphosphorylated Chk2. *J. Biol. Chem.* **282**, 29431–29440.

Haller, K., Wu, Y., Derow, E., Schmitt, I., Jeang, K. T., and Grassmann, R. (2002). Physical interaction of human T-cell leukemia virus type 1 Tax with cyclin-dependent kinase 4 stimulates the phosphorylation of retinoblastoma protein. *Mol. Cell. Biol.* **22**, 3327–3338.

Hangaishi, A., Ogawa, S., Imamura, N., Miyawaki, S., Miura, Y., Uike, N., Shimazaki, C., Emi, N., Takeyama, K., Hirosawa, S., *et al.* (1996). Inactivation of multiple tumor-suppressor genes involved in negative regulation of the cell cycle, MTS1/p16INK4A/CDKN2, MTS2/p15INK4B, p53, and Rb genes in primary lymphoid malignancies. *Blood* **87**, 4949–4958.

Hanon, E., Stinchcombe, J. C., Saito, M., Asquith, B. E., Taylor, G. P., Tanaka, Y., Weber, J. N., Griffiths, G. M., and Bangham, C. R. (2000). Fratricide among CD8(+) T lymphocytes naturally infected with human T cell lymphotropic virus type I. *Immunity* **13**, 657–664.

Haoudi, A., and Semmes, O. J. (2003). The HTLV-1 tax oncoprotein attenuates DNA damage induced G1 arrest and enhances apoptosis in p53 null cells. *Virology* **305**, 229–239.

Haoudi, A., Daniels, R. C., Wong, E., Kupfer, G., and Semmes, O. J. (2003). Human T-cell leukemia virus-I tax oncoprotein functionally targets a subnuclear complex involved in cellular DNA damage-response. *J. Biol. Chem.* **278**, 37736–37744.

Harhaj, E. W., and Sun, S. C. (1999). IKKgamma serves as a docking subunit of the IkappaB kinase (IKK) and mediates interaction of IKK with the human T-cell leukemia virus Tax protein. *J. Biol. Chem.* **274**, 22911–22914.

Harper, J. W., and Schulman, B. A. (2006). Structural complexity in ubiquitin recognition. *Cell* **124**, 1133–1136.

Harrod, R., Kuo, Y. L., Tang, Y., Yao, Y., Vassilev, A., Nakatani, Y., and Giam, C. Z. (2000). p300 and p300/cAMP-responsive element-binding protein associated factor interact with human T-cell lymphotropic virus type-1 Tax in a multi-histone acetyltransferase/activator-enhancer complex. *J. Biol. Chem.* **275**, 11852–11857.

Hasegawa, H., Sawa, H., Lewis, M. J., Orba, Y., Sheehy, N., Yamamoto, Y., Ichinohe, T., Tsunetsugu-Yokota, Y., Katano, H., Takahashi, H., *et al.* (2006). Thymus-derived leukemia-lymphoma in mice transgenic for the Tax gene of human T-lymphotropic virus type I. *Nat. Med.* **12**, 466–472.

Hatta, Y., Hirama, T., Miller, C. W., Yamada, Y., Tomonaga, M., and Koeffler, H. P. (1995). Homozygous deletions of the p15 (MTS2) and p16 (CDKN2/MTS1) genes in adult T-cell leukemia. *Blood* **85**, 2699–2704.

Hayden, M. S., and Ghosh, S. (2004). Signaling to NF-kappaB. *Genes Dev.* **18**, 2195–2224.

Hemelaar, J., Bex, F., Booth, B., Cerundolo, V., McMichael, A., and Daenke, S. (2001). Human T-cell leukemia virus type 1 Tax protein binds to assembled nuclear proteasomes and enhances their proteolytic activity. *J. Virol.* **75**, 11106–11115.

Henkel, T., Zabel, U., van Zee, K., Muller, J. M., Fanning, E., and Baeuerle, P. A. (1992). Intramolecular masking of the nuclear location signal and dimerization domain in the precursor for the p50 NF-kappa B subunit. *Cell* **68**, 1121–1133.

Hicke, L., Schubert, H. L., and Hill, C. P. (2005). Ubiquitin-binding domains. *Nat. Rev. Mol. Cell Biol.* **6**, 610–621.

Higuchi, M., Tsubata, C., Kondo, R., Yoshida, S., Takahashi, M., Oie, M., Tanaka, Y., Mahieux, R., Matsuoka, M., and Fujii, M. (2007). Cooperation of NF-kappaB2/p100 activation and the PDZ domain binding motif signal in human T-cell leukemia virus type 1 (HTLV-1) Tax1 but not HTLV-2 Tax2 is crucial for interleukin-2-independent growth transformation of a T-cell line. *J. Virol.* **81**, 11900–11907.

Higuchi, M., and Fujii, M. (2009). Distinct functions of HTLV-1 Tax1 from HTLV-2 Tax2 contribute key roles to viral pathogenesis. *Retrovirology* **17**(6), 117.

Hirai, H., Fujisawa, J., Suzuki, T., Ueda, K., Muramatsu, M., Tsuboi, A., Arai, N., and Yoshida, M. (1992). Transcriptional activator Tax of HTLV-1 binds to the NF-kappa B precursor p105. *Oncogene* **7**, 1737–1742.

Hirai, H., Suzuki, T., Fujisawa, J., Inoue, J., and Yoshida, M. (1994). Tax protein of human T-cell leukemia virus type I binds to the ankyrin motifs of inhibitory factor kappa B and induces nuclear translocation of transcription factor NF-kappa B proteins for transcriptional activation. *Proc. Natl. Acad. Sci. USA* **91**, 3584–3588.

Hirata, A., Higuchi, M., Niinuma, A., Ohashi, M., Fukushi, M., Oie, M., Akiyama, T., Tanaka, Y., Gejyo, F., and Fujii, M. (2004). PDZ domain-binding motif of human T-cell leukemia virus type 1 Tax oncoprotein augments the transforming activity in a rat fibroblast cell line. *Virology* **318**, 327–336.

Hochstrasser, M. (2004). Ubiquitin signalling: what's in a chain? *Nat. Cell Biol.* **6**, 571–572.

Huang, Y., Ohtani, K., Iwanaga, R., Matsumura, Y., and Nakamura, M. (2001). Direct transactivation of the human cyclin D2 gene by the oncogene product Tax of human T-cell leukemia virus type I. *Oncogene* **20**, 1094–1102.

Huang, T. T., Wuerzberger-Davis, S. M., Wu, Z. H., and Miyamoto, S. (2003). Sequential modification of NEMO/IKKgamma by SUMO-1 and ubiquitin mediates NF-kappaB activation by genotoxic stress. *Cell* **115**, 565–576.

Huang, J., Ren, T., Guan, H., Jiang, Y., and Cheng, H. (2009). HTLV-1 Tax is a critical lipid raft modulator that hijacks IkappaB kinases to the microdomains for persistent activation of NF-kappaB. *J. Biol. Chem.* **284**, 6208–6217.

Hurley, J. H., Lee, S., and Prag, G. (2006). Ubiquitin-binding domains. *Biochem. J.* **399**, 361–372.

Inoue, J., Kerr, L. D., Kakizuka, A., and Verma, I. M. (1992). I kappa B gamma, a 70 kd protein identical to the C-terminal half of p110 NF-kappa B: a new member of the I kappa B family. *Cell* **68**, 1109–1120.

Israel, A. (2010). The IKK complex, a central regulator of NF-kappaB activation. *Cold Spring Harb. Perspect. Biol.* **2**, a000158.

Itoyama, T., Sadamori, N., Tokunaga, S., Sasagawa, I., Nakamura, H., Yao, E., Jubashi, T., Yamada, Y., Ikeda, S., and Ichimaru, M. (1990). Cytogenetic studies of human T-cell leukemia virus type I carriers. A family study. *Cancer Genet. Cytogenet.* **49**, 157–163.

Iwanaga, R., Ohtani, K., Hayashi, T., and Nakamura, M. (2001). Molecular mechanism of cell cycle progression induced by the oncogene product Tax of human T-cell leukemia virus type I. *Oncogene* **20**, 2055–2067.

Jain, P., Mostoller, K., Flaig, K. E., Ahuja, J., Lepoutre, V., Alefantis, T., Khan, Z. K., and Wigdahl, B. (2007). Identification of human T cell leukemia virus type 1 tax amino acid signals and cellular factors involved in secretion of the viral oncoprotein. *J. Biol. Chem.* **282**, 34581–34593.

Jain, P., Manuel, S. L., Khan, Z. K., Ahuja, J., Quann, K., and Wigdahl, B. (2009). DC-SIGN mediates cell-free infection and transmission of human T-cell lymphotropic virus type 1 by dendritic cells. *J. Virol.* **83**, 10908–10921.

Jeang, K. T., Widen, S. G., Semmes, O. J., 4th, and Wilson, S. H. (1990). HTLV-I trans-activator protein, tax, is a trans-repressor of the human beta-polymerase gene. *Science* **247**, 1082–1084.

Jeong, S. J., Radonovich, M., Brady, J. N., and Pise-Masison, C. A. (2004). HTLV-I Tax induces a novel interaction between p65/RelA and p53 that results in inhibition of p53 transcriptional activity. *Blood* **104**, 1490–1497.

Jeong, S. J., Pise-Masison, C. A., Radonovich, M. F., Park, H. U., and Brady, J. N. (2005). A novel NF-kappaB pathway involving IKKbeta and p65/RelA Ser-536 phosphorylation results in p53 Inhibition in the absence of NF-kappaB transcriptional activity. *J. Biol. Chem.* **280**, 10326–10332.

Jiang, H., Lu, H., Schiltz, R. L., Pise-Masison, C. A., Ogryzko, V. V., Nakatani, Y., and Brady, J. N. (1999). PCAF interacts with tax and stimulates tax transactivation in a histone acetyltransferase-independent manner. *Mol. Cell. Biol.* **19**, 8136–8145.

Jin, D. Y., and Jeang, K. T. (1997a). HTLV-I Tax self-association in optimal trans-activation function. *Nucleic Acids Res.* **25**, 379–387.

Jin, D. Y., and Jeang, K. T. (1997b). Transcriptional activation and self-association in yeast: protein-protein dimerization as a pleiotropic mechanism of HTLV-I Tax function. *Leukemia* **11**(Suppl. 3), 3–6.

Jin, D. Y., Spencer, F., and Jeang, K. T. (1998). Human T cell leukemia virus type 1 oncoprotein Tax targets the human mitotic checkpoint protein MAD1. *Cell* **93**, 81–91.

Jin, D. Y., Giordano, V., Kibler, K. V., Nakano, H., and Jeang, K. T. (1999). Role of adapter function in oncoprotein-mediated activation of NF-kappaB. Human T-cell leukemia virus type I Tax interacts directly with IkappaB kinase gamma. *J. Biol. Chem.* **274**, 17402–17405.

Jin, Z., Nagakubo, D., Shirakawa, A. K., Nakayama, T., Shigeta, A., Hieshima, K., Yamada, Y., and Yoshie, O. (2009). CXCR7 is inducible by HTLV-1 Tax and promotes growth and survival of HTLV-1-infected T cells. *Int. J. Cancer* **125**, 2229–2235.

Jones, K. S., Petrow-Sadowski, C., Bertolette, D. C., Huang, Y., and Ruscetti, F. W. (2005). Heparan sulfate proteoglycans mediate attachment and entry of human T-cell leukemia virus type 1 virions into CD4+ T cells. *J. Virol.* **79**, 12692–12702.

Jones, K. S., Petrow-Sadowski, C., Huang, Y. K., Bertolette, D. C., and Ruscetti, F. W. (2008). Cell-free HTLV-1 infects dendritic cells leading to transmission and transformation of CD4(+) T cells. *Nat. Med.* **14**, 429–436.

Journo, C., Douceron, E., and Mahieux, R. (2009a). HTLV gene regulation: because size matters, transcription is not enough. *Future Microbiol.* **4**, 425–440.

Journo, C., Filipe, J., About, F., Chevalier, S. A., Afonso, P. V., Brady, J. N., Flynn, D., Tangy, F., Israel, A., Vidalain, P. O., et al. (2009b). NRP/optineurin cooperates with TAX1BP1 to potentiate the activation of NF-kappaB by human T-lymphotropic virus type 1 tax protein. *PLoS Pathog.* **5**, e1000521.

Kalyanaraman, V. S., Sarngadharan, M. G., Robert-Guroff, M., Miyoshi, I., Golde, D., and Gallo, R. C. (1982). A new subtype of human T-cell leukemia virus (HTLV-II) associated with a T-cell variant of hairy cell leukemia. *Science* **218**, 571–573.

Kao, S. Y., Lemoine, F. J., and Mariott, S. J. (2000). HTLV-1 Tax protein sensitizes cells to apoptotic cell death induced by DNA damaging agents. *Oncogene* **19**, 2240–2248.

Kao, S. Y., Lemoine, F. J., and Marriott, S. J. (2001). p53-independent induction of apoptosis by the HTLV-I tax protein following UV irradiation. *Virology* **291**, 292–298.

Kawakami, A., Nakashima, T., Sakai, H., Urayama, S., Yamasaki, S., Hida, A., Tsuboi, M., Nakamura, H., Ida, H., Migita, K., et al. (1999). Inhibition of caspase cascade by HTLV-I tax through induction of NF-kappaB nuclear translocation. *Blood* **94**, 3847–3854.

Kehn, K., Fuente Cde, L., Strouss, K., Berro, R., Jiang, H., Brady, J., Mahieux, R., Pumfery, A., Bottazzi, M. E., and Kashanchi, F. (2005). The HTLV-I Tax oncoprotein targets the retinoblastoma protein for proteasomal degradation. *Oncogene* **24**, 525–540.

Kerscher, O., Felberbaum, R., and Hochstrasser, M. (2006). Modification of proteins by ubiquitin and ubiquitin-like proteins. *Annu. Rev. Cell Dev. Biol.* **22**, 159–180.

Kfoury, Y., Nasr, R., Hermine, O., de The, H., and Bazarbachi, A. (2005). Proapoptotic regimes for HTLV-I-transformed cells: targeting Tax and the NF-kappaB pathway. *Cell Death Differ.* **12**(Suppl. 1), 871–877.

Kfoury, Y., Nasr, R., Favre-Bonvin, A., El-Sabban, M., Renault, N., Giron, M. L., Setterblad, N., Hajj, H. E., Chiari, E., Mikati, A. G., et al. (2008). Ubiquitylated Tax targets and binds the IKK signalosome at the centrosome. *Oncogene* **27**, 1665–1676.

Kfoury, Y., Setterblad, N., El-Sabban, M., Zamborlini, A., Dassouki, Z., El Hajj, H., Hermine, O., Pique, C., de The, H., Saib, A., and Bazarbachi, A. (2011). Tax ubiquitylation and SUMOylation control the dynamic shuttling of Tax and NEMO between Ubc9 nuclear bodies and the centrosome. *Blood* **117**, 190–199.

Kinjo, T., Ham-Terhune, J., Peloponese, J. M., Jr., and Jeang, K. T. (2010). Induction of reactive oxygen species by human T-cell leukemia virus type 1 tax correlates with DNA damage and expression of cellular senescence marker. *J. Virol.* **84**, 5431–5437.

Kirisako, T., Kamei, K., Murata, S., Kato, M., Fukumoto, H., Kanie, M., Sano, S., Tokunaga, F., Tanaka, K., and Iwai, K. (2006). A ubiquitin ligase complex assembles linear polyubiquitin chains. *EMBO J.* **25,** 4877–4887.

Koyanagi, Y., Itoyama, Y., Nakamura, N., Takamatsu, K., Kira, J., Iwamasa, T., Goto, I., and Yamamoto, N. (1993). In vivo infection of human T-cell leukemia virus type I in non-T cells. *Virology* **196,** 25–33.

Krause Boehm, A., Stawhecker, J. A., Semmes, O. J., Jankowski, P. E., Lewis, R., and Hinrichs, S. H. (1999). Analysis of potential phosphorylation sites in human T cell leukemia virus type 1 Tax. *J. Biomed. Sci.* **6,** 206–212.

Krueger, A., Fas, S. C., Giaisi, M., Bleumink, M., Merling, A., Stumpf, C., Baumann, S., Holtkotte, D., Bosch, V., Krammer, P. H., and Li-Weber, M. (2006). HTLV-1 Tax protects against CD95-mediated apoptosis by induction of the cellular FLICE-inhibitory protein (c-FLIP). *Blood* **107,** 3933–3939.

Kuo, Y. L., and Giam, C. Z. (2006). Activation of the anaphase promoting complex by HTLV-1 tax leads to senescence. *EMBO J.* **25,** 1741–1752.

Lacoste, J., Petropoulos, L., Pepin, N., and Hiscott, J. (1995). Constitutive phosphorylation and turnover of I kappa B alpha in human T-cell leukemia virus type I-infected and Tax-expressing T cells. *J. Virol.* **69,** 564–569.

Lambert, S., Bouttier, M., Vassy, R., Seigneuret, M., Petrow-Sadowski, C., Janvier, S., Heveker, N., Ruscetti, F. W., Perret, G., Jones, K. S., and Pique, C. (2009). HTLV-1 uses HSPG and neuropilin-1 for entry by molecular mimicry of VEGF165. *Blood* **113,** 5176–5185.

Lamsoul, I., Lodewick, J., Lebrun, S., Brasseur, R., Burny, A., Gaynor, R. B., and Bex, F. (2005). Exclusive ubiquitination and sumoylation on overlapping lysine residues mediate NF-kappaB activation by the human T-cell leukemia virus tax oncoprotein. *Mol. Cell. Biol.* **25,** 10391–10406.

Lanoix, J., Lacoste, J., Pepin, N., Rice, N., and Hiscott, J. (1994). Overproduction of NFKB2 (lyt-10) and c-Rel: a mechanism for HTLV-I Tax-mediated trans-activation via the NF-kappa B signalling pathway. *Oncogene* **9,** 841–852.

Lemoine, F. J., and Marriott, S. J. (2001). Accelerated G(1) phase progression induced by the human T cell leukemia virus type I (HTLV-I) Tax oncoprotein. *J. Biol. Chem.* **276,** 31851–31857.

Liang, M. H., Geisbert, T., Yao, Y., Hinrichs, S. H., and Giam, C. Z. (2002). Human T-lymphotropic virus type 1 oncoprotein tax promotes S-phase entry but blocks mitosis. *J. Virol.* **76,** 4022–4033.

Liu, B., Liang, M. H., Kuo, Y. L., Liao, W., Boros, I., Kleinberger, T., Blancato, J., and Giam, C. Z. (2003). Human T-lymphotropic virus type 1 oncoprotein tax promotes unscheduled degradation of Pds1p/securin and Clb2p/cyclin B1 and causes chromosomal instability. *Mol. Cell. Biol.* **23,** 5269–5281.

Lo, K. M., Vivier, E., Rochet, N., Dehni, G., Levine, H., Haseltine, W. A., and Anderson, P. (1992). Infection of human natural killer (NK) cells with replication-defective human T cell leukemia virus type I provirus. Increased proliferative capacity and prolonged survival of functionally competent NK cells. *J. Immunol.* **149,** 4101–4108.

Lodewick, J., Lamsoul, I., Polania, A., Lebrun, S., Burny, A., Ratner, L., and Bex, F. (2009). Acetylation of the human T-cell leukemia virus type 1 Tax oncoprotein by p300 promotes activation of the NF-kappaB pathway. *Virology* **386,** 68–78.

Mahieux, R., and Gessain, A. (2009). The human HTLV-3 and HTLV-4 retroviruses: new members of the HTLV family. *Pathol. Biol. (Paris)* **57,** 161–166.

Majone, F., and Jeang, K. T. (2000). Clastogenic effect of the human T-cell leukemia virus type I Tax oncoprotein correlates with unstabilized DNA breaks. *J. Biol. Chem.* **275,** 32906–32910.

Majone, F., Luisetto, R., Zamboni, D., Iwanaga, Y., and Jeang, K. T. (2005). Ku protein as a potential human T-cell leukemia virus type 1 (HTLV-1) Tax target in clastogenic chromosomal instability of mammalian cells. *Retrovirology* **2**, 45.

Makino, M., Shimokubo, S., Wakamatsu, S. I., Izumo, S., and Baba, M. (1999). The role of human T-lymphotropic virus type 1 (HTLV-1)-infected dendritic cells in the development of HTLV-1-associated myelopathy/tropical spastic paraparesis. *J. Virol.* **73**, 4575–4581.

Manel, N., Kim, F. J., Kinet, S., Taylor, N., Sitbon, M., and Battini, J. L. (2003). The ubiquitous glucose transporter GLUT-1 is a receptor for HTLV. *Cell* **115**, 449–459.

Mani, K. S., Mani, A. J., and Montgomery, R. D. (1969). A spastic paraplegic syndrome in South India. *J. Neurol. Sci.* **9**, 179–199.

Maruyama, M., Shibuya, H., Harada, H., Hatakeyama, M., Seiki, M., Fujita, T., Inoue, J., Yoshida, M., and Taniguchi, T. (1987). Evidence for aberrant activation of the interleukin-2 autocrine loop by HTLV 1 encoded p40x and T3/Ti complex triggering. *Cell* **48**, 343–350.

Matsumoto, K., Shibata, H., Fujisawa, J. I., Inoue, M., Hakura, A., Tsukahara, T., and Fujii, M. (1997). Human T-cell leukemia virus type 1 Tax protein transforms rat fibroblasts via two distinct pathways. *J. Virol.* **71**, 4445–4451.

Matsuoka, M., and Green, P. L. (2009). The HBZ gene, a key player in HTLV-1 pathogenesis. *Retrovirology* **6**, 71.

Meytes, D., Schochat, B., Lee, H., Nadel, G., Sidi, Y., Cerney, M., Swanson, P., Shaklai, M., Kilim, Y., Elgat, M., *et al.* (1990). Serological and molecular survey for HTLV-I infection in a high-risk Middle Eastern group. *Lancet* **336**, 1533–1535.

Miyamoto, Y., Yamaguchi, K., Nishimura, H., Takatsuki, K., Motoori, T., Morimatsu, M., Yasaka, T., Ohya, I., and Koga, T. (1985). Familial adult T-cell leukemia. *Cancer* **55**, 181–185.

Mori, N., Fujii, M., Cheng, G., Ikeda, S., Yamasaki, Y., Yamada, Y., Tomonaga, M., and Yamamoto, N. (2001). Human T-cell leukemia virus type I tax protein induces the expression of anti-apoptotic gene Bcl-xL in human T-cells through nuclear factor-kappaB and c-AMP responsive element binding protein pathways. *Virus Genes* **22**, 279–287.

Morosetti, R., Kawamata, N., Gombart, A. F., Miller, C. W., Hatta, Y., Hirama, T., Said, J. W., Tomonaga, M., and Koeffler, H. P. (1995). Alterations of the p27KIP1 gene in non-Hodgkin's lymphomas and adult T-cell leukemia/lymphoma. *Blood* **86**, 1924–1930.

Mukhopadhyay, D., and Riezman, H. (2007). Proteasome-independent functions of ubiquitin in endocytosis and signaling. *Science* **315**, 201–205.

Nagai, M., Brennan, M. B., Sakai, J. A., Mora, C. A., and Jacobson, S. (2001). CD8(+) T cells are an in vivo reservoir for human T-cell lymphotropic virus type I. *Blood* **98**, 1858–1861.

Nagatani, T., Matsuzaki, T., Iemoto, G., Kim, S., Baba, N., Miyamoto, H., and Nakajima, H. (1990). Comparative study of cutaneous T-cell lymphoma and adult T-cell leukemia/lymphoma. Clinical, histopathologic, and immunohistochemical analyses. *Cancer* **66**, 2380–2386.

Nasr, R., Rosenwald, A., El-Sabban, M. E., Arnulf, B., Zalloua, P., Lepelletier, Y., Bex, F., Hermine, O., Staudt, L., de The, H., and Bazarbachi, A. (2003). Arsenic/interferon specifically reverses 2 distinct gene networks critical for the survival of HTLV-1-infected leukemic cells. *Blood* **101**, 4576–4582.

Nasr, R., Chiari, E., El-Sabban, M., Mahieux, R., Kfoury, Y., Abdulhay, M., Yazbeck, V., Hermine, O., de The, H., Pique, C., and Bazarbachi, A. (2006). Tax ubiquitylation and sumoylation control critical cytoplasmic and nuclear steps of NF-kappaB activation. *Blood* **107**, 4021–4029.

Nejmeddine, M., Barnard, A. L., Tanaka, Y., Taylor, G. P., and Bangham, C. R. (2005). Human T-lymphotropic virus, type 1, tax protein triggers microtubule reorientation in the virological synapse. *J. Biol. Chem.* **280**, 29653–29660.

Nejmeddine, M., Negi, V. S., Mukherjee, S., Tanaka, Y., Orth, K., Taylor, G. P., and Bangham, C. R. (2009). HTLV-1-Tax and ICAM-1 act on T-cell signal pathways to polarize the microtubule-organizing center at the virological synapse. *Blood* **114**, 1016–1025.

Nicot, C., and Harrod, R. (2000). Distinct p300-responsive mechanisms promote caspase-dependent apoptosis by human T-cell lymphotropic virus type 1 Tax protein. *Mol. Cell. Biol.* **20**, 8580–8589.

Nicot, C., Harrod, R. L., Ciminale, V., and Franchini, G. (2005). Human T-cell leukemia/lymphoma virus type 1 nonstructural genes and their functions. *Oncogene* **24**, 6026–6034.

Nishikawa, H., Ooka, S., Sato, K., Arima, K., Okamoto, J., Klevit, R. E., Fukuda, M., and Ohta, T. (2004). Mass spectrometric and mutational analyses reveal Lys-6-linked polyubiquitin chains catalyzed by BRCA1-BARD1 ubiquitin ligase. *J. Biol. Chem.* **279**, 3916–3924.

Nosaka, K., Maeda, M., Tamiya, S., Sakai, T., Mitsuya, H., and Matsuoka, M. (2000). Increasing methylation of the CDKN2A gene is associated with the progression of adult T-cell leukemia. *Cancer Res.* **60**, 1043–1048.

Nyunoya, H., Akagi, T., Ogura, T., Maeda, S., and Shimotohno, K. (1988). Evidence for phosphorylation of trans-activator p40x of human T-cell leukemia virus type I produced in insect cells with a baculovirus expression vector. *Virology* **167**, 538–544.

Ozawa, T., Itoyama, T., Sadamori, N., Yamada, Y., Hata, T., Tomonaga, M., and Isobe, M. (2004). Rapid isolation of viral integration site reveals frequent integration of HTLV-1 into expressed loci. *J. Hum. Genet.* **49**, 154–165.

Park, R. E., Haseltine, W. A., and Rosen, C. A. (1988). A nuclear factor is required for transactivation of HTLV-I gene expression. *Oncogene* **3**, 275–279.

Park, H. U., Jeong, J. H., Chung, J. H., and Brady, J. N. (2004). Human T-cell leukemia virus type 1 Tax interacts with Chk1 and attenuates DNA-damage induced G2 arrest mediated by Chk1. *Oncogene* **23**, 4966–4974.

Peloponese, J. M., Jr., and Jeang, K. T. (2006). Role for Akt/protein kinase B and activator protein-1 in cellular proliferation induced by the human T-cell leukemia virus type 1 tax oncoprotein. *J. Biol. Chem.* **281**, 8927–8938.

Peloponese, J. M., Jr., Iha, H., Yedavalli, V. R., Miyazato, A., Li, Y., Haller, K., Benkirane, M., and Jeang, K. T. (2004). Ubiquitination of human T-cell leukemia virus type 1 tax modulates its activity. *J. Virol.* **78**, 11686–11695.

Peloponese, J. M., Jr., Yasunaga, J., Kinjo, T., Watashi, K., and Jeang, K. T. (2009). Peptidylproline cis-trans-isomerase Pin1 interacts with human T-cell leukemia virus type 1 tax and modulates its activation of NF-kappaB. *J. Virol.* **83**, 3238–3248.

Pepin, N., Roulston, A., Lacoste, J., Lin, R., and Hiscott, J. (1994). Subcellular redistribution of HTLV-1 Tax protein by NF-kappa B/Rel transcription factors. *Virology* **204**, 706–716.

Petropoulos, L., and Hiscott, J. (1998). Association between HTLV-1 Tax and I kappa B alpha is dependent on the I kappa B alpha phosphorylation state. *Virology* **252**, 189–199.

Pickart, C. M., and Fushman, D. (2004). Polyubiquitin chains: polymeric protein signals. *Curr. Opin. Chem. Biol.* **8**, 610–616.

Pise-Masison, C. A., Choi, K. S., Radonovich, M., Dittmer, J., Kim, S. J., and Brady, J. N. (1998). Inhibition of p53 transactivation function by the human T-cell lymphotropic virus type 1 Tax protein. *J. Virol.* **72**, 1165–1170.

Poiesz, B. J., Ruscetti, F. W., Gazdar, A. F., Bunn, P. A., Minna, J. D., and Gallo, R. C. (1980). Detection and isolation of type C retrovirus particles from fresh and cultured lymphocytes of a patient with cutaneous T-cell lymphoma. *Proc. Natl. Acad. Sci. USA* **77**, 7415–7419.

Pomerantz, J. L., and Baltimore, D. (2002). Two pathways to NF-kappaB. *Mol. Cell* **10**, 693–695.

Proietti, F. A., Carneiro-Proietti, A. B., Catalan-Soares, B. C., and Murphy, E. L. (2005). Global epidemiology of HTLV-I infection and associated diseases. *Oncogene* **24**, 6058–6068.

Richardson, J. H., Edwards, A. J., Cruickshank, J. K., Rudge, P., and Dalgleish, A. G. (1990). In vivo cellular tropism of human T-cell leukemia virus type 1. *J. Virol.* **64**, 5682–5687.

Rivera, I., Harhaj, E. W., and Sun, S. C. (1998). Involvement of NF-AT in type I human T-cell leukemia virus Tax-mediated Fas ligand promoter transactivation. *J. Biol. Chem.* **273**, 22382–22388.

Rivera-Walsh, I., Waterfield, M., Xiao, G., Fong, A., and Sun, S. C. (2001). NF-kappaB signaling pathway governs TRAIL gene expression and human T-cell leukemia virus-I Tax-induced T-cell death. *J. Biol. Chem.* **276**, 40385–40388.

Rosen, C. A., Sodroski, J. G., and Haseltine, W. A. (1985). Location of cis-acting regulatory sequences in the human T-cell leukemia virus type I long terminal repeat. *Proc. Natl. Acad. Sci. USA* **82**, 6502–6506.

Rousset, R., Desbois, C., Bantignies, F., and Jalinot, P. (1996). Effects on NF-kappa B1/p105 processing of the interaction between the HTLV-1 transactivator Tax and the proteasome. *Nature* **381**, 328–331.

Rousset, R., Fabre, S., Desbois, C., Bantignies, F., and Jalinot, P. (1998). The C-terminus of the HTLV-1 Tax oncoprotein mediates interaction with the PDZ domain of cellular proteins. *Oncogene* **16**, 643–654.

Ryseck, R. P., Bull, P., Takamiya, M., Bours, V., Siebenlist, U., Dobrzanski, P., and Bravo, R. (1992). RelB, a new Rel family transcription activator that can interact with p50-NF-kappa B. *Mol. Cell. Biol.* **12**, 674–684.

Saito, K., Saito, M., Taniura, N., Okuwa, T., and Ohara, Y. (2010). Activation of the PI3K-Akt pathway by human T cell leukemia virus type 1 (HTLV-1) oncoprotein Tax increases Bcl3 expression, which is associated with enhanced growth of HTLV-1-infected T cells. *Virology* **403**, 173–180.

Sakashita, A., Hattori, T., Miller, C. W., Suzushima, H., Asou, N., Takatsuki, K., and Koeffler, H. P. (1992). Mutations of the p53 gene in adult T-cell leukemia. *Blood* **79**, 477–480.

Santiago, F., Clark, E., Chong, S., Molina, C., Mozafari, F., Mahieux, R., Fujii, M., Azimi, N., and Kashanchi, F. (1999). Transcriptional up-regulation of the cyclin D2 gene and acquisition of new cyclin-dependent kinase partners in human T-cell leukemia virus type 1-infected cells. *J. Virol.* **73**, 9917–9927.

Semmes, O. J., and Jeang, K. T. (1992). Mutational analysis of human T-cell leukemia virus type I Tax: regions necessary for function determined with 47 mutant proteins. *J. Virol.* **66**, 7183–7192.

Semmes, O. J., and Jeang, K. T. (1996). Localization of human T-cell leukemia virus type 1 tax to subnuclear compartments that overlap with interchromatin speckles. *J. Virol.* **70**, 6347–6357.

Sen, R., and Baltimore, D. (1986). Multiple nuclear factors interact with the immunoglobulin enhancer sequences. *Cell* **46**, 705–716.

Shembade, N., and Harhaj, E. (2010). A20 inhibition of NFkappaB and inflammation: targeting E2:E3 ubiquitin enzyme complexes. *Cell Cycle* **9**, 2481–2482.

Shembade, N., Harhaj, N. S., Yamamoto, M., Akira, S., and Harhaj, E. W. (2007). The human T-cell leukemia virus type 1 Tax oncoprotein requires the ubiquitin-conjugating enzyme Ubc13 for NF-kappaB activation. *J. Virol.* **81**, 13735–13742.

Shembade, N., Harhaj, N. S., Parvatiyar, K., Copeland, N. G., Jenkins, N. A., Matesic, L. E., and Harhaj, E. W. (2008). The E3 ligase Itch negatively regulates inflammatory signaling pathways by controlling the function of the ubiquitin-editing enzyme A20. *Nat. Immunol.* **9**, 254–262.

Shembade, N., Ma, A., and Harhaj, E. W. (2010). Inhibition of NF-kappaB signaling by A20 through disruption of ubiquitin enzyme complexes. *Science* **327**, 1135–1139.

Shimizu, A., Baratchian, M., Takeuchi, Y., Escors, D., Macdonald, D., Barrett, T., Bagneris, C., Collins, M., and Noursadeghi, M. (2011). Kaposi's sarcoma-associated herpesvirus vFLIP and human T cell lymphotropic virus type 1 Tax oncogenic proteins activate I{kappa}B kinase subunit gamma by different mechanisms independent of the physiological cytokine-induced pathways. *J. Virol.* 85, 7444–7448.

Shimotohno, K., Takano, M., Teruuchi, T., and Miwa, M. (1986). Requirement of multiple copies of a 21-nucleotide sequence in the U3 regions of human T-cell leukemia virus type I and type II long terminal repeats for trans-acting activation of transcription. *Proc. Natl. Acad. Sci. USA* 83, 8112–8116.

Shimoyama, M. (1991). Diagnostic criteria and classification of clinical subtypes of adult T-cell leukaemia-lymphoma. A report from the Lymphoma Study Group (1984-87). *Br. J. Haematol.* 79, 428–437.

Siekevitz, M., Feinberg, M. B., Holbrook, N., Wong-Staal, F., and Greene, W. C. (1987). Activation of interleukin 2 and interleukin 2 receptor (Tac) promoter expression by the trans-activator (tat) gene product of human T-cell leukemia virus, type I. *Proc. Natl. Acad. Sci. USA* 84, 5389–5393.

Silverman, N., and Maniatis, T. (2001). NF-kappaB signaling pathways in mammalian and insect innate immunity. *Genes Dev.* 15, 2321–2342.

Sinha-Datta, U., Horikawa, I., Michishita, E., Datta, A., Sigler-Nicot, J. C., Brown, M., Kazanji, M., Barrett, J. C., and Nicot, C. (2004). Transcriptional activation of hTERT through the NF-kappaB pathway in HTLV-I-transformed cells. *Blood* 104, 2523–2531.

Smith, M. R., and Greene, W. C. (1992). Characterization of a novel nuclear localization signal in the HTLV-I tax transactivator protein. *Virology* 187, 316–320.

Song, G., Ouyang, G., and Bao, S. (2005). The activation of Akt/PKB signaling pathway and cell survival. *J. Cell. Mol. Med.* 9, 59–71.

Spector, D. L. (1993). Macromolecular domains within the cell nucleus. *Annu. Rev. Cell Biol.* 9, 265–315.

Sun, S. C. (2011). Non-canonical NF-kappaB signaling pathway. *Cell Res.* 21, 71–85.

Sun, S. C., and Ballard, D. W. (1999). Persistent activation of NF-kappaB by the tax transforming protein of HTLV-1: hijacking cellular IkappaB kinases. *Oncogene* 18, 6948–6958.

Sun, S. C., and Yamaoka, S. (2005). Activation of NF-kappaB by HTLV-I and implications for cell transformation. *Oncogene* 24, 5952–5964.

Sun, S. C., Elwood, J., Beraud, C., and Greene, W. C. (1994). Human T-cell leukemia virus type I Tax activation of NF-kappa B/Rel involves phosphorylation and degradation of I kappa B alpha and RelA (p65)-mediated induction of the c-rel gene. *Mol. Cell. Biol.* 14, 7377–7384.

Suzuki, T., Hirai, H., and Yoshida, M. (1994). Tax protein of HTLV-1 interacts with the Rel homology domain of NF-kappa B p65 and c-Rel proteins bound to the NF-kappa B binding site and activates transcription. *Oncogene* 9, 3099–3105.

Suzuki, T., Hirai, H., Murakami, T., and Yoshida, M. (1995). Tax protein of HTLV-1 destabilizes the complexes of NF-kappa B and I kappa B-alpha and induces nuclear translocation of NF-kappa B for transcriptional activation. *Oncogene* 10, 1199–1207.

Switzer, W. M., Salemi, M., Qari, S. H., Jia, H., Gray, R. R., Katzourakis, A., Marriott, S. J., Pryor, K. N., Wolfe, N. D., Burke, D. S., Folks, T. M., and Heneine, W. (2009). Ancient, independent evolution and distinct molecular features of the novel human T-lymphotropic virus type 4. *Retrovirology* 6, 9.

Takatsuki, K., Uchiyama, T., Sagawa, K., and Hattori, T. (1977). Lymphoma and immunoglobulin abnormalities, with special reference to M proteinemia. *Nippon Rinsho* 35, 3757–3767.

Takatsuki, K., Yamaguchi, K., Kawano, F., Hattori, T., Nishimura, H., Tsuda, H., Sanada, I., Nakada, K., and Itai, Y. (1985). Clinical diversity in adult T-cell leukemia-lymphoma. *Cancer Res.* 45, 4644s–4645s.

Takeuchi, S., Takeuchi, N., Tsukasaki, K., Fermin, A. C., De Vas, S., Seo, H., and Koeffler, H. P. (2003). Mutations in the retinoblastoma-related gene RB2/p130 in adult T-cell leukaemia/lymphoma. *Leuk. Lymphoma* **44**, 699–701.

Tamiya, S., Etoh, K., Suzushima, H., Takatsuki, K., and Matsuoka, M. (1998). Mutation of CD95 (Fas/Apo-1) gene in adult T-cell leukemia cells. *Blood* **91**, 3935–3942.

Tanaka, A., Takahashi, C., Yamaoka, S., Nosaka, T., Maki, M., and Hatanaka, M. (1990). Oncogenic transformation by the tax gene of human T-cell leukemia virus type I in vitro. *Proc. Natl. Acad. Sci. USA* **87**, 1071–1075.

Tatham, M. H., Jaffray, E., Vaughan, O. A., Desterro, J. M., Botting, C. H., Naismith, J. H., and Hay, R. T. (2001). Polymeric chains of SUMO-2 and SUMO-3 are conjugated to protein substrates by SAE1/SAE2 and Ubc9. *J. Biol. Chem.* **276**, 35368–35374.

Tie, F., Adya, N., Greene, W. C., and Giam, C. Z. (1996). Interaction of the human T-lymphotropic virus type 1 Tax dimer with CREB and the viral 21-base-pair repeat. *J. Virol.* **70**, 8368–8374.

Tsubata, C., Higuchi, M., Takahashi, M., Oie, M., Tanaka, Y., Gejyo, F., and Fujii, M. (2005). PDZ domain-binding motif of human T-cell leukemia virus type 1 Tax oncoprotein is essential for the interleukin 2 independent growth induction of a T-cell line. *Retrovirology* **2**, 46.

Tsuchiya, H., Fujii, M., Tanaka, Y., Tozawa, H., and Seiki, M. (1994). Two distinct regions form a functional activation domain of the HTLV-1 trans-activator Tax1. *Oncogene* **9**, 337–340.

Tsukahara, T., Kannagi, M., Ohashi, T., Kato, H., Arai, M., Nunez, G., Iwanaga, Y., Yamamoto, N., Ohtani, K., Nakamura, M., and Fujii, M. (1999). Induction of Bcl-x(L) expression by human T-cell leukemia virus type 1 Tax through NF-kappaB in apoptosis-resistant T-cell transfectants with Tax. *J. Virol.* **73**, 7981–7987.

Tsukasaki, K., Miller, C. W., Kubota, T., Takeuchi, S., Fujimoto, T., Ikeda, S., Tomonaga, M., and Koeffler, H. P. (2001). Tumor necrosis factor alpha polymorphism associated with increased susceptibility to development of adult T-cell leukemia/lymphoma in human T-lymphotropic virus type 1 carriers. *Cancer Res.* **61**, 3770–3774.

Uhlik, M., Good, L., Xiao, G., Harhaj, E. W., Zandi, E., Karin, M., and Sun, S. C. (1998). NF-kappaB-inducing kinase and IkappaB kinase participate in human T-cell leukemia virus I Tax-mediated NF-kappaB activation. *J. Biol. Chem.* **273**, 21132–21136.

Ulrich, H. D. (2009). The SUMO system: an overview. *Methods Mol. Biol.* **497**, 3–16.

Uy, R., and Wold, F. (1977). Posttranslational covalent modification of proteins. *Science* **198**, 890–896.

Vidali, G., Gershey, E. L., and Allfrey, V. G. (1968). Chemical studies of histone acetylation. The distribution of epsilon-N-acetyllysine in calf thymus histones. *J. Biol. Chem.* **243**, 6361–6366.

Waldele, K., Silbermann, K., Schneider, G., Ruckes, T., Cullen, B. R., and Grassmann, R. (2006). Requirement of the human T-cell leukemia virus (HTLV-1) tax-stimulated HIAP-1 gene for the survival of transformed lymphocytes. *Blood* **107**, 4491–4499.

Welchman, R. L., Gordon, C., and Mayer, R. J. (2005). Ubiquitin and ubiquitin-like proteins as multifunctional signals. *Nat. Rev. Mol. Cell Biol.* **6**, 599–609.

Whang-Peng, J., Bunn, P. A., Knutsen, T., Kao-Shan, C. S., Broder, S., Jaffe, E. S., Gelmann, E., Blattner, W., Lofters, W., Young, R. C., et al. (1985). Cytogenetic studies in human T-cell lymphoma virus (HTLV)-positive leukemia-lymphoma in the United States. *J. Natl. Cancer Inst.* **74**, 357–369.

Whiteside, S. T., and Israel, A. (1997). I kappa B proteins: structure, function and regulation. *Semin. Cancer Biol.* **8**, 75–82.

Wolfe, N. D., Heneine, W., Carr, J. K., Garcia, A. D., Shanmugam, V., Tamoufe, U., Torimiro, J. N., Prosser, A. T., Lebreton, M., Mpoudi-Ngole, E., et al. (2005). Emergence of unique primate T-lymphotropic viruses among central African bushmeat hunters. *Proc. Natl. Acad. Sci. USA* **102**, 7994–7999.

Wu, X., and Sun, S. C. (2007). Retroviral oncoprotein Tax deregulates NF-kappaB by activating Tak1 and mediating the physical association of Tak1-IKK. *EMBO Rep.* **8**, 510–515.

Wu, Z. H., Shi, Y., Tibbetts, R. S., and Miyamoto, S. (2006). Molecular linkage between the kinase ATM and NF-kappaB signaling in response to genotoxic stimuli. *Science* **311**, 1141–1146.

Xiao, G., Harhaj, E. W., and Sun, S. C. (2000). Domain-specific interaction with the I kappa B kinase (IKK)regulatory subunit IKK gamma is an essential step in tax-mediated activation of IKK. *J. Biol. Chem.* **275**, 34060–34067.

Xiao, G., Cvijic, M. E., Fong, A., Harhaj, E. W., Uhlik, M. T., Waterfield, M., and Sun, S. C. (2001). Retroviral oncoprotein Tax induces processing of NF-kappaB2/p100 in T cells: evidence for the involvement of IKKalpha. *EMBO J.* **20**, 6805–6815.

Xie, L., Yamamoto, B., Haoudi, A., Semmes, O. J., and Green, P. L. (2006). PDZ binding motif of HTLV-1 Tax promotes virus-mediated T-cell proliferation in vitro and persistence in vivo. *Blood* **107**, 1980–1988.

Xu, P., Duong, D. M., Seyfried, N. T., Cheng, D., Xie, Y., Robert, J., Rush, J., Hochstrasser, M., Finley, D., and Peng, J. (2009). Quantitative proteomics reveals the function of unconventional ubiquitin chains in proteasomal degradation. *Cell* **137**, 133–145.

Yamaoka, S., Courtois, G., Bessia, C., Whiteside, S. T., Weil, R., Agou, F., Kirk, H. E., Kay, R. J., and Israel, A. (1998). Complementation cloning of NEMO, a component of the IkappaB kinase complex essential for NF-kappaB activation. *Cell* **93**, 1231–1240.

Yan, P., Fu, J., Qu, Z., Li, S., Tanaka, T., Grusby, M. J., and Xiao, G. (2009). PDLIM2 suppresses human T-cell leukemia virus type I Tax-mediated tumorigenesis by targeting Tax into the nuclear matrix for proteasomal degradation. *Blood* **113**, 4370–4380.

Yashiki, S., Fujiyoshi, T., Arima, N., Osame, M., Yoshinaga, M., Nagata, Y., Tara, M., Nomura, K., Utsunomiya, A., Hanada, S., *et al.* (2001). HLA-A*26, HLA-B*4002, HLA-B*4006, and HLA-B*4801 alleles predispose to adult T cell leukemia: the limited recognition of HTLV type 1 tax peptide anchor motifs and epitopes to generate anti-HTLV type 1 tax CD8(+) cytotoxic T lymphocytes. *AIDS Res. Hum. Retroviruses* **17**, 1047–1061.

Yasunaga, J., Lin, F. C., Lu, X., and Jeang, K. T. (2011). Ubiquitin-specific peptidase 20 targets TRAF6 and human T cell leukemia virus type 1 Tax to negatively regulate NF-{kappa}B signaling. *J. Virol.* **85**, 6212–6219.

Yin, M. J., Christerson, L. B., Yamamoto, Y., Kwak, Y. T., Xu, S., Mercurio, F., Barbosa, M., Cobb, M. H., and Gaynor, R. B. (1998). HTLV-I Tax protein binds to MEKK1 to stimulate IkappaB kinase activity and NF-kappaB activation. *Cell* **93**, 875–884.

Yoshida, M., Miyoshi, I., and Hinuma, Y. (1982). Isolation and characterization of retrovirus from cell lines of human adult T-cell leukemia and its implication in the disease. *Proc. Natl. Acad. Sci. USA* **79**, 2031–2035.

Yu, Q., Minoda, Y., Yoshida, R., Yoshida, H., Iha, H., Kobayashi, T., Yoshimura, A., and Takaesu, G. (2008). HTLV-1 Tax-mediated TAK1 activation involves TAB2 adapter protein. *Biochem. Biophys. Res. Commun.* **365**, 189–194.

Zabel, U., and Baeuerle, P. A. (1990). Purified human I kappa B can rapidly dissociate the complex of the NF-kappa B transcription factor with its cognate DNA. *Cell* **61**, 255–265.

Zane, L., Sibon, D., Mortreux, F., and Wattel, E. (2009). Clonal expansion of HTLV-1 infected cells depends on the CD4 versus CD8 phenotype. *Front. Biosci.* **14**, 3935–3941.

Zhao, J. (2007). Sumoylation regulates diverse biological processes. *Cell. Mol. Life Sci.* **64**, 3017–3033.

Zhi, H., Yang, L., Kuo, Y. L., Ho, Y. K., Shih, H. M., and Giam, C. Z. (2011). NF-kappaB hyper-activation by HTLV-1 tax induces cellular senescence, but can be alleviated by the viral anti-sense protein HBZ. *PLoS Pathog.* **7**, e1002025.

Lynch or Not Lynch? Is that Always a Question?

Chrystelle Colas,[*,†,‡] Florence Coulet,[†,‡] Magali Svrcek,[*,†,§] Ada Collura,[*,†] Jean-François Fléjou,[*,†,§] Alex Duval,[*,†] and Richard Hamelin[*,†]

*INSERM, UMRS 938, Centre de Recherche Saint-Antoine, Equipe "Instabilité des Microsatellites et Cancers," Paris, France
†Université Pierre et Marie Curie-Paris6, Paris, France
‡APHP, Laboratoire d'Oncogénétique et d'Angiogénétique, Groupe hospitalier Pitié-Salpêtrière, Paris, France
§AP-HP, Hôpital Saint-Antoine, Service d'Anatomie et Cytologie Pathologiques, Paris, France

I. Introduction
II. What is HNPCC?
III. What is MSI?
IV. What is an MMR Defect?
V. What is Lynch Syndrome?
   A. HNPCC: Lynch or Not Lynch?
   B. MSI: Lynch or Sporadic?
VI. Unusual Variants of Lynch Syndrome
   A. Muir Torre Syndrome
   B. Turcot Syndrome
   C. Constitutional MMR-Deficiency Syndrome
VII. Genetic Alterations Responsible for MSI Tumor Progression in Sporadic and Lynch Syndrome Cases
VIII. Current Approaches to the Detection of Lynch Syndrome Patients
IX. Surveillance of Individuals with Lynch Syndrome
X. Conclusions
   Acknowledgment
   References

The familial cancer syndrome referred to as Lynch I and II was renamed hereditary nonpolyposis colorectal cancer (HNPCC) only to revert later to Lynch syndrome (LS). LS is the most frequent human predisposition for the development of colorectal cancer (CRC), and probably also for endometrial and gastric cancers, although it has yet to acquire a consensus name. Its estimated prevalence ranges widely from 2% to 7% of all CRCs due to the fact that tumors from patients with LS are difficult to recognize at both the clinical and molecular level. This review is based on two assumptions. First, all LS patients inherit a predisposition to develop CRC (without polyposis) and/or other tumors from the Lynch spectrum. Second, all LS patients have a germline defect in one of the DNA mismatch repair (MMR) genes. When a somatic second hit inactivates the relevant MMR gene, the consequence is instability of DNA repeat sequences such as microsatellites and the tumors are referred to as having the microsatellite instability (MSI) phenotype. However, some of the inherited predisposition to develop CRC without concurrent polyposis, termed HNPCC, is found in non-LS patients, while not all MSI tumors are from LS cases. LS tumors are therefore at the junction of inherited and MSI cases. We describe here the defining characteristics of LS tumors that differentiate them from inherited non-MSI tumors and from non-inherited MSI tumors. © 2012 Elsevier Inc.

I. INTRODUCTION

Colorectal cancer (CRC) is the second cause of cancer-related death (behind lung cancer) in developed countries. It is estimated that 20–30% of CRC have a familial or hereditary component, but germline mutations of characterized genes have been reported in only about 5% of cases (Rustgi, 2007). The two main forms are categorized based on the presence of a more or less large number of adenomatous polyps and are called familial adenomatous polyposis (FAP) and hereditary nonpolyposis colorectal cancer (HNPCC), respectively. In these highly penetrant inherited syndromes, one copy of a gene is constitutionally inactivated in all cells of the individual. The second allele is somatically inactivated in the tumor following the two-hit model of cancer progression originally proposed by Knudson for tumor suppressor genes (Knudson, 2001).

FAP is easy to recognize since patients who inherit an *APC* gene germline mutation present with hundreds if not thousands of polyps in their colon. As a consequence, the prevalence of FAP is consistently estimated amongst authors to be about 1% of all CRC cases (Bisgaard *et al.*, 1994). Attenuated polyposis (AFAP) has less numerous polyps and is due to germline mutations in the 5′ and 3′ regions of the *APC* gene (Soravia *et al.*, 1998). This condition can sometimes be confused with MAP, or MUTYH associated polyposis, caused by biallelic mutation of the *MUTYH* gene (Sieber *et al.*, 2003). AFAP and MAP are rare predisposition events, as are hamartomatous polyposis syndromes such as Peutz-Jeghers syndrome (PJS) Cowden disease and juvenile polyposis syndrome (JPS).

HNPCC is a more frequent inherited predisposition to develop CRC, but as it will later be discussed, the term is incorrectly used since it corresponds to at least two different entities named Lynch syndrome (LS) and type X familial CRC. It is common nowadays to read in many reviews and research articles the following phrase: "LS also known as HNPCC." These terms are, however, not synonymous. LS is characterized by a germline defect of the mismatch repair (MMR) system, but its true prevalence is unknown and Lynch himself wrote in 2004 that "it accounts for about 2 to 7% of the total colorectal cancer burden" (Lynch and Lynch, 2004). In a number of reviews, H. Lynch acknowledged the pivotal role of A. Warthin who first described in

1913 an inherited predisposition to develop cancers of the colon and other sites such as the endometrium in a family called "G" (Warthin, 1913). In the early 1960s, Lynch *et al.* studied other two large families (N and M) with many early-onset CRCs, in association with carcinomas of the endometrium and ovary (Lynch *et al.*, 1966). This led them to hypothesize the existence of a novel hereditary cancer syndrome. Even with the description of other affected families, including the G family of Warthin, considerable time elapsed before the possibility of this new hereditary syndrome was widely accepted. Finally, it was acknowledged that Lynch I syndrome was a predisposition to develop CRC, while Lynch II syndrome predisposed to the development of CRC as well as extracolonic cancers including cancer of the endometrium, ovary, stomach, urinary tract, biliary tract, pancreas, small bowel, brain, and skin (Lynch *et al.*, 1988). Moreover patients with LS may develop multiple synchronous or metachronous cancers.

In the early 1990s, the genetic defect responsible for LS was identified as a germline mutation in one of the DNA MMR genes (Bronner *et al.*, 1994; Fishel *et al.*, 1993; Leach *et al.*, 1993; Papadopoulos *et al.*, 1994), with the consequence of a microsatellite instability (MSI) phenotype (Ionov *et al.*, 1993; Thibodeau *et al.*, 1993). Because these carcinomas were observed to develop in the absence of polyposis, the term HNPCC was used instead of LS. However, "Hereditary Non-Polyposis Colorectal Cancer" includes the word colorectal, whereas it has been known for many years that LS also predisposes to carcinoma of the endometrium and ovary, for example. This fact alone should have been sufficient to prevent HNPCC and LS being used in parallel (Jass, 2006). Clinical and molecular advances have allowed much more to become known about LS, and definitions such as the Amsterdam (Vasen *et al.*, 1991) or Bethesda criteria (Rodriguez-Bigas *et al.*, 1997) have been proposed to identify HNPCC patients. However, a number of tumors from these patients are not MSI and show an intact MMR system and therefore, cannot be considered as coming from LS patients. On the other hand, there are colorectal as well as gastric and endometrial tumors that are MSI, but arise sporadically because of a somatic MMR gene defect.

In conclusion, not all HNPCC cases and not all MSI cases are due to LS, but LS tumors are always due to an inherited predisposition, without polyposis, with an MSI phenotype (Fig. 1).

II. WHAT IS HNPCC?

The definition is in the name itself: HNPCC. Given that, there is no polyposis in these cases and FAP patients are excluded. The first clinical and familial criteria to define HNPCC (and LS when these terms were

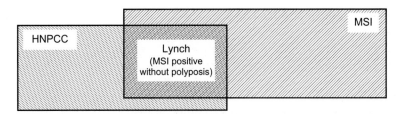

Fig. 1 Tumors from LS patients are at the junction of inherited cases without polyposis (HNPCC) and MSI cases with a MMR defect.

considered synonymous) were published in 1991 and are called Amsterdam criteria (Vasen et al., 1991). There were three criteria and all must be fulfilled: three relatives with CRC, one of them being a first degree relative of the other two; CRC in at least two successive generations; one CRC before the age of 50 years. Initially, these criteria only related to CRC, but in 1998 the Amsterdam II criteria replaced CRC with HNPCC-associated cancers (colorectum, endometrium, small bowel, ureter, and renal pelvis cancers) (Vasen et al., 1999). In three large consecutive series of CRC patients for which personal and familial history of cancer was available, 2% fulfilled Amsterdam I or II criteria (177/8,668 and 321/15,776 for Amsterdam I and II, respectively) (Bapat et al., 2009; Chen et al., 2008; Pinol et al., 2004) (Table I). This percentage corresponds approximately to the prevalence of LS, but it quickly became obvious that Amsterdam I and II criteria were too stringent, and not sensitive enough to recognize all patients

Table I Amsterdam I/II and/or Bethesda I/II Cases in Consecutive Unselected CRC Series

	Pinol 2004	Chen 2007	Bapat 2009
Consecutive unselected	1872	7108	6796
Amsterdam I/II	32/46	?/83	145/192[a]
Bethesda I/II	360/?	?	?/2503[a]
Familial clustering (non-Amsterdam)	504	?	?
Below 60 years	?	?	3025[a]
Below 50 years	129	?	1144[a]
Below 40 years	?	?	230[a]
MSI/MMR loss (n)	ND	ND	961[a]
Germline MMR mutation (n)	ND	ND	ND
Lynch (%)	?	?	?

ND, not determined.
[a]Estimation due to partial analysis.
? information not available.

with LS. Small families, for example, could not be positive for all the criteria. A number of studies that screened for MMR germline mutations in consecutive series of unselected patients showed that only 43% (54/125) of demonstrated LS patients fulfilled Amsterdam I or II criteria (Aaltonen et al., 1998; Cunningham et al., 2001; Green et al., 2009; Hampel et al., 2008, 2005a; Julie et al., 2008; Pinol et al., 2005; Salovaara et al., 2000; Trano et al., 2010) (Table II). These non-sensitive criteria were, however, thought to be specific, because most cases fulfilling Amsterdam I or II criteria were considered to be true LS patients. However, when large cohorts of patients meeting Amsterdam criteria were analyzed for germline MMR mutations, this was found to be inaccurate. Only 46% (274/594) of patients positive for the Amsterdam I or II criteria showed a germline MMR mutation or at least a tumor with an MSI phenotype (Bisgaard et al., 2002; Caldes et al., 2002; Dove-Edwin et al., 2006; Katballe et al., 2002; Lindor et al., 2005; Llor et al., 2005; Scott et al., 2001; Syngal et al., 2000; Valle et al., 2007; Wijnen et al., 1998; Wolf et al., 2006) (Table III). These observations indicate that within HNPCC cases, there is probably another hereditary syndrome different from LS. Consequently, Amsterdam criteria are considered neither sensitive nor specific for LS (Fig. 2).

Bethesda and revised Bethesda guidelines were published in 1997 and 2004, respectively (Rodriguez-Bigas et al., 1997; Umar et al., 2004). There are seven Bethesda criteria and five revised Bethesda criteria, but only one must be fulfilled to classify a patient as potentially having LS. However, these criteria are very broad with 360/1872 (19%) and 2503/6796 (37%) patients with CRC and a known personal and family history fulfilling the Bethesda and revised Bethesda criteria, respectively (Bapat et al., 2009; Pinol et al., 2004) (Table I). Needless to say, only a fraction of these are likely to be LS patients. MSI analysis or immunohistochemical (IHC) investigation of MMR protein expression is recommended to further discriminate potential LS patients from other patients. Bethesda and revised Bethesda criteria cannot be considered specific for LS since only 81 patients from a series of 274 patients (29.5%) that fulfilled Bethesda criteria had a germline MMR defect (Caldes et al., 2002; Scott et al., 2001; Syngal et al., 2000; Wolf et al., 2006) (Table IV). These criteria were considered sensitive, but it was also shown that some LS patients were not positive for Bethesda criteria (20/100, 20%) (Aaltonen et al., 1998; Green et al., 2009; Hampel et al., 2008, 2005a; Julie et al., 2008; Pinol et al., 2005; Trano et al., 2010) (Table II). Consequently, Bethesda criteria are not considered to be specific or sensitive for LS (Fig. 2). Preselection of patients according to age at diagnosis (Schofield et al., 2009; Southey et al., 2005), or to non-Amsterdam familial clustering (Bisgaard et al., 2002; Dove-Edwin et al., 2006; Katballe et al., 2002; Wijnen et al., 1998) did not contribute to a higher specificity (6.5–17%) (Table IV).

Table II Lynch Syndrome Cases in Consecutive Unselected CRC Series

	Aaltonen	Salavaara	Cunningham	Pinol	Hampel	Julié	Trano	Hampel	Green	Total
	1998	2000	2001	2005	2005	2008	2010	2008	2009	
Consecutive unselected	509	535	225	1222	1066	214	336	500	725	
Amsterdam I/II	?/7	?/12	7/9	18/22	?	?/5	?/8	?	28/31	
Bethesda I/II	?	?	?	224/287	?	39/90	?/87	?	?/364	
Below 50 years	40	45	24	58	?	37	?	?	93	
MSI/MMR loss (n)	63	66	47	91	135	21	57	64	69	
Germline MMR mutation (n)	10	18	7	11	23	8	12[a]	18	18	
Lynch (%)	2	3.4	3.1	0.9	2.2	3.8	3.6	3.6	2.5	
Amsterdam I or II/Lynch	7/10	12/18	3/7	4/11	3/23	3/8	3/12	7/18	12/18	54/125
Bethesda I or II/Lynch	10/10	?	?	10/11	18/23	6/8	6/12	13/18	17/18	80/100

The number of LS cases fulfilling Amsterdam I/II and/or Bethesda I/II criteria is indicated.
[a] Putative germline MMR mutations due to the absence of *BRAF* mutation and of *MLH1* promoter methylation.
? information not available.

Table III Lynch Syndrome Cases in Patients Fulfilling Amsterdam I or II Criteria

	Wijnen 1998	Syngal 2000	Scott 2001	Bisgaard 2002	Katballe 2002	Caldes 2002	Lindor 2005	Llor 2005	Wolf 2006	Dove-Edwin 2006	Valle 2007	Total
Amsterdam I or II (n)	92/?	28/34	?/33	31/?	11/18	24/30	161/?	17/25	?/35	?/71	64/?	594
MSI/MMR loss (n)	–	14	–	–	11	–	90	10	–	26	38	126[a]
Mutation MMR (n)	41	41	20	19	8	10	–	–	15	–	23	148
Lynch (%)	44.5	41	60.5	61	44	33	56	40	37	36.5	59.5	46
Not Lynch (%)	55.5	59	39.5	39	56	67	44	60	63	63.5	40.5	54

[a]Total of MSI or MMR defective cases for which MMR mutation was not searched for.
? information not available.

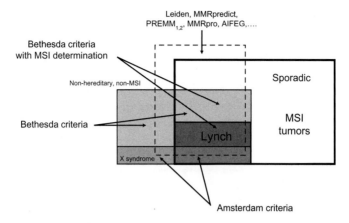

Fig. 2 Schematic representation of the sensitivity and specificity of Amsterdam criteria, Bethesda criteria, and multivariate models to detect LS patients.

A number of models have been proposed to estimate the occurrence of LS amongst CRC patients (Castells *et al.*, 2007; Green *et al.*, 2009). All are complex algorithms and multivariate models that combine data on personal and familial history of cancer. They include the Leiden model (Wijnen *et al.*, 1998), the PREMM$_{1,2}$ model (Balmana *et al.*, 2006), the MMR predict model (Barnetson *et al.*, 2006), the MMRpro model (Chen *et al.*, 2006), and the AIFEG model (Marroni *et al.*, 2006). None have yet been demonstrated to be fully specific and/or sensitive in their prediction (Tresallet *et al.*, 2011) (Fig. 2).

III. WHAT IS MSI?

MSI is defined by the appearance of new alleles of repeated sequences in tumor DNA that are absent in matching normal DNA, and this can be found in both hereditary and sporadic cancers. The first report on genetic instability at the nucleotide level described "ubiquitous somatic mutations in simple repeated sequences" as revealing a new mechanism for colonic carcinogenesis (Ionov *et al.*, 1993). This report was followed by others showing MSI in "HNPCC" and in sporadic CRC (Aaltonen *et al.*, 1993; Thibodeau *et al.*, 1993). A genetic locus predisposing to "HNPCC" was first localized on chromosome 2 (Peltomaki *et al.*, 1993), and a second predisposing locus on chromosome 3 was identified soon after (Lindblom *et al.*, 1993). The observed MSI in these tumors is reminiscent of the consequences in *Escherichia coli* of defects in the mutHLS system implicated in verifying the fidelity of DNA replication. Soon after, it was shown that human

Table IV Lynch Syndrome Cases in Patients Fulfilling Bethesda or Modified Bethesda Criteria, or Selected Because of Non-Amsterdam Familial Clustering or Age of Onset

	Syngal	Scott	Caldes	Wolf	Wijnen	Bisgaard	Katballe	Dove-Edwin	Scoffield	Southey
	2000	2001	2002	2006	1998	2002	2002	2006	2009	2008
Bethesda I/II	56/?	95/?	42/?	72/81	?	?	?	?	?	?
Familial clustering (non-Amsterdam)	?	?	?	?	92	54	23	26	?	(32)
Below 60 years	?	?	?	?	(63)	?	?	?	1344	NA
Below 50 years	?	?	?	?	(37)	?	?	?	(454)	NA
Below 45 years	?	?	?	?	(21)	?	?	?	(235)	105
Amsterdam I/II	(28/34)	(?/33)	(24/30)	(?/35)	0	0	0	0	?	(?/12)
MSI/MMR loss (n)	?	?	?	?	?	?	4	3	105	27
Germline MMR mutation (n)	17	32	13	19	6	8	2	–	48[a]	18
Lynch (%)	30.3	33.7	31	23.5	6.5	15	8.7	11.5	3.6	17
Amsterdam I or II/Lynch	14/17	20/32	10/13	13/19	NA	NA	NA	NA	?	9/18

The number of LS cases fulfilling Amsterdam I/II is indicated when known.
[a]Estimation due to partial analysis.
? information not available. Numbers in parenthesis indicate subsets of larger series.

homologs of the genes of the *E. coli* mutHLS system were localized on chromosomes 2 and 3 and were constitutionally mutated in some "HNPCC" patients (Bronner *et al.*, 1994; Fishel *et al.*, 1993; Leach *et al.*, 1993; Papadopoulos *et al.*, 1994). It is now known that most LS patients (but not all HNPCC patients) have a germline mutation of the *MLH1* or *MSH2* genes, with *MSH6* and *PMS2* also being implicated but less frequently. These genes belong to the MMR system whose function is to check DNA replication fidelity and to repair DNA mismatches that arises due to replication errors. Since replication errors occur more frequently in DNA repeated sequences, the consequence of MMR inactivation in tumors is the accumulation of numerous deletions and insertions in microsatellites. This event was first called RER for replication errors but is now more commonly referred to as MSI (or MIN) for MSI in hereditary as well as in sporadic cases (Boland *et al.*, 1998).

Many studies were published reporting MSI in various types of human tumors. The number and nature of microsatellites analyzed in these studies together with the threshold of instability required to classify tumors as MSI were highly variable (Perucho, 1999). As a consequence, MSI was reported to occur in almost all human cancer types and guidelines to define MSI were soon deemed necessary. In an analysis of 90 CRC and matching normal DNA with an average of 65 microsatellites each, our group found that tumors could divided in two clear subgroups: MSI tumors showing instability at more than 50% of the markers and non-MSI, or MSS (MicroSatellite Stable) tumors showing instability at less than 10% of the markers. We also observed that the BAT-26 microsatellite containing a 26A mononucleotide repeat was unstable in all but one MSI tumor defined using the 65 microsatellites panel, and in none of the MSS tumors. We thus proposed BAT-26 as an excellent marker for the detection of MSI (Hoang *et al.*, 1997). Moreover, BAT-26 was almost monomorphic in normal DNA from Caucasian populations, meaning it was not necessary to compare allelic size profiles between tumor DNA and matching normal DNA in order to determine MSI status. We also showed that BAT-25 and BAT-34C4, two other mononucleotide repeats, shared similar properties to BAT-26 and could thus be analyzed contemporaneously (Zhou *et al.*, 1997, 1998).

The first consensus meeting on the definition of MSI tumors was held in Bethesda in December 1996 and the guidelines emanating from this meeting were published in 1998 (Boland *et al.*, 1998). The use of BAT-26 alone, or together with BAT-25 and/or BAT-34C4 was not recognized, mainly because of the lack of information on ethnic variability of these markers at the time. Instead, a panel of five microsatellites was recommended, comprising two mononucleotide repeats (BAT-26 and BAT-25), and three dinucleotide repeats (D2S123, D5S346, and D17S250) (Boland *et al.*, 1998). Following PCR amplification the allelic size profiles obtained with these markers is compared between tumor and matching normal DNA. A tumor is considered to be MSI-H if it shows instability for at least two of the five markers. It is termed MSI-L or MSS if it shows instability at only one or none of the markers, respectively. The widespread use of the standardized Bethesda panel of markers, or of BAT-26 alone, clarified the frequency of MSI tumors. This phenotype occurs in about 10–15% of colorectal, endometrial, and gastric cancers; less frequently in other cancers such as pancreatic, kidney, or bladder; and almost never in breast or lung, for example.

Despite this progress in the detection of true MSI using BAT-26 alone or the Bethesda panel, some caveats became apparent. For example, ethnicity-related size variants of the BAT-26 locus were found, particularly in populations from Africa (Samowitz et al., 1999). Moreover, some LS germline mutations comprised deletion of all or part of the *MSH2* gene containing the BAT-26 locus. When the second *MSH2* allele was inactivated in tumor tissue by LOH, this led to the absence of both *MSH2* alleles. In such rare cases, BAT-26 was not amplified from tumor DNA since this repeat is localized in intron 5 of the *MSH2* gene.

The Bethesda panel also has some weaknesses. First, dinucleotide amplification profiles may be difficult to interpret and lead to discrepancies between readers (Loukola et al., 2001). There is also some background instability of dinucleotide repeats in MMR proficient tumors. This can be observed in MSI-L tumors which generally show instability at only one of the three dinucleotide repeats. To date, no genetic or clinical particularities have been shown to differentiate MSI-L tumors from MSS tumors, causing the biological relevance of MSI-L tumors to be questioned (Tomlinson et al., 2002). Finally, tumors can be classified as MSI-H because of observed instability at two dinucleotide repeats, but this may be due to background instability at the two loci rather than to defective MMR (Laiho et al., 2002).

Our group proposed a novel microsatellite panel for MSI comprised of five mononucleotide repeats (BAT-26, BAT-25, NR-21, NR-22, and NR-24) having the following properties. All markers are highly unstable in MSI-H tumors and highly stable in MSI-L and MSS tumors (Suraweera et al., 2002). They are quasi-monomorphic in Caucasian populations, meaning that concomitant analysis of allelic size profiles in normal DNA is not mandatory. Moreover, the conditions for amplification of all five markers in a single pentaplex PCR have been optimized (Suraweera et al., 2002). These were established using fluorescent primers labeled with different dyes. To avoid potential interference of one dye by another during laser scanning, we designed a novel pentaplex assay such that PCR products of different sizes were produced for each marker (Buhard et al., 2004). The most recent pentaplex assay version comprises of the BAT-26, BAT-25, NR-21, NR-24, and NR-27 markers. Analysis of these markers in 1206 healthy individuals from 55 different ethnic groups worldwide revealed the presence of some rare variant alleles (Buhard et al.,

2006). For this reason, and in the absence of comparison with normal DNA, we proposed the stringent condition that instability of at least three markers should be met in order to classify a tumor as MSI. A commercially available MSI detection kit has been developed based on the new pentaplex panel in which NR-27 is replaced with the mononucleotide repeat Mono-27 (Murphy et al., 2006).

The most recent consensus meeting on MSI tumors acknowledged the limitations of the Bethesda panel of markers and proposed the use of additional mononucleotide repeats in place of the dinucleotide repeats (Umar et al., 2004). Indeed, recent published studies by independent groups established that the pentaplex assay was at least as sensitive and specific as the Bethesda panel of markers, and far easier to analyze (Goel et al., 2010; Laghi et al., 2008; Xicola et al., 2007). It was also shown that analysis of MSI status of the same tumors with the pentaplex in several different laboratories gave the same results without discrepancies and without the need to discuss data in order to reach a consensus between sites (Nardon et al., 2010). We have also shown the pentaplex assay is effective in determining the MSI status of endometrial tumors (Wong et al., 2006) and of tumors with *MSH6* MMR gene mutations (You et al., 2010). These two conditions have been associated with a low level of instability. The pentaplex markers are therefore progressively becoming the standard used to establish the MSI status of hereditary as well as sporadic tumors from different tissues.

IV. WHAT IS AN MMR DEFECT?

MSI is the direct consequence of an MMR defect. The implicated proteins of the MMR system are mainly MSH2 or MLH1, and less frequently MSH6 or PMS2, affected by a genetic defect in LS cases, and MLH1 affected by an epigenetic defect in sporadic MSI cases. When functional, the role of the MMR system is to repair bases mismatches occurring during DNA replication due to DNA polymerase errors. Alternative to the determination of MSI by microsatellite genotyping is to analyze the expression of these MMR proteins by immunohistochemistry (IHC). This technique is relatively inexpensive and can be performed in most clinical pathology departments. IHC uses monoclonal antibodies to MMR proteins (MLH1, MSH2, MSH6, and PMS2) that are widely available commercially. MSI is strongly associated with the loss of reactivity of at least one MMR protein. The deleterious MMR mutation usually results in loss of reactivity of the corresponding protein. Most initial studies used a two antibody panel (MLH1 and MSH2) due to the limited availability of antibodies to the other proteins and the predominance of *MLH1* and *MSH2* gene mutations in LS (Leach et al.,

1996; Salahshor *et al.*, 2001). More recently, the use of a four antibody panel has shown that all MSH2 negative tumors were also MSH6 negative, while MLH1 loss was always associated with PMS2 loss (Shia, 2008). This pattern is explained by the fact that MMR proteins form functional heterodimers (MLH1 and PMS2 form MutLα, MSH2 and MSH6 form MutSα). Since MLH1 and MSH2 are obligatory partners in their respective dimers, abnormalities in them lead to their partner proteins becoming degraded (Boland *et al.*, 2008). However, rare cases occur with isolated loss of MSH6 or PMS2 without loss of their corresponding partner, MSH2 or MLH1, respectively. Based on this observation, the investigation of tumor expression of MSH6 and PMS2 has been proposed as a first-screening method. MSH2 and MLH1 staining would then be restricted to cases showing loss of MSH6 or PMS2 in order to define the primary gene defect (Hall *et al.*, 2010; Shia *et al.*, 2009). In most studies IHC has been performed on surgical specimens. However, it can also be performed on endoscopic biopsies, with similar results obtained in terms of the diagnosis of MSI (Kumarasinghe *et al.*, 2010; Shia *et al.*, 2011). The criteria for interpreting IHC results as a tool for identifying MSI are a major issue. Most studies use the binary classification of either positive (no loss of expression) or negative (complete loss of expression). Our experience from the study of a large consecutive series of CRCs indicates that only the completely negative pattern (total loss of staining) is indicative of MSI, assuming that internal positive controls (lymphocytes, normal glands) retain their positivity. The significance of partial loss of expression and/or heterogeneous expression is still debated, especially for MLH1 (Mangold *et al.*, 2005; Watson *et al.*, 2007). It may be largely due to technical artifacts such as inadequate tissue fixation. Neoadjuvant treatments could also affect the expression pattern, especially for MSH6 expression (Bao *et al.*, 2010). To overcome these problems, some authors have proposed that IHC for MMR proteins should only be performed in expert laboratories (Overbeek *et al.*, 2008) (Klarskov *et al.*, 2010) and that semiquantitative scoring systems should be used to interpret the staining pattern (Barrow *et al.*, 2010).

V. WHAT IS LYNCH SYNDROME?

A. HNPCC: Lynch or Not Lynch?

Several studies have analyzed series of HNPCC families defined as fulfilling Amsterdam I or II criteria for MMR defects in one proband from each family (Table III). This was performed by direct sequencing of the *MLH1* and *MSH2* genes or by analyzing MSI and/or the absence of MMR gene expression in the tumors by IHC. Concordant results were obtained in these studies, with the percentage of Amsterdam families with a MMR defect ranging from 33% to 61%. The average was 46% for a total of 594 families analyzed. These patients belong to LS families and can be compared with the remaining patients (54%) from Amsterdam positive families who have no discernible MMR defect. The latter patients are likely to have an inherited predisposition to develop CRC in the absence of polyposis. They are HNPCC cases but do not belong to LS families. They have been referred to as familial CRC type X cases (Lindor et al., 2005). Tumors from this new entity arise at a later age and are more often left-sided CRC as compared to LS cases which are mainly right-sided. They are also less often poorly differentiated and mucinous compared to LS tumors. Patients from familial CRC type X do not present multiple cancers (Valle et al., 2007). They probably represent a heterogeneous group involving mutations of as yet undiscovered gene(s), or of less penetrant mutations in genes known to cause CRC predisposition (Dove-Edwin et al., 2006).

B. MSI: Lynch or Sporadic?

1. FAMILIAL AND CLINICAL FEATURES THAT DIFFERENTIATE LYNCH CANCERS FROM SPORADIC MSI CANCERS

The major features are

(a). family history: this is considered the most useful clinical indicator of LS in individual patients
(b). age: the mean age of cancer onset in patients with sporadic MSI tumors is considerably older compared to LS patients (Young et al., 2001)
(c). gender: sporadic MSI CRC are found predominantly in older females, whereas LS CRC often occurs in younger males (Malkhosyan et al., 2000; Young et al., 2001)
(d). tumor location: although LS CRC have long been known to show a predilection for the proximal colon, up to 40% of LS CRC are present in the distal bowel and rectum. In contrast, 90% of sporadic MSI CRC occurs in the proximal colon (Kim et al., 1994)
(e). histopathology: several histological features are more commonly seen in MSI CRCs compared to MSS tumors, including poor differentiation, mucinous phenotype, lack of "dirty" necrosis, increased number of tumor intraepithelial lymphocytes, circumscribed/expansive growth pattern, and a prominent inflammatory reaction at the edge of the tumor (so-called Crohn's like reaction). These phenotypic particularities have been used to create various morphological scores aimed at discriminating MSI and MSS tumors (Greenson et al., 2009; Hyde et al., 2010; Jenkins et al., 2007; Roman et al.,

2010). The scores generally have a high negative predictive value to diagnose MSI, but only a moderate to low positive predictive value. Moreover, they cannot be applied to discriminate sporadic and hereditary MSI cases. For this purpose, it has been shown that sporadic MSI cancers are more frequently heterogeneous, poorly differentiated, and mucinous than LS MSI CRC. In sporadic MSI cancers, contiguous adenomas are likely to be serrated whereas traditional adenomas are dominant in LS cases (Jass, 2004; Young et al., 2001).

The above criteria are insufficient to completely discriminate hereditary from sporadic MSI cases. Patients of intermediate age (50–65 years) and uncertain family history who present with an MSI CRC represent a clinical dilemma because it is not immediately apparent whether their cancer has occurred sporadically or because of an underlying germline mutation.

2. MOLECULAR PROPERTIES DIFFERENTIATING LYNCH CANCERS FROM SPORADIC MSI CANCERS

The mechanism for inactivation of the MMR system is clearly different between hereditary and sporadic MSI cases. Patients from LS families have in most cases a germline mutation of one of the MMR genes (see below), while most sporadic MSI cases arise because of the absence of *MLH1* gene expression due to the methylation-induced silencing of its promoter (Fig. 3).

In mammalian cells, DNA methylation is found mainly at the level of CpG dinucleotides, and isolated CpG are globally hypomethylated in cancer cells compared to normal cells. CpG clusters are called CpG islands and these are

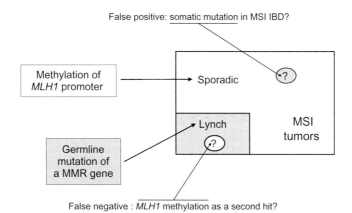

Fig. 3 Molecular differences between sporadic and inherited MSI cases according to their respective MMR defects.

frequently observed in promoter regions of up to 60% of human genes. CpG islands are generally unmethylated in normal cells, regardless of whether or not the gene is expressed. CpG island methylation is a physiological process related to aging and is also known to be more frequent in women. An increasing number of gene promoters have been found to be specifically methylated in human cancers, representing a potent mechanism to inhibit gene expression. The first report of *MLH1* promoter methylation in CRC was published in 1997 (Kane *et al.*, 1997). Since then it has been shown that most, if not all, sporadic MSI tumors arise due to *MLH1* silencing of expression following biallelic promoter methylation. Of note, methylation of the proximal region only of the *MLH1* promoter is linked to *MLH1* silencing (Deng *et al.*, 1999). In a review of the literature performed in 2006, we showed that 60% of the 161 articles published in this area analyzed methylation of nonspecific regions of the *MLH1* promoter (Capel *et al.*, 2007). Consequently, it is important to examine specific regions when investigating *MLH1* promoter methylation in putative sporadic MSI cases.

Sporadic MSI cases usually also show frequent methylation at many other promoters, thus constituting the CpG Island Methylation Phenotype (CIMP) (Toyota *et al.*, 1999). Depending on the methylation markers analyzed, this CRC subgroup is now quite well defined (Weisenberger *et al.*, 2006). Most sporadic MSI cases are CIMP+ type as opposed to hereditary MSI cases that are generally CIMP−. The CIMP+ cases that are not MSI appear to have distinctive and specific clinical and molecular characteristics (Goel *et al.*, 2007).

Also associated with *MLH1* promoter methylation and/or with the CIMP + type is the presence of *BRAF* mutations. *BRAF* is a kinase encoding gene from the RAS/RAF/MAPK pathway (Rajagopalan *et al.*, 2002). *BRAF* mutations were first demonstrated in melanomas and other types of cancers (Davies *et al.*, 2002). Approximately 10% of CRC have mutations in *BRAF*, almost always the V600E hotspot, representing about 40% of sporadic MSI cases, but interestingly this mutation is not found in LS MSI tumors (Domingo *et al.*, 2004; Wang *et al.*, 2003a).

Genetic testing for LS has been available since the mid-1990s after DNA linkage analysis identified mutations in the DNA MMR genes *MSH2* and *MLH1* in families who met the classic Amsterdam criteria (Dunlop *et al.*, 1997; Liu *et al.*, 1996). Initial studies identified mutations in only 40–60% of these families and in a small percentage of patients from families who did not fulfill these criteria (Giardiello *et al.*, 2001). However, improvements in molecular testing techniques and the additional testing of other MMR genes have increased the yield of germline mutations. During the past 15 years, screening techniques for mutations in MMR genes have been evaluated and optimized. Most early studies used various exon-by-exon polymerase chain reaction-based methods to screen subjects with a family history of CRC.

Different prescreening methods based on the detection of heteroduplexes have been proposed, such as denaturing high-performance liquid chromatography (dHPLC) (Holinski-Feder et al., 2001), denaturing gradient gel electrophoresis (DGGE) (Liu, 2010; Wijnen et al., 1995), and single-strand conformational polymorphism (SSCP) (Beck et al., 1997). The presence of larger rearrangements such as exonic deletions or duplications is usually investigated by multiplex ligation probe amplification (MLPA) (Grabowski et al., 2005). More recent technology combines the screening of point mutations and large rearrangements with the qPCR-high resolution melting (qPCR-HRM) technology (Rouleau et al., 2009). Direct DNA sequencing can of course be used instead of prescreening methods (Wahlberg et al., 1999).

Although at least eight genes have been implicated in MMR (*MSH2* (Fishel et al., 1993; Leach et al., 1993), *MLH1* (Bronner et al., 1994; Papadopoulos et al., 1994), *MSH6* (Palombo et al., 1995), *PMS1* (Nicolaides et al., 1994), *PMS2* (Nicolaides et al., 1994), *MSH3* (Watanabe et al., 1996), *MLH3* (Wu et al., 2001b) and *EXO1* (Wu et al., 2001a)), mutations in only four of these have been shown to cause LS: *MLH1* (chromosome 3p21.3), *MSH2* (chromosome 2p21), *MSH6* (chromosome 2p16), and *PMS2* (chromosome 7p22). Mutations in *MSH2* and *MLH1* account for up to 90% of LS cases (Peltomaki, 2003; Peltomaki and Vasen, 2004). The mutation spectrum of the MMR genes is spread throughout the exons and intronic splice regions. There are no hotspot mutations, which makes the screening process very time consuming. Only some mutations occur regularly, such as c.942+3A>T in *MSH2* (Desai et al., 2000; Froggatt et al., 1999) or Ashkenasi Jewish founder mutations in *MSH2* (Foulkes et al., 2002) and *MSH6* (Raskin et al., 2010). According to the InSIGHT Database (International Society for Gastrointestinal Hereditary Tumors) (www.insight-group.org), more than 500 different MMR gene mutations have been described. Germline mutations in MMR genes are mainly point mutations (nonsense and missense), small deletions/insertions of a few nucleotides or splice-site changes. Studies that have used a combination of other mutation detection methods to detect larger genomic rearrangements show higher mutation frequencies than was first reported (Charbonnier et al., 2000; Kohonen-Corish et al., 1996; van der Klift et al., 2005). Large rearrangements account for up to 15% of all pathogenic mutations in *MSH2* and *MLH1* with the relative frequencies of these varying according to the study population (Charbonnier et al., 2000; Wang et al., 2003c).

Many families with suspected LS who initially test negative for mutations in *MLH1* and *MSH2* have subsequently been found to have mutations in *MSH6* or *PMS2*, thus confirming the clinical diagnosis of a hereditary cancer syndrome. Mutations in *MSH6* account for approximately 7–10% of LS cases (Peltomaki, 2003), while mutations in *PMS2* are detected only rarely (De Rosa et al., 2000). Mutations in *MSH6* and *PMS2* often occur in less typical families probably because the risk estimates for CRC are substantially lower for these genes compared to *MLH1* and *MSH2* (Aarnio *et al.*, 1999; Baglietto *et al.*, 2010; Kariola *et al.*, 2004; Talseth-Palmer *et al.*, 2010; Vasen *et al.*, 1996).

Although most genetic test results for LS are either positive or negative for a pathogenic mutation, a significant proportion yield a "variant of uncertain significance." These are also referred as "unclassified variants" (UVs). UVs in LS represent 32%, 18%, and 38% for *MLH1*, *MSH2*, and *MSH6* genes, respectively (Peltomaki and Vasen, 2004). This type of variant consists mostly of missense and intronic mutations, in-frame deletions or insertions, or mutations that do not lead to premature termination codon (Anczukow *et al.*, 2008). Whether or not an UV contributes to the disease or merely represents a neutral mutation or polymorphism is a major problem with clinical consequences. In practice, several classification methods may be used to evaluate the clinical significance of UVs. This is done using a multidisciplinary approach involving cosegregation with disease within pedigrees, population genetics, co-occurrence with deleterious mutation, RNA analysis and studies of protein function (Gammie *et al.*, 2007; Raevaara *et al.*, 2005), comparisons of allele frequencies, evolutionary conservation of the amino acids, and combinatorial approaches (Arnold *et al.*, 2009). Mutation databases can help to classify variants by providing a unique source of frequency information and co-occurrence data. Several MMR UV databases already exist, including the INSIGHT database (www.insight-group.org), which collates variants that are identified worldwide(Peltomaki and Vasen, 2004); the MMR Gene Unclassified Variants Database (www.mmruv.info); as well as a French database of *MMR* variants developed with the UMD software (Beroud *et al.*, 2005) (www.umd.be/MMR.html).

A novel cause of LS was recently described. Germline deletions in the epithelial cell adhesion molecule gene (*EpCAM*), also known as *TACSTD1*, were found in a subset of families with LS (Kovacs *et al.*, 2009; Ligtenberg *et al.*, 2009). These deletions provoke the synthesis of a fusion mRNA transcript between the *TACSTD1* and *MSH2* genes, resulting in epigenetic silencing of *MSH2* gene by promoter methylation. These families displayed an early onset LS phenotype. Varying deletions involving the 3′end of EPCAM have been described and represent at least 1–3% of confirmed LS families (Kuiper *et al.*, 2011), indicating that screening for these deletions should be implemented in routine LS diagnostics. *EPCAM* deletion carriers have a high risk of CRC, but only those with deletions extending close to the *MSH2* promoter have an increased risk of endometrial cancer (Kempers *et al.*, 2011).

Finally, constitutional epimutation of the *MLH1* promoter has been described in some cases (Gazzoli *et al.*, 2002; Hitchins, 2010). These are *de novo* events and are not transmitted from carriers to offspring. As a consequence, probands do not have a family history, but only a personal history of cancers with early onset. To date, only one case of non-Mendelian transmission between a mother and one of her children has been described (Hitchins *et al.*, 2007), with another possible case showing mosaic transmission (Morak *et al.*, 2008b).

In conclusion, currently available genetic testing for LS includes options for screening of *MLH1*, *MSH2*, *MSH6*, and *PMS2*, with additional testing available for large rearrangements in these genes as well as deletions in *TACSTD1/EpCAM*. The number of LS families whose genetic background has been clarified has gradually increased over time due to improvements in diagnostic procedures. Approximately 95% of MSI tumors can be accounted for by mutation or epigenetic inactivation of MMR genes (Hampel *et al.*, 2005a), while the remaining 5% are poorly understood. Other mechanisms could involve microRNAs that might regulate components of the MMR machinery (Valeri *et al.*, 2010). With the development of next generation sequencing technologies, new candidate CRC genes may be discovered. Deep sequencing of tumor exomes could identify mutations with potential clinical relevance (Timmermann *et al.*, 2010).

3. CLINICAL CONSEQUENCES OF DIFFERENT MMR DEFECTS

LS is an autosomal dominant predisposition to develop cancers. LS patients with a germline MMR mutation have a lifetime CRC risk of 69% in men and 52% in women (Hampel *et al.*, 2005b). The cumulative lifetime risk of endometrial cancer is equal to or greater than the cumulative risk of CRC (Hampel *et al.*, 2005b; Quehenberger *et al.*, 2005).

The specific risk of cancer in individuals with LS depends upon the MMR gene that is mutated. For example, males with *MLH1* mutations exhibit a significantly higher CRC risk than females, whereas the risk is similar in *MSH2* carriers (Choi *et al.*, 2009). *MLH1* carriers have a higher prevalence of CRC and younger age at diagnosis when compared to *MSH2* carriers (Goecke *et al.*, 2006; Kastrinos *et al.*, 2008). While the prevalence of endometrial cancer in women is similar for *MLH1* and *MSH2* mutation carriers, other extracolonic cancers such as renal and urethral cancers are more frequent in *MSH2* carriers (Geary *et al.*, 2008; Watson *et al.*, 2008).

MSH6 and *PMS2* germline mutation carriers present with specific clinical phenotypes. Compared to *MLH1* and *MSH2* mutation carriers, *MSH6* carriers are reported to show a lower cumulative risk for LS-associated tumors, including CRC, and an older age of onset (Hendriks *et al.*, 2004; Watson *et al.*, 2008). Baglietto *et al.* reported in a large series of 113 *MSH6* families that the lifetime risk of CRC was 22% and 10% for men and women, respectively, and 26% for endometrial cancer in women (Baglietto *et al.*, 2010). Compared to the general population, there was a 25-fold increase in the risk for endometrial cancer in *MSH6* mutation carriers.

The phenotype of *PMS2* germline mutations is usually attenuated. They are associated with older age of CRC (mean age of onset: 59 years) and a generally lower risk of LS-associated cancers. Senter *et al.* performed an analysis of *PMS2* mutations in 99 probands diagnosed with LS-associated tumors that showed isolated loss of PMS2 by IHC (Senter *et al.*, 2008). Germline *PMS2* mutations were detected in 62% of probands with a cumulative cancer risk by the age of 70 of only 15–20% for CRC, 15% for endometrial cancer, and 25–32% for any LS-associated cancers.

4. POTENTIAL FALSE NEGATIVE OR FALSE POSITIVE LYNCH SYNDROME CASES AS DETERMINED BY MOLECULAR ANALYSIS

False Negative: Second Hit in LS Tumors

According to the Knudson two-hit hypothesis, both alleles of classical tumor suppressor genes (TSG) are inactivated in tumors (Knudson, 2001). The second hit is generally loss of heterozygosity (LOH) of a relatively large chromosomal fragment containing the TSG. Alternatively, the wild-type TSG allele can be inactivated by mutation or transcriptional silencing by methylation of its promoter. Only a few studies have analyzed the nature of the second inactivating event ("hit") of the wild-type MMR gene allele in relatively large series of tumors from LS patients with a known germline MMR defect. In a series of 45 LS tumors carrying a pathogenic or nonpathogenic *MLH1* or *MSH2* germline mutation, Sanchez de Abajo *et al.* reported LOH at the MMR loci in 56% of the cases (Sanchez de Abajo *et al.*, 2006). In 40% of the cases with a pathogenic mutation and LOH, this LOH event targeted the mutant MMR gene allele. This unexpected finding was explained by the fact that the wild-type MMR allele could be inactivated by another mechanism such as mutation or promoter methylation, although this was not demonstrated. These authors postulated the observed LOH had a dual role, perhaps targeting other nearby cancer genes with relevance to tumor progression. In another study of 25 tumors from LS patients with a known *MSH2* or *MLH1* germline mutation, such a dual phenomenon was not reported since LOH was observed in 14 cases (56%) and in each of these, the wild-type MMR gene allele was targeted (Tuupanen *et al.*, 2007). A third study was published on a series of 57 colorectal and endometrial cancers from patients carrying one of the three Finnish *MLH1* founding germline mutations (Ollikainen *et al.*, 2007). These workers reported intermediate results with LOH of the *MLH1* locus observed in 31 (54.4%) cases and involving both the wild-type ($n = 23$) and mutant ($n = 8$) alleles.

Whether LOH has dual roles in the progression of LS tumors as a second hit of the MMR gene and/or the inactivation of other nearby cancer gene(s) remains to be determined, but certainly LOH would appear to be a common mechanism in these tumors.

A related MMR gene inactivation mechanism has been reported by Zhang *et al*. These authors analyzed 16 putative Swiss LS patients with an MSI tumor and showing loss of MLH1 or MSH2 expression by IHC but in which no germline mutation was detected by conventional DNA sequencing (Zhang *et al*., 2006). They looked for large genomic rearrangements of *MLH1* and *MSH2* in the germline DNA from these patients and found exon deletions in the *MLH1* and *MSH2* genes in 2/8 and 3/8 cases, respectively. A series of 11 formalin-fixed cancers from six genomic deletion carriers was then analyzed for genomic rearrangement. In four cases, the somatic MMR gene inactivation was a deletion similar to the one identified in the germline. A second series of seven Finnish patients was analyzed and the same phenomenon of somatic genomic rearrangement similar to that occurring in the germline was observed in two cases. In all cases, both MMR alleles were thus inactivated by the same deletion and to the same extent. This could not happen by chance and the proposed and likely explanation is that of gene conversion, that is, a locus-restricted recombination event, whereby the wild-type MMR allele in the tumor DNA is replaced by the mutated allele.

Apart from LOH, there are very few examples of other mechanisms that could account for the second inactivating hit of the MMR genes in tumors from LS patients. To the best of our knowledge, there is no described case of inactivation of the wild-type MMR gene by somatic mutation. Promoter methylation of the wild-type allele was described in 4/55 cases from the Ollikainen series (Ollikainen *et al*., 2007) and in 2/98 cases of another series with *MSH2* or *MLH1* germline mutation (Rahner *et al*., 2008).

For rare LS cases, it is possible that the second MMR allele is inactivated by promoter methylation. This could then lead to false negative determination of LS if the presence of *MLH1* promoter methylation was being used to differentiate sporadic MSI cases from LS MSI cases (Fig. 3). Constitutional epimutation of the *MLH1* promoter would not contribute as a negative determinant of LS cases according to this differentiation procedure, since the epigenetic defect generally appears *de novo* and is not transmitted to offspring (Hitchins, 2010). This epigenetic defect leads to a "sporadic case of Lynch syndrome."

False Positive: MSI and Inflammatory Bowel Disease

Inflammatory bowel diseases (IBDs) include Crohn's disease (CD) and ulcerative colitis (UC). Patients with UC have an increased risk of developing CRC compared with the general population (Ekbom *et al*., 1990). Despite the initial belief that CD carries a lower risk of CRC than UC, there is now growing evidence that both diseases are associated with an increased risk of

CRC (Gillen et al., 1994a,b). An increased risk for the development of small bowel adenocarcinoma (SBA) is also reported in patients with CD (Palascak-Juif et al., 2005). Risk factors for the development of CRC in the setting of IBD include disease duration, anatomic extent of disease, age at time of diagnosis, severity of inflammation, family history of CRC, and concomitant primary sclerosing cholangitis (Itzkowitz and Harpaz, 2004).

CRCs that develop in the setting of IBD are mainly adenocarcinomas. Like sporadic CRC, IBD-associated adenocarcinomas develop from the precursor lesion of dysplasia (or intraepithelial neoplasia). However, the pathogenesis and biological significance of dysplasia are different in these two types of carcinogenesis: in sporadic CRC the dysplastic precursor lesion is an adenomatous polyp, whereas in IBD the dysplasia may be polypoid or flat, and is often widespread and multifocal. Although the molecular events underlying IBD-associated CRC are similar to those described in sporadic colorectal carcinogenesis, the frequency and timing of the genetic alterations are different.

MSI has been reported to be a feature of some-IBD-associated intestinal cancers, but with variable frequencies ranging from less than 1% to 45% (Fleisher et al., 2000; Fujiwara et al., 2008; Lyda et al., 2000; Schulmann et al., 2005; Suzuki et al., 1994; Willenbucher et al., 1999). The main reasons for such a discrepancy are probably the use of nonstandardized microsatellite markers to evaluate MSI, as well as the small number of cases evaluated due to their rarity. In a study led by our laboratory on a large series of intestinal neoplasic lesions arising in the context of IBD (total of 277 neoplastic lesions in 205 patients), we observed the MSI phenotype in 8.3% of cases (Svrcek et al., 2007). The incidence of MSI was approximately the same in CD- and UC-related tumors. It was observed in both the dysplastic lesions (including low-grade dysplasia) and the CRCs of three patients, suggesting it was an early event during tumor progression of IBD-associated CRCs. The emergence of an MSI phenotype in IBD-associated tumors might be related to chronic inflammation and/or immunosuppression, as proposed recently in immunodeficiency-related non-Hodgkin's lymphomas (Borie et al., 2009; Duval et al., 2004).

The mechanisms underlying MMR deficiency in MSI IBDs are different to those in sporadic MSI tumors and appear to be more closely related to those observed in hereditary MSI tumors. Compared with sporadic MSI CRCs, patients showed a younger age at diagnosis and there was no female or right-sided predominance. Unlike sporadic MSI CRCs, MSI IBD-associated neoplasias presented with heterogeneous MMR defects involving MLH1, MSH2, MSH6, or PMS2 (as judged by IHC), and a low frequency of *MLH1* promoter methylation.

These cases, presenting a putative somatic mutation in MSI IBDs, would lead to false LS determination in instances where LS cases were differentiated from sporadic MSI cases by evidence of an MMR mutation (Fig. 3).

5. DE NOVO MUTATIONS

Constitutional *de novo* mutations are rare events and require a special set of circumstances in order to be detected. Absolute proof of a constitutional *de novo* mutation requires analysis of parental samples to demonstrate they are not affected by the germline MMR defect. There are only a few published examples of this for LS, all of them involving *MLH1* and *MSH2* (Kraus *et al.*, 1999; Morak *et al.*, 2008a; Stulp *et al.*, 2006). These patients do not have a family history of cancer in ascendant relatives but may have one in descendant ones. Constitutional epimutation of the *MLH1* gene is also a *de novo* event but affected patients generally do not have a family history of cancer in either ascendant or descendant relatives (Hitchins, 2010).

VI. UNUSUAL VARIANTS OF LYNCH SYNDROME

A. Muir Torre Syndrome

Muir Torre Syndrome (MTS) was first described by Muir *et al.* (1967) and subsequently by Torre (1968). MTS is usually diagnosed clinically by the synchronous or metachronous manifestation of a sebaceous neoplasm with or without kerathoacantomas and visceral malignancy, most commonly CRC. Later, the discovery of several patients with sebaceous neoplasms, CRC and a familial cancer predisposition led to the consideration of MTS as an uncommon phenotypic variant of LS (Lynch *et al.*, 1981). Although MTS is rare, the serious but curable nature of its associated cancers makes early identification of the syndrome important. Cutaneous neoplasms can serve as visible markers of internal malignancy. The most important is sebaceous adenoma, a relatively rare but benign tumor that presents as yellow papules or nodules. Whenever multiple sebaceous tumors are identified involving any site or any sebaceous tumor is discovered outside of the head and neck region and especially in young patients (<50 years), it is important to consider MTS as a possible cause. The initial test should be IHC in order to determine whether MMR proteins are expressed abnormally in the dermatological lesion. Germline mutations in *MSH2* and *MSH6* are the most common cause of MTS and contribute about 90% of cases, with the remainder due to *MLH1* mutation (Mercader, 2010).

B. Turcot Syndrome

Turcot syndrome is characterized clinically by the occurrence of primary brain tumors and CRCs (Turcot *et al.*, 1959). TS has been associated with germline mutations in the *APC* gene (associated with adenomatous polyposis), but also in the *MLH1*, *MHS6*, and *PMS2* genes. The type of brain tumor tends to differ according to the underlying gene defect. *APC* mutations are more commonly associated with medulloblastoma, while MMR genes mutations are more commonly associated with glioblastoma (Hamilton *et al.*, 1995). Biallelic mutations of MMR genes have been described in such families with MSI positive brain tumors (Miyaki *et al.*, 1997), thus it is unclear whether TS should be considered an uncommon phenotypic variant of LS or as CMMRD syndrome (see below) (Agostini *et al.*, 2005; Felton *et al.*, 2007; Hegde *et al.*, 2005; Sjursen *et al.*, 2009).

C. Constitutional MMR-Deficiency Syndrome

Patients with homozygous or compound heterozygous mutations in MMR genes have already been reported. They show a distinctive tumor spectrum with childhood cancers that are typically hematological malignancies, brain tumors, and early Lynch-related tumors. Almost all of these patients show café-au-lait spots (CLS) reminiscent of neurofibromatosis. This syndrome has been named constitutional MMR-deficiency syndrome (CMMR-D) or Lynch III Syndrome (Felton *et al.*, 2007). More than 78 cases from at least 47 families of homozygous or compound heterozygous mutations in the four MMR genes have been reported to date. Most of these families present biallelic mutation of the *PMS2* gene. *MSH6* and *MLH1* biallelic mutations have been found in 9 and 10 families, respectively. Surprisingly, only a few patients from a smaller number of families have so far been reported with biallelic mutation of *MSH2* (Durno *et al.*, 2010).

The preponderance of *PMS2* mutation may in part be explained by the presence of a *PMS2* founder mutation in consanguineous Pakistani families living in the United Kingdom. Nevertheless, even when these patients are excluded, mutant *PMS2* alleles still constitute the largest group of patients with biallelic mutations. A speculative hypothesis is that certain *MSH2* mutations are not viable in a homozygous state, whereas this is less likely to be the case for *PMS2*, considering the reported low penetrance of heterozygous *PMS2* mutations.

Multiple CLS in young children are highly suggestive of NF1 although they are not pathognomonic. It is therefore not surprising that most of these patients were initially misdiagnosed with NF1. It is tempting to speculate that NF1 mutations are responsible for the formation of CLS and other signs reminiscent of NF1 as Lish nodules. The NF1 gene, a large gene containing microsatellite sequences, is known to be a mutational target of MMR deficiency (Wang *et al.*, 2003b). However, CLS in MMR deficient patients differs from typical NF1-associated CLS in that they often have a variable degree of pigmentation and irregular borders (De Vos *et al.*, 2006; Kruger *et al.*, 2008; Scott *et al.*, 2007). Patients also present with hypochromic spots which may be part of this syndrome rather than coincidental. Hypopigmentation is not a feature of NF1 and its presence should be carefully investigated. If present, it could be a useful means to differentiate CMMR-D syndrome from NF1. It is, however, conceivable that different genetic alterations account for the cutaneous phenotype of this syndrome and that it cannot be reduced to a NF1 phenotype. CMMRD syndrome should be considered in any individual presenting with skin pigmentation abnormalities and childhood malignancy, even in the absence of a strong family history of tumors and especially in a context of consanguinity.

VII. GENETIC ALTERATIONS RESPONSIBLE FOR MSI TUMOR PROGRESSION IN SPORADIC AND LYNCH SYNDROME CASES

Many genetic and epigenetic alterations have been shown to occur in CRC. The main ones are compatible with the model of colorectal tumor progression proposed by Fearon and Vogelstein (1990). They include biallelic inactivation of the tumor suppressor genes *APC*, *TP53*, and *DPC4*, and activating mutations of the *KRAS* or *BRAF* oncogenes. However, MSI tumors do not follow this model which is more relevant to MSS tumors showing frequent chromosomal instability.

MSI tumors do not show frequent alterations of the *APC*, *TP53*, and *DPC4* genes and are generally diploid with few chromosomal abnormalities (Cottu *et al.*, 1996; Olschwang *et al.*, 1997). Indeed, in both hereditary and sporadic MSI tumors the first event is inactivation of an MMR gene that is not in itself a transforming event. Because of this MMR defect, MSI tumors accumulate numerous non-repaired mutations (mainly deletions or insertions) in genes containing small mononucleotide repeats within their coding sequence. These genes have been termed target genes for instability. The first to be identified was the *TGFβRII* gene which is mutated in about 90% of MSI CRC (Markowitz *et al.*, 1995). Following this, frequent mutations were reported in the *BAX*, *MSH3*, *MSH6*, and *TCF4* genes, for example (Duval *et al.*, 1999; Malkhosyan *et al.*, 1996; Rampino *et al.*, 1997). About 20% of human genes contain a coding repeat tract $\geq 7N$ (El-Bchiri *et al.*, 2008). All are potential target genes for instability, depending on their function and the length of their coding repeat. A number of these repeats have been analyzed in MSI tumors in "Instabilotyping" experiments, with a large range of mutation frequencies identified (Mori *et al.*, 2001; Woerner *et al.*, 2001). The challenge is to differentiate real target genes for instability from bystander events. Different statistical analyses have been proposed for this purpose (Duval and Hamelin, 2002; Woerner et al., 2003). Target genes for instability are involved in DNA repair, cell cycle regulation, and apoptosis. Functional studies are required to confirm that candidate genes are targets for instability. Recently, genes with very specific functions such as *Exportin-5* or *TARBP2* involved in miRNA nuclear export and processing, respectively, have been described (Melo et al., 2009, 2010).

There are also several reports concerning mutations in noncoding regulatory sequences believed to be important for gene expression. For example, the *MRE11* gene is mutated in an intronic repeat sequence, leading to exon skipping and synthesis of a truncated protein (Giannini et al., 2004). The *MYB* oncogene appears to be overexpressed in MSI tumors due to mutations of an intronic repeat within a regulatory transcriptional elongation sequence (Hugo et al., 2006).

There are differences in the repertoire of target genes for instability between gastrointestinal and endometrial MSI tumors (Duval et al., 2002), but very few differences, if any, between sporadic and hereditary cases.

Besides mutations in coding repeats due to their MMR defect, MSI CRC tumors frequently display activating mutations of the *BRAF* oncogene whereas *KRAS* mutations predominate in MSS CRC tumors (Oliveira et al., 2003, 2004). Interestingly, *BRAF* mutations are never observed in hereditary MSI cases but are present in 40% of sporadic MSI cases (Domingo et al., 2004). It appears thus that *BRAF* mutations are associated with *MLH1* promoter methylation and the CIMP phenotype generally, rather than specifically with MMR mutation. The reason for this association remains unknown.

VIII. CURRENT APPROACHES TO THE DETECTION OF LYNCH SYNDROME PATIENTS

The diagnosis of LS is based on the identification of germline MMR mutations. However, because of practical and financial constraints, it is first necessary to select patients prior to testing for mutations in the DNA MMR genes.

Besides the fact that clinical and familial criteria and/or predictive models to determine LS lack specificity and sensitivity, it has been reported that patients fulfilling these criteria are not frequently identified by clinicians, even in specialized cancer medical centers. Two retrospective studies conducted in the Netherlands (Van Lier et al., 2009) and in the United States (Mukherjee et al., 2010) reached similar conclusions about patients who should have been suspected as having LS. Among 1905 Dutch CRC patients, 169 met at least one of the revised Bethesda criteria, but MSI analysis was performed in only 23 (14%) of these. From 380 Lynch-associated tumors in the U.S. study, 41 met at least one of the revised Bethesda guidelines, but only 8 (19.5%) had been referred for genetic counselling and just 2 were seen by a medical geneticist. This underutilization of clinical and familial criteria can be explained by the frequent lack of knowledge by the patient of their own personal and family history of cancer, as well as inadequate knowledge of LS characteristics by clinicians.

For these reasons, many LS families remain undetected based on the current diagnosis of this syndrome which relies on personal features and family data. It also argues against the recently proposed strategy of genetic testing for a MMR gene defect in healthy patients selected by predictive models of LS (Dinh *et al.*, 2011). As highlighted by Hampel and de la Chapelle (2011), this strategy is also limited by the fact that sequencing may present challenges due to the frequent presence of large rearrangements in the MMR genes that are not detected using classical sequencing approaches.

In the absence of specific clinical and familial features to characterize LS patients, screening is now often performed using molecular analysis (Aaltonen *et al.*, 1998; Cunningham *et al.*, 2001; Hampel *et al.*, 2005a, 2008; Julie *et al.*, 2008; Pinol *et al.*, 2005; Salovaara *et al.*, 2000; Trano *et al.*, 2010). This is done by a systematic screen of all colorectal and endometrial tumors for MMR defects and/or MSI. These test results are then followed up to determine which of the positive cases are potentially due to an inherited predisposition. MSI status can be determined by microsatellite analysis or by IHC with antibodies against the four MMR proteins (MSH, MLH1 and MSH6 and PMS2). As discussed earlier, when correctly performed, these methods give identical results in the large majority of cases (Shia, 2008; Zhang, 2008). IHC has the advantage of indicating which MMR gene is defective. On the other hand, some missense MMR mutations may functionally inactivate the corresponding MMR protein (leading to instability detected by microsatellite genotyping), without obviously altering its expression as detected by IHC. We and others have proposed the use of both methods, with each serving as a positive control for the other. MSI classification as hereditary or sporadic is dependent on the fact that inherited MSI tumors are due to germline MMR mutation, while sporadic MSI tumors arise because of *MLH1* promoter methylation (Fig. 3).

There are four possibilities in the decision tree (Fig. 4):

A. No instability and IHC detection of all MMR proteins. This is an MSS tumor and cannot be from a LS patient. If the patient belongs to a family with a high incidence of CRC or LS-associated tumors, it could be a "phenocopy," or a patient with familial CRC type X.

B. MSI and IHC detection of all MMR proteins. This is likely to be an MSI tumor harboring a missense MMR gene mutation that inactivates its activity but does not affect its expression. The patient probably belongs to an LS

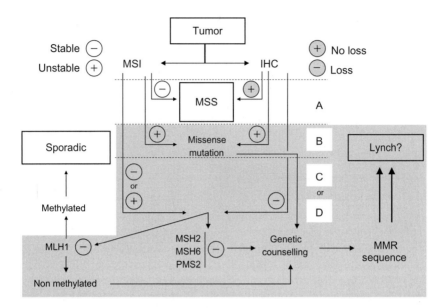

Fig. 4 Decision tree to detect LS among unselected CRC patients. (A) Absence of MSI and no MMR gene loss: MSS tumor. (B) MSI and no MMR gene loss: LS tumor with germline missense mutation. (C) Absence of MSI and loss of a MMR gene: putative LS tumor. (D) MSI and loss of a MMR gene: putative LS tumor according to the defective MMR gene

family, but there is no indication of which MMR gene is mutated, and therefore all must be sequenced until the germline mutation is found.

C. No instability and one of the MMR proteins is not detected by IHC. This is an MSI tumor in which instability was not detected, probably due to the low content of tumor cells in the analyzed sample (Brennetot et al., 2005). The presence of instability should be investigated in another sample after tumor enrichment by microdissection in order to confirm the IHC result. If instability is found in the new sample, the patient may belong to a LS family and would belong to the next category.

D. MSI and one of the MMR proteins is not detected by IHC. This is an MSI tumor. If the absent MMR protein is MSH2, MSH6 (alone), or PMS2 (alone), the patient probably belongs to a LS family and the MMR gene corresponding to the absent protein should be sequenced to look for a germline mutation. If the absent protein is MLH1, additional analysis of the *MLH1* promoter is necessary. If it is methylated, the tumor is likely to be sporadic and this can be confirmed in 40–50% of cases by the presence of a *BRAF* mutation. If the *MLH1* promoter is not methylated, it is indicative of the presence of a germline *MLH1* mutation and this gene should be sequenced to demonstrate LS.

All together, LS cases are in the gray part of Fig. 4. By following such a decision tree, LS patients not fulfilling modified Bethesda criteria and/or predictive models can be detected. The number of patients eligible for DNA sequencing to detect germline mutations is limited when the decision tree described above is used and the MMR gene to sequence is known in most cases.

IX. SURVEILLANCE OF INDIVIDUALS WITH LYNCH SYNDROME

The analysis for germline mutations in MMR genes can sometimes be difficult due to the absence of mutational hot spots and the frequent occurrence of large deletions. However, once a pathogenic mutation has been identified in an affected individual, it is much easier to offer genetic testing and counseling to relatives in order to identify those with a germline MMR gene mutation. In families in whom no mutation is found, the physician must make a judgment as to whether the risk of LS is sufficiently likely to justify intensive surveillance.

Clinical management can thus be directed only toward at-risk patients who are proven mutation carriers (Vasen *et al.*, 2007). Several studies have shown that surveillance of LS families reduces the development of CRC by 60% and also decreases mortality (de Jong *et al.*, 2006; Jarvinen *et al.*, 2000). LS family individuals who are at-risk are recommended to begin surveillance from the age of 20 to 25 years and to continue with intervals of 1–2 years (Lindor *et al.*, 2006; Vasen *et al.*, 2007).

Relative to conventional colonoscopy, high-resolution colonoscopy with chromoendoscopy markedly improves the detection of adenomas in individuals with LS and might help to prevent the development of CRC in these patients who are at very high risk (Lecomte *et al.*, 2005).

Annual gynecological surveillance is also recommended beginning at the age of 30, but the modalities are not well established. They are mainly based on clinical examination, ultrasonography, and endometrial biopsy. However, it is unclear whether screening reduces morbidity and mortality for endometrial cancer. Abnormal bleeding must be systematically followed up with further exploration. Increased risks of ovarian cancer have also led some to propose annual transvaginal ultrasound (Gerritzen *et al.*, 2009; Renkonen-Sinisalo *et al.*, 2007; Vasen *et al.*, 2007). Prophylactic hysterectomy with salpingo-oophorectomy can be offered to women in whom abdominal surgery is performed for other reasons, to those with particularly increased risk such as *MSH6* mutation carriers, and/or to women with multiple relatives affected by endometrial carcinoma (Pistorius *et al.*, 2006; Schmeler *et al.*, 2006).

Surveillance of the upper digestive tract by endoscopy can be considered every 2 years starting from age 30 to 35 years, or 5 years earlier than the youngest onset of gastric cancer. According to a German study, only 26% of gastric cancer cases had a family history for this cancer type, with the majority of these diagnosed after the age of 35 years (Goecke et al., 2006). There is no definitive evidence that supports a benefit in LS individuals from surveillance for early gastric cancers (Goecke et al., 2006; Renkonen-Sinisalo et al., 2002) or for small bowel cancers (Koornstra et al., 2008).

Because of the increased risk of urinary system cancers, annual renal ultrasound, urinalysis, and urine cytology should be considered in families with a history of urinary tract tumors. Lindor et al. recommend annual urinalysis with cytology because of the low cost and non-invasiveness (Lindor et al., 2006).

Screening for other LS-associated malignancies is not standardized and most decisions are based upon the specific pattern of tumors observed to occur within the family.

X. CONCLUSIONS

The best approach to detect LS patients is currently to analyze MSI and MMR expression by IHC. When the aberrant MMR gene expression is MLH1, discrimination between sporadic and LS is achieved by analyzing the methylation status of the MLH1 promoter. These analyses should be performed on all CRC and endometrial cancers, regardless of the age of diagnosis or the personal and family history of cancer (de la Chapelle and Hampel, 2010; Canard et al., 2011). This recommendation increases the number of analysis to be performed but should allow to detect most, if not all LS patients, avoiding cases that escape detection after any clinical and/or familial preselection. Moreover, such analyses are not only useful for detecting LS patients but are also of broader interest in the case of sporadic MSI. Indeed, it is known that MSI tumors have a more favorable outcome than MSS tumors (Popat et al., 2005). Moreover, patients with MSI tumors may not benefit as much from commonly used fluorouracil-based adjuvant chemotherapy as patients with MSS tumors (Ribic et al., 2003). On the other hand, Irinotecan-based chemotherapy appears to be more efficient for treating patients with MSI tumors (Fallik et al., 2003). Therefore, knowledge of the tumor MSI status is very useful for the clinician. These observations suggest that chemotherapy treatments could in future and with further validation be adapted according to the MSI status of tumors (Vilar and Gruber, 2010).

IHC and microsatellite analysis are becoming routine in cancer centers and are relatively low cost approaches. The cost of microsatellite genotyping may be reduced if the Bethesda panel is replaced by a panel of five mononucleotide repeats. This is because only a single pentaplex PCR is necessary, and extraction and analysis of matching normal DNA is not mandatory. The cost of IHC may also be reduced by using MSH6 and PMS2 antibodies as a first screening step, with MLH1 and MSH2 antibodies used only for positive cases from the first screen in order to define the exact MMR gene defect.

Sequencing of MMR genes is more expensive and time consuming but cannot be avoided since it brings definitive proof that a patient has LS. Current developments with next generation sequencing technologies will reduce this cost and facilitate the detection of LS patients. It is likely that the true prevalence of LS amongst CRCs and Lynch-associated tumors will soon be determined, thus allowing all at-risk patients to benefit from intensive surveillance.

ACKNOWLEDGMENT

We thank Dr. Barry Iacopetta for critical reading of the chapter.

REFERENCES

Aaltonen, L. A., Peltomaki, P., Leach, F. S., Sistonen, P., Pylkkanen, L., Mecklin, J. P., Jarvinen, H., Powell, S. M., Jen, J., Hamilton, S. R., *et al.* (1993). Clues to the pathogenesis of familial colorectal cancer. *Science (New York, NY)* **260**, 812–816.

Aaltonen, L. A., Salovaara, R., Kristo, P., Canzian, F., Hemminki, A., Peltomaki, P., Chadwick, R. B., Kaariainen, H., Eskelinen, M., Jarvinen, H., Mecklin, J. P., and de la Chapelle, A. (1998). Incidence of hereditary nonpolyposis colorectal cancer and the feasibility of molecular screening for the disease. *N. Engl. J. Med.* **338**, 1481–1487.

Aarnio, M., Sankila, R., Pukkala, E., Salovaara, R., Aaltonen, L. A., de la Chapelle, A., Peltomaki, P., Mecklin, J. P., and Jarvinen, H. J. (1999). Cancer risk in mutation carriers of DNA-mismatch-repair genes. *Int. J. Cancer* **81**, 214–218.

Agostini, M., Tibiletti, M. G., Lucci-Cordisco, E., Chiaravalli, A., Morreau, H., Furlan, D., Boccuto, L., Pucciarelli, S., Capella, C., Boiocchi, M., and Viel, A. (2005). Two PMS2 mutations in a Turcot syndrome family with small bowel cancers. *Am. J. Gastroenterol.* **100**, 1886–1891.

Anczukow, O., Ware, M. D., Buisson, M., Zetoune, A. B., Stoppa-Lyonnet, D., Sinilnikova, O. M., and Mazoyer, S. (2008). Does the nonsense-mediated mRNA decay mechanism prevent the synthesis of truncated BRCA1, CHK2, and p53 proteins? *Hum. Mutat.* **29**, 65–73.

Arnold, S., Buchanan, D. D., Barker, M., Jaskowski, L., Walsh, M. D., Birney, G., Woods, M. O., Hopper, J. L., Jenkins, M. A., Brown, M. A., Tavtigian, S. V., Goldgar, D. E., *et al.* (2009). Classifying MLH1 and MSH2 variants using bioinformatic prediction, splicing assays, segregation, and tumor characteristics. *Hum. Mutat.* **30**, 757–770.

Baglietto, L., Lindor, N. M., Dowty, J. G., White, D. M., Wagner, A., Gomez Garcia, E. B., Vriends, A. H., Cartwright, N. R., Barnetson, R. A., Farrington, S. M., Tenesa, A., Hampel, H., et al. (2010). Risks of Lynch syndrome cancers for MSH6 mutation carriers. *J. Natl. Cancer Inst.* **102**, 193–201.

Balmana, J., Stockwell, D. H., Steyerberg, E. W., Stoffel, E. M., Deffenbaugh, A. M., Reid, J. E., Ward, B., Scholl, T., Hendrickson, B., Tazelaar, J., Burbidge, L. A., and Syngal, S. (2006). Prediction of MLH1 and MSH2 mutations in Lynch syndrome. *JAMA* **296**, 1469–1478.

Bao, F., Panarelli, N. C., Rennert, H., Sherr, D. L., and Yantiss, R. K. (2010). Neoadjuvant therapy induces loss of MSH6 expression in colorectal carcinoma. *Am. J. Surg. Pathol.* **34**, 1798–1804.

Bapat, B., Lindor, N. M., Baron, J., Siegmund, K., Li, L., Zheng, Y., Haile, R., Gallinger, S., Jass, J. R., Young, J. P., Cotterchio, M., Jenkins, M., et al. (2009). The association of tumor microsatellite instability phenotype with family history of colorectal cancer. *Cancer Epidemiol. Biomarkers Prev.* **18**, 967–975.

Barnetson, R. A., Tenesa, A., Farrington, S. M., Nicholl, I. D., Cetnarskyj, R., Porteous, M. E., Campbell, H., and Dunlop, M. G. (2006). Identification and survival of carriers of mutations in DNA mismatch-repair genes in colon cancer. *N. Engl. J. Med.* **354**, 2751–2763.

Barrow, E., Jagger, E., Brierley, J., Wallace, A., Evans, G., Hill, J., and McMahon, R. (2010). Semiquantitative assessment of immunohistochemistry for mismatch repair proteins in Lynch syndrome. *Histopathology* **56**, 331–344.

Beck, N. E., Tomlinson, I. P., Homfray, T., Frayling, I., Hodgson, S. V., Harocopos, C., and Bodmer, W. F. (1997). Use of SSCP analysis to identify germline mutations in HNPCC families fulfilling the Amsterdam criteria. *Hum. Genet.* **99**, 219–224.

Beroud, C., Hamroun, D., Collod-Beroud, G., Boileau, C., Soussi, T., and Claustres, M. (2005). UMD (Universal Mutation Database): 2005 update. *Hum. Mutat.* **26**, 184–191.

Bisgaard, M. L., Fenger, K., Bulow, S., Niebuhr, E., and Mohr, J. (1994). Familial adenomatous polyposis (FAP): frequency, penetrance, and mutation rate. *Hum. Mutat.* **3**, 121–125.

Bisgaard, M. L., Jager, A. C., Myrhoj, T., Bernstein, I., and Nielsen, F. C. (2002). Hereditary non-polyposis colorectal cancer (HNPCC): phenotype-genotype correlation between patients with and without identified mutation. *Hum. Mutat.* **20**, 20–27.

Boland, C. R., Thibodeau, S. N., Hamilton, S. R., Sidransky, D., Eshleman, J. R., Burt, R. W., Meltzer, S. J., Rodriguez-Bigas, M. A., Fodde, R., Ranzani, G. N., and Srivastava, S. (1998). A National Cancer Institute Workshop on Microsatellite Instability for cancer detection and familial predisposition: development of international criteria for the determination of microsatellite instability in colorectal cancer. *Cancer Res.* **58**, 5248–5257.

Boland, C. R., Koi, M., Chang, D. K., and Carethers, J. M. (2008). The biochemical basis of microsatellite instability and abnormal immunohistochemistry and clinical behavior in Lynch syndrome: from bench to bedside. *Fam. Cancer* **7**, 41–52.

Borie, C., Colas, C., Dartigues, P., Lazure, T., Rince, P., Buhard, O., Folliot, P., Chalastanis, A., Muleris, M., Hamelin, R., Mercier, D., Oliveira, C., et al. (2009). The mechanisms underlying MMR deficiency in immunodeficiency-related non-Hodgkin lymphomas are different from those in other sporadic microsatellite instable neoplasms. *Int. J. Cancer* **125**, 2360–2366.

Brennetot, C., Buhard, O., Jourdan, F., Flejou, J. F., Duval, A., and Hamelin, R. (2005). Mononucleotide repeats BAT-26 and BAT-25 accurately detect MSI-H tumors and predict tumor content: implications for population screening. *Int. J. Cancer* **113**, 446–450.

Bronner, C. E., Baker, S. M., Morrison, P. T., Warren, G., Smith, L. G., Lescoe, M. K., Kane, M., Earabino, C., Lipford, J., Lindblom, A., et al. (1994). Mutation in the DNA mismatch repair gene homologue hMLH1 is associated with hereditary non-polyposis colon cancer. *Nature* **368**, 258–261.

Buhard, O., Suraweera, N., Lectard, A., Duval, A., and Hamelin, R. (2004). Quasimonomorphic mononucleotide repeats for high-level microsatellite instability analysis. *Dis. Markers* **20**, 251–257.

Buhard, O., Cattaneo, F., Wong, Y. F., Yim, S. F., Friedman, E., Flejou, J. F., Duval, A., and Hamelin, R. (2006). Multipopulation analysis of polymorphisms in five mononucleotide repeats used to determine the microsatellite instability status of human tumors. *J. Clin. Oncol.* **24**, 241–251.

Caldes, T., Godino, J., de la Hoya, M., Garcia Carbonero, I., Perez Segura, P., Eng, C., Benito, M., and Diaz-Rubio, E. (2002). Prevalence of germline mutations of MLH1 and MSH2 in hereditary nonpolyposis colorectal cancer families from Spain. *Int. J. Cancer* **98**, 774–779.

Canard, G., Lefevre, J. H., Colas, C., Coulet, F., Svrcek, M., Lascols, O., Hamelin, R., Shields, C., Duval, A., Fléjou, J. F., Soubrier, F., and Tiret, E. (2011). Screening for Lynch syndrome in colorectal cancer: are we doing enough? *Ann. Surg. Oncol.* Epub ahead of print.

Capel, E., Flejou, J. F., and Hamelin, R. (2007). Assessment of MLH1 promoter methylation in relation to gene expression requires specific analysis. *Oncogene* **26**, 7596–7600.

Castells, A., Balaguer, F., Castellvi-Bel, S., Gonzalo, V., and Ocana, T. (2007). Identification of Lynch syndrome: how should we proceed in the 21st century? *World J. Gastroenterol.* **13**, 4413–4416.

Charbonnier, F., Raux, G., Wang, Q., Drouot, N., Cordier, F., Limacher, J. M., Saurin, J. C., Puisieux, A., Olschwang, S., and Frebourg, T. (2000). Detection of exon deletions and duplications of the mismatch repair genes in hereditary nonpolyposis colorectal cancer families using multiplex polymerase chain reaction of short fluorescent fragments. *Cancer Res.* **60**, 2760–2763.

Chen, S., Wang, W., Lee, S., Nafa, K., Lee, J., Romans, K., Watson, P., Gruber, S. B., Euhus, D., Kinzler, K. W., Jass, J., Gallinger, S., *et al.* (2006). Prediction of germline mutations and cancer risk in the Lynch syndrome. *JAMA* **296**, 1479–1487.

Chen, J. R., Chiang, J. M., Changchien, C. R., Chen, J. S., Tang, R. P., and Wang, J. Y. (2008). Mismatch repair protein expression in Amsterdam II criteria-positive patients in Taiwan. *Br. J. Surg.* **95**, 102–110.

Choi, Y. H., Cotterchio, M., McKeown-Eyssen, G., Neerav, M., Bapat, B., Boyd, K., Gallinger, S., McLaughlin, J., Aronson, M., and Briollais, L. (2009). Penetrance of colorectal cancer among MLH1/MSH2 carriers participating in the colorectal cancer familial registry in Ontario. *Hered. Cancer Clin. Pract.* **7**, 14.

Cottu, P. H., Muzeau, F., Estreicher, A., Flejou, J. F., Iggo, R., Thomas, G., and Hamelin, R. (1996). Inverse correlation between RER+ status and p53 mutation in colorectal cancer cell lines. *Oncogene* **13**, 2727–2730.

Cunningham, J. M., Kim, C. Y., Christensen, E. R., Tester, D. J., Parc, Y., Burgart, L. J., Halling, K. C., McDonnell, S. K., Schaid, D. J., Walsh Vockley, C., Kubly, V., Nelson, H., *et al.* (2001). The frequency of hereditary defective mismatch repair in a prospective series of unselected colorectal carcinomas. *Am. J. Hum. Genet.* **69**, 780–790.

Davies, H., Bignell, G. R., Cox, C., Stephens, P., Edkins, S., Clegg, S., Teague, J., Woffendin, H., Garnett, M. J., Bottomley, W., Davis, N., Dicks, E., *et al.* (2002). Mutations of the BRAF gene in human cancer. *Nature* **417**, 949–954.

de Jong, A. E., Hendriks, Y. M., Kleibeuker, J. H., de Boer, S. Y., Cats, A., Griffioen, G., Nagengast, F. M., Nelis, F. G., Rookus, M. A., and Vasen, H. F. (2006). Decrease in mortality in Lynch syndrome families because of surveillance. *Gastroenterology* **130**, 665–671.

de la Chapelle, A., and Hampel, H. (2010). Clinical relevance of microsatellite instability in colorectal cancer. *J. Clin. Oncol.* **28**, 3380–3387.

De Rosa, M., Fasano, C., Panariello, L., Scarano, M. I., Belli, G., Iannelli, A., Ciciliano, F., and Izzo, P. (2000). Evidence for a recessive inheritance of Turcot's syndrome caused by compound heterozygous mutations within the PMS2 gene. *Oncogene* **19**, 1719–1723.

De Vos, M., Hayward, B. E., Charlton, R., Taylor, G. R., Glaser, A. W., Picton, S., Cole, T. R., Maher, E. R., McKeown, C. M., Mann, J. R., Yates, J. R., Baralle, D., et al. (2006). PMS2 mutations in childhood cancer. *J. Natl. Cancer Inst.* **98**, 358–361.

Deng, G., Chen, A., Hong, J., Chae, H. S., and Kim, Y. S. (1999). Methylation of CpG in a small region of the hMLH1 promoter invariably correlates with the absence of gene expression. *Cancer Res.* **59**, 2029–2033.

Desai, D. C., Lockman, J. C., Chadwick, R. B., Gao, X., Percesepe, A., Evans, D. G., Miyaki, M., Yuen, S. T., Radice, P., Maher, E. R., Wright, F. A., and de La Chapelle, A. (2000). Recurrent germline mutation in MSH2 arises frequently de novo. *J. Med. Genet.* **37**, 646–652.

Dinh, T. A., Rosner, B. I., Atwood, J. C., Boland, C. R., Syngal, S., Vasen, H. F., Gruber, S. B., and Burt, R. W. (2011). Health benefits and cost-effectiveness of primary genetic screening for Lynch syndrome in the general population. *Cancer Prev. Res. (Phila.)* **4**, 9–22.

Domingo, E., Laiho, P., Ollikainen, M., Pinto, M., Wang, L., French, A. J., Westra, J., Frebourg, T., Espin, E., Armengol, M., Hamelin, R., Yamamoto, H., et al. (2004). BRAF screening as a low-cost effective strategy for simplifying HNPCC genetic testing. *J. Med. Genet.* **41**, 664–668.

Dove-Edwin, I., de Jong, A. E., Adams, J., Mesher, D., Lipton, L., Sasieni, P., Vasen, H. F., and Thomas, H. J. (2006). Prospective results of surveillance colonoscopy in dominant familial colorectal cancer with and without Lynch syndrome. *Gastroenterology* **130**, 1995–2000.

Dunlop, M. G., Farrington, S. M., Carothers, A. D., Wyllie, A. H., Sharp, L., Burn, J., Liu, B., Kinzler, K. W., and Vogelstein, B. (1997). Cancer risk associated with germline DNA mismatch repair gene mutations. *Hum. Mol. Genet.* **6**, 105–110.

Durno, C. A., Holter, S., Sherman, P. M., and Gallinger, S. (2010). The gastrointestinal phenotype of germline biallelic mismatch repair gene mutations. *Am. J. Gastroenterol.* **105**, 2449–2456.

Duval, A., and Hamelin, R. (2002). Mutations at coding repeat sequences in mismatch repair-deficient human cancers: toward a new concept of target genes for instability. *Cancer Res.* **62**, 2447–2454.

Duval, A., Gayet, J., Zhou, X. P., Iacopetta, B., Thomas, G., and Hamelin, R. (1999). Frequent frameshift mutations of the TCF-4 gene in colorectal cancers with microsatellite instability. *Cancer Res.* **59**, 4213–4215.

Duval, A., Reperant, M., Compoint, A., Seruca, R., Ranzani, G. N., Iacopetta, B., and Hamelin, R. (2002). Target gene mutation profile differs between gastrointestinal and endometrial tumors with mismatch repair deficiency. *Cancer Res.* **62**, 1609–1612.

Duval, A., Raphael, M., Brennetot, C., Poirel, H., Buhard, O., Aubry, A., Martin, A., Krimi, A., Leblond, V., Gabarre, J., Davi, F., Charlotte, F., et al. (2004). The mutator pathway is a feature of immunodeficiency-related lymphomas. *Proc. Natl. Acad. Sci. U.S.A.* **101**, 5002–5007.

Ekbom, A., Helmick, C., Zack, M., and Adami, H. O. (1990). Ulcerative colitis and colorectal cancer. A population-based study. *N. Engl. J. Med.* **323**, 1228–1233.

El-Bchiri, J., Guilloux, A., Dartigues, P., Loire, E., Mercier, D., Buhard, O., Sobhani, I., de la Grange, P., Auboeuf, D., Praz, F., Flejou, J. F., and Duval, A. (2008). Nonsense-mediated mRNA decay impacts MSI-driven carcinogenesis and anti-tumor immunity in colorectal cancers. *PLoS One* **3**, e2583.

Fallik, D., Borrini, F., Boige, V., Viguier, J., Jacob, S., Miquel, C., Sabourin, J. C., Ducreux, M., and Praz, F. (2003). Microsatellite instability is a predictive factor of the tumor response to irinotecan in patients with advanced colorectal cancer. *Cancer Res.* **63**, 5738–5744.

Fearon, E. R., and Vogelstein, B. (1990). A genetic model for colorectal tumorigenesis. *Cell* **61**, 759–767.

Felton, K. E., Gilchrist, D. M., and Andrew, S. E. (2007). Constitutive deficiency in DNA mismatch repair. *Clin. Genet.* **71**, 483–498.

Fishel, R., Lescoe, M. K., Rao, M. R., Copeland, N. G., Jenkins, N. A., Garber, J., Kane, M., and Kolodner, R. (1993). The human mutator gene homolog MSH2 and its association with hereditary nonpolyposis colon cancer. *Cell* **75**, 1027–1038.

Fleisher, A. S., Esteller, M., Harpaz, N., Leytin, A., Rashid, A., Xu, Y., Liang, J., Stine, O. C., Yin, J., Zou, T. T., Abraham, J. M., Kong, D., *et al.* (2000). Microsatellite instability in inflammatory bowel disease-associated neoplastic lesions is associated with hypermethylation and diminished expression of the DNA mismatch repair gene, hMLH1. *Cancer Res.* **60**, 4864–4868.

Foulkes, W. D., Thiffault, I., Gruber, S. B., Horwitz, M., Hamel, N., Lee, C., Shia, J., Markowitz, A., Figer, A., Friedman, E., Farber, D., Greenwood, C. M., *et al.* (2002). The founder mutation MSH2*1906G–>C is an important cause of hereditary nonpolyposis colorectal cancer in the Ashkenazi Jewish population. *Am. J. Hum. Genet.* **71**, 1395–1412.

Froggatt, N. J., Green, J., Brassett, C., Evans, D. G., Bishop, D. T., Kolodner, R., and Maher, E. R. (1999). A common MSH2 mutation in English and North American HNPCC families: origin, phenotypic expression, and sex specific differences in colorectal cancer. *J. Med. Genet.* **36**, 97–102.

Fujiwara, I., Yashiro, M., Kubo, N., Maeda, K., and Hirakawa, K. (2008). Ulcerative colitis-associated colorectal cancer is frequently associated with the microsatellite instability pathway. *Dis. Colon Rectum* **51**, 1387–1394.

Gammie, A. E., Erdeniz, N., Beaver, J., Devlin, B., Nanji, A., and Rose, M. D. (2007). Functional characterization of pathogenic human MSH2 missense mutations in Saccharomyces cerevisiae. *Genetics* **177**, 707–721.

Gazzoli, I., Loda, M., Garber, J., Syngal, S., and Kolodner, R. D. (2002). A hereditary nonpolyposis colorectal carcinoma case associated with hypermethylation of the MLH1 gene in normal tissue and loss of heterozygosity of the unmethylated allele in the resulting microsatellite instability-high tumor. *Cancer Res.* **62**, 3925–3928.

Geary, J., Sasieni, P., Houlston, R., Izatt, L., Eeles, R., Payne, S. J., Fisher, S., and Hodgson, S. V. (2008). Gene-related cancer spectrum in families with hereditary non-polyposis colorectal cancer (HNPCC). *Fam. Cancer* **7**, 163–172.

Gerritzen, L. H., Hoogerbrugge, N., Oei, A. L., Nagengast, F. M., van Ham, M. A., Massuger, L. F., and de Hullu, J. A. (2009). Improvement of endometrial biopsy over transvaginal ultrasound alone for endometrial surveillance in women with Lynch syndrome. *Fam. Cancer* **8**, 391–397.

Giannini, G., Rinaldi, C., Ristori, E., Ambrosini, M. I., Cerignoli, F., Viel, A., Bidoli, E., Berni, S., D'Amati, G., Scambia, G., Frati, L., Screpanti, I., *et al.* (2004). Mutations of an intronic repeat induce impaired MRE11 expression in primary human cancer with microsatellite instability. *Oncogene* **23**, 2640–2647.

Giardiello, F. M., Brensinger, J. D., and Petersen, G. M. (2001). AGA technical review on hereditary colorectal cancer and genetic testing. *Gastroenterology* **121**, 198–213.

Gillen, C. D., Andrews, H. A., Prior, P., and Allan, R. N. (1994a). Crohn's disease and colorectal cancer. *Gut* **35**, 651–655.

Gillen, C. D., Walmsley, R. S., Prior, P., Andrews, H. A., and Allan, R. N. (1994b). Ulcerative colitis and Crohn's disease: a comparison of the colorectal cancer risk in extensive colitis. *Gut* **35**, 1590–1592.

Goecke, T., Schulmann, K., Engel, C., Holinski-Feder, E., Pagenstecher, C., Schackert, H. K., Kloor, M., Kunstmann, E., Vogelsang, H., Keller, G., Dietmaier, W., Mangold, E., *et al.* (2006). Genotype-phenotype comparison of German MLH1 and MSH2 mutation carriers clinically affected with Lynch syndrome: a report by the German HNPCC Consortium. *J. Clin. Oncol.* **24**, 4285–4292.

Goel, A., Nagasaka, T., Arnold, C. N., Inoue, T., Hamilton, C., Niedzwiecki, D., Compton, C., Mayer, R. J., Goldberg, R., Bertagnolli, M. M., and Boland, C. R. (2007). The CpG island methylator phenotype and chromosomal instability are inversely correlated in sporadic colorectal cancer. *Gastroenterology* **132**, 127–138.

Goel, A., Nagasaka, T., Hamelin, R., and Boland, C. R. (2010). An optimized pentaplex PCR for detecting DNA mismatch repair-deficient colorectal cancers. *PLoS One* 5, e9393.

Grabowski, M., Mueller-Koch, Y., Grasbon-Frodl, E., Koehler, U., Keller, G., Vogelsang, H., Dietmaier, W., Kopp, R., Siebers, U., Schmitt, W., Neitzel, B., Gruber, M., et al. (2005). Deletions account for 17% of pathogenic germline alterations in MLH1 and MSH2 in hereditary nonpolyposis colorectal cancer (HNPCC) families. *Genet. Test.* 9, 138–146.

Green, R. C., Parfrey, P. S., Woods, M. O., and Younghusband, H. B. (2009). Prediction of Lynch syndrome in consecutive patients with colorectal cancer. *J. Natl. Cancer Inst.* 101, 331–340.

Greenson, J. K., Huang, S. C., Herron, C., Moreno, V., Bonner, J. D., Tomsho, L. P., Ben-Izhak, O., Cohen, H. I., Trougouboff, P., Bejhar, J., Sova, Y., Pinchev, M., et al. (2009). Pathologic predictors of microsatellite instability in colorectal cancer. *Am. J. Surg. Pathol.* 33, 126–133.

Hall, G., Clarkson, A., Shi, A., Langford, E., Leung, H., Eckstein, R. P., and Gill, A. J. (2010). Immunohistochemistry for PMS2 and MSH6 alone can replace a four antibody panel for mismatch repair deficiency screening in colorectal adenocarcinoma. *Pathology* 42, 409–413.

Hamilton, S. R., Liu, B., Parsons, R. E., Papadopoulos, N., Jen, J., Powell, S. M., Krush, A. J., Berk, T., Cohen, Z., Tetu, B., et al. (1995). The molecular basis of Turcot's syndrome. *N. Engl. J. Med.* 332, 839–847.

Hampel, H., and de la Chapelle, A. (2011). The search for unaffected individuals with Lynch syndrome: do the ends justify the means? *Cancer Prev. Res. (Phila.)* 4, 1–5.

Hampel, H., Frankel, W. L., Martin, E., Arnold, M., Khanduja, K., Kuebler, P., Nakagawa, H., Sotamaa, K., Prior, T. W., Westman, J., Panescu, J., Fix, D., et al. (2005a). Screening for the Lynch syndrome (hereditary nonpolyposis colorectal cancer). *N. Engl. J. Med.* 352, 1851–1860.

Hampel, H., Stephens, J. A., Pukkala, E., Sankila, R., Aaltonen, L. A., Mecklin, J. P., and de la Chapelle, A. (2005b). Cancer risk in hereditary nonpolyposis colorectal cancer syndrome: later age of onset. *Gastroenterology* 129, 415–421.

Hampel, H., Frankel, W. L., Martin, E., Arnold, M., Khanduja, K., Kuebler, P., Clendenning, M., Sotamaa, K., Prior, T., Westman, J. A., Panescu, J., Fix, D., et al. (2008). Feasibility of screening for Lynch syndrome among patients with colorectal cancer. *J. Clin. Oncol.* 26, 5783–5788.

Hegde, M. R., Chong, B., Blazo, M. E., Chin, L. H., Ward, P. A., Chintagumpala, M. M., Kim, J. Y., Plon, S. E., and Richards, C. S. (2005). A homozygous mutation in MSH6 causes Turcot syndrome. *Clin. Cancer Res.* 11, 4689–4693.

Hendriks, Y. M., Wagner, A., Morreau, H., Menko, F., Stormorken, A., Quehenberger, F., Sandkuijl, L., Moller, P., Genuardi, M., Van Houwelingen, H., Tops, C., Van Puijenbroek, M., et al. (2004). Cancer risk in hereditary nonpolyposis colorectal cancer due to MSH6 mutations: impact on counseling and surveillance. *Gastroenterology* 127, 17–25.

Hitchins, M. P. (2010). Inheritance of epigenetic aberrations (constitutional epimutations) in cancer susceptibility. *Adv. Genet.* 70, 201–243.

Hitchins, M. P., Wong, J. J., Suthers, G., Suter, C. M., Martin, D. I., Hawkins, N. J., and Ward, R. L. (2007). Inheritance of a cancer-associated MLH1 germ-line epimutation. *N. Engl. J. Med.* 356, 697–705.

Hoang, J. M., Cottu, P. H., Thuille, B., Salmon, R. J., Thomas, G., and Hamelin, R. (1997). BAT-26, an indicator of the replication error phenotype in colorectal cancers and cell lines. *Cancer Res.* 57, 300–303.

Holinski-Feder, E., Muller-Koch, Y., Friedl, W., Moeslein, G., Keller, G., Plaschke, J., Ballhausen, W., Gross, M., Baldwin-Jedele, K., Jungck, M., Mangold, E., Vogelsang, H., et al. (2001). DHPLC mutation analysis of the hereditary nonpolyposis colon cancer (HNPCC) genes hMLH1 and hMSH2. *J. Biochem. Biophys. Methods* 47, 21–32.

Hugo, H., Cures, A., Suraweera, N., Drabsch, Y., Purcell, D., Mantamadiotis, T., Phillips, W., Dobrovic, A., Zupi, G., Gonda, T. J., Iacopetta, B., and Ramsay, R. G. (2006). Mutations in the MYB intron I regulatory sequence increase transcription in colon cancers. *Genes Chromosomes Cancer* **45**, 1143–1154.

Hyde, A., Fontaine, D., Stuckless, S., Green, R., Pollett, A., Simms, M., Sipahimalani, P., Parfrey, P., and Younghusband, B. (2010). A histology-based model for predicting microsatellite instability in colorectal cancers. *Am. J. Surg. Pathol.* **34**, 1820–1829.

Ionov, Y., Peinado, M. A., Malkhosyan, S., Shibata, D., and Perucho, M. (1993). Ubiquitous somatic mutations in simple repeated sequences reveal a new mechanism for colonic carcinogenesis. *Nature* **363**, 558–561.

Itzkowitz, S. H., and Harpaz, N. (2004). Diagnosis and management of dysplasia in patients with inflammatory bowel diseases. *Gastroenterology* **126**, 1634–1648.

Jarvinen, H. J., Aarnio, M., Mustonen, H., Aktan-Collan, K., Aaltonen, L. A., Peltomaki, P., De La Chapelle, A., and Mecklin, J. P. (2000). Controlled 15-year trial on screening for colorectal cancer in families with hereditary nonpolyposis colorectal cancer. *Gastroenterology* **118**, 829–834.

Jass, J. R. (2004). HNPCC and sporadic MSI-H colorectal cancer: a review of the morphological similarities and differences. *Fam. Cancer* **3**, 93–100.

Jass, J. R. (2006). Hereditary Non-Polyposis Colorectal Cancer: the rise and fall of a confusing term. *World J. Gastroenterol.* **12**, 4943–4950.

Jenkins, M. A., Hayashi, S., O'Shea, A. M., Burgart, L. J., Smyrk, T. C., Shimizu, D., Waring, P. M., Ruszkiewicz, A. R., Pollett, A. F., Redston, M., Barker, M. A., Baron, J. A., et al. (2007). Pathology features in Bethesda guidelines predict colorectal cancer microsatellite instability: a population-based study. *Gastroenterology* **133**, 48–56.

Julie, C., Tresallet, C., Brouquet, A., Vallot, C., Zimmermann, U., Mitry, E., Radvanyi, F., Rouleau, E., Lidereau, R., Coulet, F., Olschwang, S., Frebourg, T., et al. (2008). Identification in daily practice of patients with Lynch syndrome (hereditary nonpolyposis colorectal cancer): revised Bethesda guidelines-based approach versus molecular screening. *Am. J. Gastroenterol.* **103**, 2825–2835quiz 2836.

Kane, M. F., Loda, M., Gaida, G. M., Lipman, J., Mishra, R., Goldman, H., Jessup, J. M., and Kolodner, R. (1997). Methylation of the hMLH1 promoter correlates with lack of expression of hMLH1 in sporadic colon tumors and mismatch repair-defective human tumor cell lines. *Cancer Res.* **57**, 808–811.

Kariola, R., Hampel, H., Frankel, W. L., Raevaara, T. E., de la Chapelle, A., and Nystrom-Lahti, M. (2004). MSH6 missense mutations are often associated with no or low cancer susceptibility. *Br. J. Cancer* **91**, 1287–1292.

Kastrinos, F., Stoffel, E. M., Balmana, J., Steyerberg, E. W., Mercado, R., and Syngal, S. (2008). Phenotype comparison of MLH1 and MSH2 mutation carriers in a cohort of 1,914 individuals undergoing clinical genetic testing in the United States. *Cancer Epidemiol. Biomarkers Prev.* **17**, 2044–2051.

Katballe, N., Christensen, M., Wikman, F. P., Orntoft, T. F., and Laurberg, S. (2002). Frequency of hereditary non-polyposis colorectal cancer in Danish colorectal cancer patients. *Gut* **50**, 43–51.

Kempers, M. J., Kuiper, R. P., Ockeloen, C. W., Chappuis, P. O., Hutter, P., Rahner, N., Schackert, H. K., Steinke, V., Holinski-Feder, E., Morak, M., Kloor, M., Buttner, R., et al. (2011). Risk of colorectal and endometrial cancers in EPCAM deletion-positive Lynch syndrome: a cohort study. *Lancet Oncol.* **12**, 49–55.

Kim, H., Jen, J., Vogelstein, B., and Hamilton, S. R. (1994). Clinical and pathological characteristics of sporadic colorectal carcinomas with DNA replication errors in microsatellite sequences. *Am. J. Pathol.* **145**, 148–156.

Klarskov, L., Ladelund, S., Holck, S., Roenlund, K., Lindebjerg, J., Elebro, J., Halvarsson, B., von Salome, J., Bernstein, I., and Nilbert, M. (2010). Interobserver variability in the evaluation of mismatch repair protein immunostaining. *Hum. Pathol.* **41**, 1387–1396.

Knudson, A. G. (2001). Two genetic hits (more or less) to cancer. *Nat. Rev. Cancer* **1**, 157–162.

Kohonen-Corish, M., Ross, V. L., Doe, W. F., Kool, D. A., Edkins, E., Faragher, I., Wijnen, J., Khan, P. M., Macrae, F., and St John, D. J. (1996). RNA-based mutation screening in hereditary nonpolyposis colorectal cancer. *Am. J. Hum. Genet.* **59**, 818–824.

Koornstra, J. J., Kleibeuker, J. H., and Vasen, H. F. (2008). Small-bowel cancer in Lynch syndrome: is it time for surveillance? *Lancet Oncol.* **9**, 901–905.

Kovacs, M. E., Papp, J., Szentirmay, Z., Otto, S., and Olah, E. (2009). Deletions removing the last exon of TACSTD1 constitute a distinct class of mutations predisposing to Lynch syndrome. *Hum. Mutat.* **30**, 197–203.

Kraus, C., Kastl, S., Gunther, K., Klessinger, S., Hohenberger, W., and Ballhausen, W. G. (1999). A proven de novo germline mutation in HNPCC. *J. Med. Genet.* **36**, 919–921.

Kruger, S., Kinzel, M., Walldorf, C., Gottschling, S., Bier, A., Tinschert, S., von Stackelberg, A., Henn, W., Gorgens, H., Boue, S., Kolble, K., Buttner, R., *et al.* (2008). Homozygous PMS2 germline mutations in two families with early-onset haematological malignancy, brain tumours, HNPCC-associated tumours, and signs of neurofibromatosis type 1. *Eur. J. Hum. Genet.* **16**, 62–72.

Kuiper, R. P., Vissers, L. E., Venkatachalam, R., Bodmer, D., Hoenselaar, E., Goossens, M., Haufe, A., Kamping, E., Niessen, R. C., Hogervorst, F. B., Gille, J. J., Redeker, B., *et al.* (2011). Recurrence and variability of germline EPCAM deletions in Lynch syndrome. *Hum. Mutat.* **32**, 407–414.

Kumarasinghe, A. P., de Boer, B., Bateman, A. C., and Kumarasinghe, M. P. (2010). DNA mismatch repair enzyme immunohistochemistry in colorectal cancer: a comparison of biopsy and resection material. *Pathology* **42**, 414–420.

Laghi, L., Bianchi, P., and Malesci, A. (2008). Differences and evolution of the methods for the assessment of microsatellite instability. *Oncogene* **27**, 6313–6321.

Laiho, P., Launonen, V., Lahermo, P., Esteller, M., Guo, M., Herman, J. G., Mecklin, J. P., Jarvinen, H., Sistonen, P., Kim, K. M., Shibata, D., Houlston, R. S., *et al.* (2002). Low-level microsatellite instability in most colorectal carcinomas. *Cancer Res.* **62**, 1166–1170.

Leach, F. S., Nicolaides, N. C., Papadopoulos, N., Liu, B., Jen, J., Parsons, R., Peltomaki, P., Sistonen, P., Aaltonen, L. A., Nystrom-Lahti, M., *et al.* (1993). Mutations of a mutS homolog in hereditary nonpolyposis colorectal cancer. *Cell* **75**, 1215–1225.

Leach, F. S., Polyak, K., Burrell, M., Johnson, K. A., Hill, D., Dunlop, M. G., Wyllie, A. H., Peltomaki, P., de la Chapelle, A., Hamilton, S. R., Kinzler, K. W., and Vogelstein, B. (1996). Expression of the human mismatch repair gene hMSH2 in normal and neoplastic tissues. *Cancer Res.* **56**, 235–240.

Lecomte, T., Cellier, C., Meatchi, T., Barbier, J. P., Cugnenc, P. H., Jian, R., Laurent-Puig, P., and Landi, B. (2005). Chromoendoscopic colonoscopy for detecting preneoplastic lesions in hereditary nonpolyposis colorectal cancer syndrome. *Clin. Gastroenterol. Hepatol.* **3**, 897–902.

Ligtenberg, M. J., Kuiper, R. P., Chan, T. L., Goossens, M., Hebeda, K. M., Voorendt, M., Lee, T. Y., Bodmer, D., Hoenselaar, E., Hendriks-Cornelissen, S. J., Tsui, W. Y., Kong, C. K., *et al.* (2009). Heritable somatic methylation and inactivation of MSH2 in families with Lynch syndrome due to deletion of the 3′ exons of TACSTD1. *Nat. Genet.* **41**, 112–117.

Lindblom, A., Tannergard, P., Werelius, B., and Nordenskjold, M. (1993). Genetic mapping of a second locus predisposing to hereditary non-polyposis colon cancer. *Nat. Genet.* **5**, 279–282.

Lindor, N. M., Rabe, K., Petersen, G. M., Haile, R., Casey, G., Baron, J., Gallinger, S., Bapat, B., Aronson, M., Hopper, J., Jass, J., LeMarchand, L., *et al.* (2005). Lower cancer incidence in Amsterdam-I criteria families without mismatch repair deficiency: familial colorectal cancer type X. *JAMA* **293**, 1979–1985.

Lindor, N. M., Petersen, G. M., Hadley, D. W., Kinney, A. Y., Miesfeldt, S., Lu, K. H., Lynch, P., Burke, W., and Press, N. (2006). Recommendations for the care of individuals with an inherited predisposition to Lynch syndrome: a systematic review. *JAMA* **296**, 1507–1517.

Liu, T. (2010). Mutational screening of hMLH1 and hMSH2 that confer inherited colorectal cancer susceptibility using denature gradient gel electrophoresis (DGGE). *Methods Mol. Biol. (Clifton, NJ)* **653**, 193–205.

Liu, B., Parsons, R., Papadopoulos, N., Nicolaides, N. C., Lynch, H. T., Watson, P., Jass, J. R., Dunlop, M., Wyllie, A., Peltomaki, P., de la Chapelle, A., Hamilton, S. R., et al. (1996). Analysis of mismatch repair genes in hereditary non-polyposis colorectal cancer patients. *Nat. Med.* **2**, 169–174.

Llor, X., Pons, E., Xicola, R. M., Castells, A., Alenda, C., Pinol, V., Andreu, M., Castellvi-Bel, S., Paya, A., Jover, R., Bessa, X., Giros, A., et al. (2005). Differential features of colorectal cancers fulfilling Amsterdam criteria without involvement of the mutator pathway. *Clin. Cancer Res.* **11**, 7304–7310.

Loukola, A., Eklin, K., Laiho, P., Salovaara, R., Kristo, P., Jarvinen, H., Mecklin, J. P., Launonen, V., and Aaltonen, L. A. (2001). Microsatellite marker analysis in screening for hereditary nonpolyposis colorectal cancer (HNPCC). *Cancer Res.* **61**, 4545–4549.

Lyda, M. H., Noffsinger, A., Belli, J., and Fenoglio-Preiser, C. M. (2000). Microsatellite instability and K-ras mutations in patients with ulcerative colitis. *Hum. Pathol.* **31**, 665–671.

Lynch, H. T., and Lynch, J. F. (2004). Lynch syndrome: history and current status. *Dis. Markers* **20**, 181–198.

Lynch, H. T., Shaw, M. W., Magnuson, C. W., Larsen, A. L., and Krush, A. J. (1966). Hereditary factors in cancer. Study of two large midwestern kindreds. *Arch. Intern. Med.* **117**, 206–212.

Lynch, H. T., Lynch, P. M., Pester, J., and Fusaro, R. M. (1981). The cancer family syndrome. Rare cutaneous phenotypic linkage of Torre's syndrome. *Arch. Intern. Med.* **141**, 607–611.

Lynch, H. T., Watson, P., Lanspa, S. J., Marcus, J., Smyrk, T., Fitzgibbons, R. J., Jr., Kriegler, M., and Lynch, J. F. (1988). Natural history of colorectal cancer in hereditary nonpolyposis colorectal cancer (Lynch syndromes I and II). *Dis. Colon Rectum* **31**, 439–444.

Malkhosyan, S., Rampino, N., Yamamoto, H., and Perucho, M. (1996). Frameshift mutator mutations. *Nature* **382**, 499–500.

Malkhosyan, S. R., Yamamoto, H., Piao, Z., and Perucho, M. (2000). Late onset and high incidence of colon cancer of the mutator phenotype with hypermethylated hMLH1 gene in women. *Gastroenterology* **119**, 598.

Mangold, E., Pagenstecher, C., Friedl, W., Fischer, H. P., Merkelbach-Bruse, S., Ohlendorf, M., Friedrichs, N., Aretz, S., Buettner, R., Propping, P., and Mathiak, M. (2005). Tumours from MSH2 mutation carriers show loss of MSH2 expression but many tumours from MLH1 mutation carriers exhibit weak positive MLH1 staining. *J. Pathol.* **207**, 385–395.

Markowitz, S., Wang, J., Myeroff, L., Parsons, R., Sun, L., Lutterbaugh, J., Fan, R. S., Zborowska, E., Kinzler, K. W., Vogelstein, B., et al. (1995). Inactivation of the type II TGF-beta receptor in colon cancer cells with microsatellite instability. *Science (New York, NY)* **268**, 1336–1338.

Marroni, F., Pastrello, C., Benatti, P., Torrini, M., Barana, D., Cordisco, E. L., Viel, A., Mareni, C., Oliani, C., Genuardi, M., Bailey-Wilson, J. E., Ponz de Leon, M., et al. (2006). A genetic model for determining MSH2 and MLH1 carrier probabilities based on family history and tumor microsatellite instability. *Clin. Genet.* **69**, 254–262.

Melo, S. A., Ropero, S., Moutinho, C., Aaltonen, L. A., Yamamoto, H., Calin, G. A., Rossi, S., Fernandez, A. F., Carneiro, F., Oliveira, C., Ferreira, B., Liu, C. G., et al. (2009). A TARBP2 mutation in human cancer impairs microRNA processing and DICER1 function. *Nat. Genet.* **41**, 365–370.

Melo, S. A., Moutinho, C., Ropero, S., Calin, G. A., Rossi, S., Spizzo, R., Fernandez, A. F., Davalos, V., Villanueva, A., Montoya, G., Yamamoto, H., Schwartz, S., Jr., et al. (2010). A genetic defect in exportin-5 traps precursor microRNAs in the nucleus of cancer cells. *Cancer Cell* **18**, 303–315.

Mercader, P. (2010). Muir-Torre syndrome. *Adv. Exp. Med. Biol.* **685**, 186–195.

Miyaki, M., Nishio, J., Konishi, M., Kikuchi-Yanoshita, R., Tanaka, K., Muraoka, M., Nagato, M., Chong, J. M., Koike, M., Terada, T., Kawahara, Y., Fukutome, A., et al. (1997). Drastic genetic instability of tumors and normal tissues in Turcot syndrome. *Oncogene* **15**, 2877–2881.

Morak, M., Laner, A., Scholz, M., Madorf, T., and Holinski-Feder, E. (2008a). Report on denovo mutation in the MSH2 gene as a rare event in hereditary nonpolyposis colorectal cancer. *Eur. J. Gastroenterol. Hepatol.* **20**, 1101–1105.

Morak, M., Schackert, H. K., Rahner, N., Betz, B., Ebert, M., Walldorf, C., Royer-Pokora, B., Schulmann, K., von Knebel-Doeberitz, M., Dietmaier, W., Keller, G., Kerker, B., et al. (2008b). Further evidence for heritability of an epimutation in one of 12 cases with MLH1 promoter methylation in blood cells clinically displaying HNPCC. *Eur. J. Hum. Genet.* **16**, 804–811.

Mori, Y., Yin, J., Rashid, A., Leggett, B. A., Young, J., Simms, L., Kuehl, P. M., Langenberg, P., Meltzer, S. J., and Stine, O. C. (2001). Instabilotyping: comprehensive identification of frameshift mutations caused by coding region microsatellite instability. *Cancer Res.* **61**, 6046–6049.

Muir, E. G., Bell, A. J., and Barlow, K. A. (1967). Multiple primary carcinomata of the colon, duodenum, and larynx associated with kerato-acanthomata of the face. *Br. J. Surg.* **54**, 191–195.

Mukherjee, A., McGarrity, T. J., Ruggiero, F., Koltun, W., McKenna, K., Poritz, L., and Baker, M. J. (2010). The revised Bethesda guidelines: extent of utilization in a university hospital medical center with a cancer genetics program. *Hered. Cancer Clin. Pract.* **8**, 9.

Murphy, K. M., Zhang, S., Geiger, T., Hafez, M. J., Bacher, J., Berg, K. D., and Eshleman, J. R. (2006). Comparison of the microsatellite instability analysis system and the Bethesda panel for the determination of microsatellite instability in colorectal cancers. *J. Mol. Diagn.* **8**, 305–311.

Nardon, E., Glavac, D., Benhattar, J., Groenen, P. J., Hofler, G., Hofler, H., Jung, A., Keller, G., Kirchner, T., Lessi, F., Ligtenberg, M. J., Mazzanti, C. M., et al. (2010). A multicenter study to validate the reproducibility of MSI testing with a panel of 5 quasimonomorphic mononucleotide repeats. *Diagn. Mol. Pathol.* **19**, 236–242.

Nicolaides, N. C., Papadopoulos, N., Liu, B., Wei, Y. F., Carter, K. C., Ruben, S. M., Rosen, C. A., Haseltine, W. A., Fleischmann, R. D., Fraser, C. M., et al. (1994). Mutations of two PMS homologues in hereditary nonpolyposis colon cancer. *Nature* **371**, 75–80.

Oliveira, C., Pinto, M., Duval, A., Brennetot, C., Domingo, E., Espin, E., Armengol, M., Yamamoto, H., Hamelin, R., Seruca, R., and Schwartz, S., Jr. (2003). BRAF mutations characterize colon but not gastric cancer with mismatch repair deficiency. *Oncogene* **22**, 9192–9196.

Oliveira, C., Westra, J. L., Arango, D., Ollikainen, M., Domingo, E., Ferreira, A., Velho, S., Niessen, R., Lagerstedt, K., Alhopuro, P., Laiho, P., Veiga, I., et al. (2004). Distinct patterns of KRAS mutations in colorectal carcinomas according to germline mismatch repair defects and hMLH1 methylation status. *Hum. Mol. Genet.* **13**, 2303–2311.

Ollikainen, M., Hannelius, U., Lindgren, C. M., Abdel-Rahman, W. M., Kere, J., and Peltomaki, P. (2007). Mechanisms of inactivation of MLH1 in hereditary nonpolyposis colorectal carcinoma: a novel approach. *Oncogene* **26**, 4541–4549.

Olschwang, S., Hamelin, R., Laurent-Puig, P., Thuille, B., De Rycke, Y., Li, Y. J., Muzeau, F., Girodet, J., Salmon, R. J., and Thomas, G. (1997). Alternative genetic pathways in colorectal carcinogenesis. *Proc. Natl. Acad. Sci. USA* **94**, 12122–12127.

Overbeek, L. I., Ligtenberg, M. J., Willems, R. W., Hermens, R. P., Blokx, W. A., Dubois, S. V., van der Linden, H., Meijer, J. W., Mlynek-Kersjes, M. L., Hoogerbrugge, N., Hebeda, K. M., and van Krieken, J. H. (2008). Interpretation of immunohistochemistry for mismatch repair proteins is only reliable in a specialized setting. *Am. J. Surg. Pathol.* **32**, 1246–1251.

Palascak-Juif, V., Bouvier, A. M., Cosnes, J., Flourie, B., Bouche, O., Cadiot, G., Lemann, M., Bonaz, B., Denet, C., Marteau, P., Gambiez, L., Beaugerie, L., *et al.* (2005). Small bowel adenocarcinoma in patients with Crohn's disease compared with small bowel adenocarcinoma de novo. *Inflamm. Bowel Dis.* **11**, 828–832.

Palombo, F., Gallinari, P., Iaccarino, I., Lettieri, T., Hughes, M., D'Arrigo, A., Truong, O., Hsuan, J. J., and Jiricny, J. (1995). GTBP, a 160-kilodalton protein essential for mismatch-binding activity in human cells. *Science (New York, NY)* **268**, 1912–1914.

Papadopoulos, N., Nicolaides, N. C., Wei, Y. F., Ruben, S. M., Carter, K. C., Rosen, C. A., Haseltine, W. A., Fleischmann, R. D., Fraser, C. M., Adams, M. D., *et al.* (1994). Mutation of a mutL homolog in hereditary colon cancer. *Science (New York, NY)* **263**, 1625–1629.

Peltomaki, P. (2003). Role of DNA mismatch repair defects in the pathogenesis of human cancer. *J. Clin. Oncol.* **21**, 1174–1179.

Peltomaki, P., and Vasen, H. (2004). Mutations associated with HNPCC predisposition—update of ICG-HNPCC/INSiGHT mutation database. *Dis. Markers* **20**, 269–276.

Peltomaki, P., Aaltonen, L. A., Sistonen, P., Pylkkanen, L., Mecklin, J. P., Jarvinen, H., Green, J. S., Jass, J. R., Weber, J. L., Leach, F. S., *et al.* (1993). Genetic mapping of a locus predisposing to human colorectal cancer. *Science (New York, NY)* **260**, 810–812.

Perucho, M. (1999). Correspondence re: C.R. Boland et al., A National Cancer Institute workshop on microsatellite instability for cancer detection and familial predisposition: development of international criteria for the determination of microsatellite instability in colorectal cancer. *Cancer Res.*, 58: 5248–5257, 1998. *Cancer Res.* **59**, 249–256.

Pinol, V., Andreu, M., Castells, A., Paya, A., Bessa, X., and Rodrigo, J. (2004). Frequency of hereditary non-polyposis colorectal cancer and other colorectal cancer familial forms in Spain: a multicentre, prospective, nationwide study. *Eur. J. Gastroenterol. Hepatol.* **16**, 39–45.

Pinol, V., Castells, A., Andreu, M., Castellvi-Bel, S., Alenda, C., Llor, X., Xicola, R. M., Rodriguez-Moranta, F., Paya, A., Jover, R., and Bessa, X. (2005). Accuracy of revised Bethesda guidelines, microsatellite instability, and immunohistochemistry for the identification of patients with hereditary nonpolyposis colorectal cancer. *JAMA* **293**, 1986–1994.

Pistorius, S., Kruger, S., Hohl, R., Plaschke, J., Distler, W., Saeger, H. D., and Schackert, H. K. (2006). Occult endometrial cancer and decision making for prophylactic hysterectomy in hereditary nonpolyposis colorectal cancer patients. *Gynecol. Oncol.* **102**, 189–194.

Popat, S., Hubner, R., and Houlston, R. S. (2005). Systematic review of microsatellite instability and colorectal cancer prognosis. *J. Clin. Oncol.* **23**, 609–618.

Quehenberger, F., Vasen, H. F., and van Houwelingen, H. C. (2005). Risk of colorectal and endometrial cancer for carriers of mutations of the hMLH1 and hMSH2 gene: correction for ascertainment. *J. Med. Genet.* **42**, 491–496.

Raevaara, T. E., Korhonen, M. K., Lohi, H., Hampel, H., Lynch, E., Lonnqvist, K. E., Holinski-Feder, E., Sutter, C., McKinnon, W., Duraisamy, S., Gerdes, A. M., Peltomaki, P., *et al.* (2005). Functional significance and clinical phenotype of nontruncating mismatch repair variants of MLH1. *Gastroenterology* **129**, 537–549.

Rahner, N., Friedrichs, N., Steinke, V., Aretz, S., Friedl, W., Buettner, R., Mangold, E., Propping, P., and Walldorf, C. (2008). Coexisting somatic promoter hypermethylation and pathogenic MLH1 germline mutation in Lynch syndrome. *J. Pathol.* **214**, 10–16.

Rajagopalan, H., Bardelli, A., Lengauer, C., Kinzler, K. W., Vogelstein, B., and Velculescu, V. E. (2002). Tumorigenesis: RAF/RAS oncogenes and mismatch-repair status. *Nature* **418**, 934.

Rampino, N., Yamamoto, H., Ionov, Y., Li, Y., Sawai, H., Reed, J. C., and Perucho, M. (1997). Somatic frameshift mutations in the BAX gene in colon cancers of the microsatellite mutator phenotype. *Science (New York, NY)* **275**, 967–969.

Raskin, L., Schwenter, F., Freytsis, M., Tischkowitz, M., Wong, N., Chong, G., Narod, S., Levine, D., Bogomolniy, F., Aronson, M., Thibodeau, S., Hunt, K., *et al.* (2010). Characterization of two Ashkenazi Jewish founder mutations in MSH6 gene causing Lynch syndrome. *Clin. Genet.* **79**, 512–522.

Renkonen-Sinisalo, L., Sipponen, P., Aarnio, M., Julkunen, R., Aaltonen, L. A., Sarna, S., Jarvinen, H. J., and Mecklin, J. P. (2002). No support for endoscopic surveillance for gastric cancer in hereditary non-polyposis colorectal cancer. *Scand. J. Gastroenterol.* **37**, 574–577.

Renkonen-Sinisalo, L., Butzow, R., Leminen, A., Lehtovirta, P., Mecklin, J. P., and Jarvinen, H. J. (2007). Surveillance for endometrial cancer in hereditary nonpolyposis colorectal cancer syndrome. *Int. J. Cancer* **120**, 821–824.

Ribic, C. M., Sargent, D. J., Moore, M. J., Thibodeau, S. N., French, A. J., Goldberg, R. M., Hamilton, S. R., Laurent-Puig, P., Gryfe, R., Shepherd, L. E., Tu, D., Redston, M., *et al.* (2003). Tumor microsatellite-instability status as a predictor of benefit from fluorouracil-based adjuvant chemotherapy for colon cancer. *N. Engl. J. Med.* **349**, 247–257.

Rodriguez-Bigas, M. A., Boland, C. R., Hamilton, S. R., Henson, D. E., Jass, J. R., Khan, P. M., Lynch, H., Perucho, M., Smyrk, T., Sobin, L., and Srivastava, S. (1997). A National Cancer Institute Workshop on Hereditary Nonpolyposis Colorectal Cancer Syndrome: meeting highlights and Bethesda guidelines. *J. Natl. Cancer Inst.* **89**, 1758–1762.

Roman, R., Verdu, M., Calvo, M., Vidal, A., Sanjuan, X., Jimeno, M., Salas, A., Autonell, J., Trias, I., Gonzalez, M., Garcia, B., Rodon, N., *et al.* (2010). Microsatellite instability of the colorectal carcinoma can be predicted in the conventional pathologic examination. A prospective multicentric study and the statistical analysis of 615 cases consolidate our previously proposed logistic regression model. *Virchows Arch.* **456**, 533–541.

Rouleau, E., Lefol, C., Bourdon, V., Coulet, F., Noguchi, T., Soubrier, F., Bieche, I., Olschwang, S., Sobol, H., and Lidereau, R. (2009). Quantitative PCR high-resolution melting (qPCR-HRM) curve analysis, a new approach to simultaneously screen point mutations and large rearrangements: application to MLH1 germline mutations in Lynch syndrome. *Hum. Mutat.* **30**, 867–875.

Rustgi, A. K. (2007). The genetics of hereditary colon cancer. *Genes Dev.* **21**, 2525–2538.

Salahshor, S., Koelble, K., Rubio, C., and Lindblom, A. (2001). Microsatellite Instability and hMLH1 and hMSH2 expression analysis in familial and sporadic colorectal cancer. *Lab. Invest.* **81**, 535–541.

Salovaara, R., Loukola, A., Kristo, P., Kaariainen, H., Ahtola, H., Eskelinen, M., Harkonen, N., Julkunen, R., Kangas, E., Ojala, S., Tulikoura, J., Valkamo, E., *et al.* (2000). Population-based molecular detection of hereditary nonpolyposis colorectal cancer. *J. Clin. Oncol.* **18**, 2193–2200.

Samowitz, W. S., Slattery, M. L., Potter, J. D., and Leppert, M. F. (1999). BAT-26 and BAT-40 instability in colorectal adenomas and carcinomas and germline polymorphisms. *Am. J. Pathol.* **154**, 1637–1641.

Sanchez de Abajo, A., de la Hoya, M., van Puijenbroek, M., Godino, J., Diaz-Rubio, E., Morreau, H., and Caldes, T. (2006). Dual role of LOH at MMR loci in hereditary non-polyposis colorectal cancer? *Oncogene* **25**, 2124–2130.

Schmeler, K. M., Lynch, H. T., Chen, L. M., Munsell, M. F., Soliman, P. T., Clark, M. B., Daniels, M. S., White, K. G., Boyd-Rogers, S. G., Conrad, P. G., Yang, K. Y., Rubin, M. M., *et al.* (2006). Prophylactic surgery to reduce the risk of gynecologic cancers in the Lynch syndrome. *N. Engl. J. Med.* **354**, 261–269.

Schofield, L., Watson, N., Grieu, F., Li, W. Q., Zeps, N., Harvey, J., Stewart, C., Abdo, M., Goldblatt, J., and Iacopetta, B. (2009). Population-based detection of Lynch syndrome in young colorectal cancer patients using microsatellite instability as the initial test. *Int. J. Cancer* **124**, 1097–1102.

Schulmann, K., Mori, Y., Croog, V., Yin, J., Olaru, A., Sterian, A., Sato, F., Wang, S., Xu, Y., Deacu, E., Berki, A. T., Hamilton, J. P., *et al.* (2005). Molecular phenotype of inflammatory bowel disease-associated neoplasms with microsatellite instability. *Gastroenterology* **129**, 74–85.

Scott, R. J., McPhillips, M., Meldrum, C. J., Fitzgerald, P. E., Adams, K., Spigelman, A. D., du Sart, D., Tucker, K., and Kirk, J. (2001). Hereditary nonpolyposis colorectal cancer in 95 families: differences and similarities between mutation-positive and mutation-negative kindreds. *Am. J. Hum. Genet.* **68**, 118–127.

Scott, R. H., Homfray, T., Huxter, N. L., Mitton, S. G., Nash, R., Potter, M. N., Lancaster, D., and Rahman, N. (2007). Familial T-cell non-Hodgkin lymphoma caused by biallelic MSH2 mutations. *J. Med. Genet.* **44**, e83.

Senter, L., Clendenning, M., Sotamaa, K., Hampel, H., Green, J., Potter, J. D., Lindblom, A., Lagerstedt, K., Thibodeau, S. N., Lindor, N. M., Young, J., Winship, I., *et al.* (2008). The clinical phenotype of Lynch syndrome due to germ-line PMS2 mutations. *Gastroenterology* **135**, 419–428.

Shia, J. (2008). Immunohistochemistry versus microsatellite instability testing for screening colorectal cancer patients at risk for hereditary nonpolyposis colorectal cancer syndrome. Part I. The utility of immunohistochemistry. *J. Mol. Diagn.* **10**, 293–300.

Shia, J., Tang, L. H., Vakiani, E., Guillem, J. G., Stadler, Z. K., Soslow, R. A., Katabi, N., Weiser, M. R., Paty, P. B., Temple, L. K., Nash, G. M., Wong, W. D., *et al.* (2009). Immunohistochemistry as first-line screening for detecting colorectal cancer patients at risk for hereditary nonpolyposis colorectal cancer syndrome: a 2-antibody panel may be as predictive as a 4-antibody panel. *Am. J. Surg. Pathol.* **33**, 1639–1645.

Shia, J., Stadler, Z., Weiser, M. R., Rentz, M., Gonen, M., Tang, L. H., Vakiani, E., Katabi, N., Xiong, X., Markowitz, A. J., Shike, M., Guillem, J., *et al.* (2011). Immunohistochemical staining for DNA mismatch repair proteins in intestinal tract carcinoma: how reliable are biopsy samples? *Am. J. Surg. Pathol.* **35**, 447–454.

Sieber, O. M., Lipton, L., Crabtree, M., Heinimann, K., Fidalgo, P., Phillips, R. K., Bisgaard, M. L., Orntoft, T. F., Aaltonen, L. A., Hodgson, S. V., Thomas, H. J., and Tomlinson, I. P. (2003). Multiple colorectal adenomas, classic adenomatous polyposis, and germ-line mutations in MYH. *N. Engl. J. Med.* **348**, 791–799.

Sjursen, W., Bjornevoll, I., Engebretsen, L. F., Fjelland, K., Halvorsen, T., and Myrvold, H. E. (2009). A homozygote splice site PMS2 mutation as cause of Turcot syndrome gives rise to two different abnormal transcripts. *Fam. Cancer* **8**, 179–186.

Soravia, C., Berk, T., Madlensky, L., Mitri, A., Cheng, H., Gallinger, S., Cohen, Z., and Bapat, B. (1998). Genotype-phenotype correlations in attenuated adenomatous polyposis coli. *Am. J. Hum. Genet.* **62**, 1290–1301.

Southey, M. C., Jenkins, M. A., Mead, L., Whitty, J., Trivett, M., Tesoriero, A. A., Smith, L. D., Jennings, K., Grubb, G., Royce, S. G., Walsh, M. D., Barker, M. A., *et al.* (2005). Use of molecular tumor characteristics to prioritize mismatch repair gene testing in early-onset colorectal cancer. *J. Clin. Oncol.* **23**, 6524–6532.

Stulp, R. P., Vos, Y. J., Mol, B., Karrenbeld, A., de Raad, M., van der Mijle, H. J., and Sijmons, R. H. (2006). First report of a de novo germline mutation in the MLH1 gene. *World J. Gastroenterol.* **12**, 809–811.

Suraweera, N., Duval, A., Reperant, M., Vaury, C., Furlan, D., Leroy, K., Seruca, R., Iacopetta, B., and Hamelin, R. (2002). Evaluation of tumor microsatellite instability using five quasimonomorphic mononucleotide repeats and pentaplex PCR. *Gastroenterology* **123**, 1804–1811.

Suzuki, H., Harpaz, N., Tarmin, L., Yin, J., Jiang, H. Y., Bell, J. D., Hontanosas, M., Groisman, G. M., Abraham, J. M., and Meltzer, S. J. (1994). Microsatellite instability in ulcerative colitis-associated colorectal dysplasias and cancers. *Cancer Res.* **54**, 4841–4844.

Svrcek, M., El-Bchiri, J., Chalastanis, A., Capel, E., Dumont, S., Buhard, O., Oliveira, C., Seruca, R., Bossard, C., Mosnier, J. F., Berger, F., Leteurtre, E., et al. (2007). Specific clinical and biological features characterize inflammatory bowel disease associated colorectal cancers showing microsatellite instability. *J. Clin. Oncol.* **25**, 4231–4238.

Syngal, S., Fox, E. A., Eng, C., Kolodner, R. D., and Garber, J. E. (2000). Sensitivity and specificity of clinical criteria for hereditary non-polyposis colorectal cancer associated mutations in MSH2 and MLH1. *J. Med. Genet.* **37**, 641–645.

Talseth-Palmer, B. A., McPhillips, M., Groombridge, C., Spigelman, A., and Scott, R. J. (2010). MSH6 and PMS2 mutation positive Australian Lynch syndrome families: novel mutations, cancer risk and age of diagnosis of colorectal cancer. *Hered. Cancer Clin. Pract.* **8**, 5.

Thibodeau, S. N., Bren, G., and Schaid, D. (1993). Microsatellite instability in cancer of the proximal colon. *Science (New York, NY)* **260**, 816–819.

Timmermann, B., Kerick, M., Roehr, C., Fischer, A., Isau, M., Boerno, S. T., Wunderlich, A., Barmeyer, C., Seemann, P., Koenig, J., Lappe, M., Kuss, A. W., et al. (2010). Somatic mutation profiles of MSI and MSS colorectal cancer identified by whole exome next generation sequencing and bioinformatics analysis. *PLoS One* **5**, e15661.

Tomlinson, I., Halford, S., Aaltonen, L., Hawkins, N., and Ward, R. (2002). Does MSI-low exist? *J. Pathol.* **197**, 6–13.

Torre, D. (1968). Multiple sebaceous tumors. *Arch. Dermatol.* **98**, 549–551.

Toyota, M., Ahuja, N., Ohe-Toyota, M., Herman, J. G., Baylin, S. B., and Issa, J. P. (1999). CpG island methylator phenotype in colorectal cancer. *Proc. Natl. Acad. Sci. USA* **96**, 8681–8686.

Trano, G., Sjursen, W., Wasmuth, H. H., Hofsli, E., and Vatten, L. J. (2010). Performance of clinical guidelines compared with molecular tumour screening methods in identifying possible Lynch syndrome among colorectal cancer patients: a Norwegian population-based study. *Br. J. Cancer* **102**, 482–488.

Tresallet, C., Brouquet, A., Julie, C., Beauchet, A., Vallot, C., Menegaux, F., Mitry, E., Radvanyi, F., Malafosse, R., Rougier, P., Nordlinger, B., Laurent-Puig, P., et al. (2012). Evaluation of predictive models for the identification in daily practice of patients with Lynch syndrome. *Int. J. Cancer* **130**, 1367–1377.

Turcot, J., Despres, J. P., and St Pierre, F. (1959). Malignant tumors of the central nervous system associated with familial polyposis of the colon: report of two cases. *Dis. Colon Rectum* **2**, 465–468.

Tuupanen, S., Karhu, A., Jarvinen, H., Mecklin, J. P., Launonen, V., and Aaltonen, L. A. (2007). No evidence for dual role of loss of heterozygosity in hereditary non-polyposis colorectal cancer. *Oncogene* **26**, 2513–2517.

Umar, A., Boland, C. R., Terdiman, J. P., Syngal, S., de la Chapelle, A., Ruschoff, J., Fishel, R., Lindor, N. M., Burgart, L. J., Hamelin, R., Hamilton, S. R., Hiatt, R. A., et al. (2004). Revised Bethesda Guidelines for hereditary nonpolyposis colorectal cancer (Lynch syndrome) and microsatellite instability. *J. Natl. Cancer Inst.* **96**, 261–268.

Valeri, N., Gasparini, P., Fabbri, M., Braconi, C., Veronese, A., Lovat, F., Adair, B., Vannini, I., Fanini, F., Bottoni, A., Costinean, S., Sandhu, S. K., et al. (2010). Modulation of mismatch repair and genomic stability by miR-155. *Proc. Natl. Acad. Sci. USA* **107**, 6982–6987.

Valle, L., Perea, J., Carbonell, P., Fernandez, V., Dotor, A. M., Benitez, J., and Urioste, M. (2007). Clinicopathologic and pedigree differences in amsterdam I-positive hereditary nonpolyposis colorectal cancer families according to tumor microsatellite instability status. *J. Clin. Oncol.* **25**, 781–786.

van der Klift, H., Wijnen, J., Wagner, A., Verkuilen, P., Tops, C., Otway, R., Kohonen-Corish, M., Vasen, H., Oliani, C., Barana, D., Moller, P., Delozier-Blanchet, C., et al. (2005). Molecular characterization of the spectrum of genomic deletions in the mismatch repair genes MSH2, MLH1, MSH6, and PMS2 responsible for hereditary nonpolyposis colorectal cancer (HNPCC). *Genes Chromosomes Cancer* **44**, 123–138.

Van Lier, M. G., De Wilt, J. H., Wagemakers, J. J., Dinjens, W. N., Damhuis, R. A., Wagner, A., Kuipers, E. J., and Van Leerdam, M. E. (2009). Underutilization of microsatellite instability analysis in colorectal cancer patients at high risk for Lynch syndrome. *Scand. J. Gastroenterol.* **44**, 600–604.

Vasen, H. F., Mecklin, J. P., Khan, P. M., and Lynch, H. T. (1991). The International Collaborative Group on Hereditary Non-Polyposis Colorectal Cancer (ICG-HNPCC). *Dis. Colon Rectum* **34**, 424–425.

Vasen, H. F., Wijnen, J. T., Menko, F. H., Kleibeuker, J. H., Taal, B. G., Griffioen, G., Nagengast, F. M., Meijers-Heijboer, E. H., Bertario, L., Varesco, L., Bisgaard, M. L., Mohr, J., *et al.* (1996). Cancer risk in families with hereditary nonpolyposis colorectal cancer diagnosed by mutation analysis. *Gastroenterology* **110**, 1020–1027.

Vasen, H. F., Watson, P., Mecklin, J. P., and Lynch, H. T. (1999). New clinical criteria for hereditary nonpolyposis colorectal cancer (HNPCC, Lynch syndrome) proposed by the International Collaborative group on HNPCC. *Gastroenterology* **116**, 1453–1456.

Vasen, H. F., Moslein, G., Alonso, A., Bernstein, I., Bertario, L., Blanco, I., Burn, J., Capella, G., Engel, C., Frayling, I., Friedl, W., Hes, F. J., *et al.* (2007). Guidelines for the clinical management of Lynch syndrome (hereditary non-polyposis cancer). *J. Med. Genet.* **44**, 353–362.

Vilar, E., and Gruber, S. B. (2010). Microsatellite instability in colorectal cancer-the stable evidence. *Nat. Rev.* **7**, 153–162.

Wahlberg, S., Liu, T., Lindblom, P., and Lindblom, A. (1999). Various mutation screening techniques in the DNA mismatch repair genes hMSH2 and hMLH1. *Genet. Test.* **3**, 259–264.

Wang, L., Cunningham, J. M., Winters, J. L., Guenther, J. C., French, A. J., Boardman, L. A., Burgart, L. J., McDonnell, S. K., Schaid, D. J., and Thibodeau, S. N. (2003a). BRAF mutations in colon cancer are not likely attributable to defective DNA mismatch repair. *Cancer Res.* **63**, 5209–5212.

Wang, Q., Montmain, G., Ruano, E., Upadhyaya, M., Dudley, S., Liskay, R. M., Thibodeau, S. N., and Puisieux, A. (2003b). Neurofibromatosis type 1 gene as a mutational target in a mismatch repair-deficient cell type. *Hum. Genet.* **112**, 117–123.

Wang, Y., Friedl, W., Lamberti, C., Jungck, M., Mathiak, M., Pagenstecher, C., Propping, P., and Mangold, E. (2003c). Hereditary nonpolyposis colorectal cancer: frequent occurrence of large genomic deletions in MSH2 and MLH1 genes. *Int. J. Cancer* **103**, 636–641.

Warthin, A. S. (1913). Heredity with reference to carcinoma as shown by the study of the cases examined in the pathological laboratory of the University of Michigan. *Arch. Intern. Med.* **12**, 546–555.

Watanabe, A., Ikejima, M., Suzuki, N., and Shimada, T. (1996). Genomic organization and expression of the human MSH3 gene. *Genomics* **31**, 311–318.

Watson, N., Grieu, F., Morris, M., Harvey, J., Stewart, C., Schofield, L., Goldblatt, J., and Iacopetta, B. (2007). Heterogeneous staining for mismatch repair proteins during population-based prescreening for hereditary nonpolyposis colorectal cancer. *J. Mol. Diagn.* **9**, 472–478.

Watson, P., Vasen, H. F., Mecklin, J. P., Bernstein, I., Aarnio, M., Jarvinen, H. J., Myrhoj, T., Sunde, L., Wijnen, J. T., and Lynch, H. T. (2008). The risk of extra-colonic, extra-endometrial cancer in the Lynch syndrome. *Int. J. Cancer* **123**, 444–449.

Weisenberger, D. J., Siegmund, K. D., Campan, M., Young, J., Long, T. I., Faasse, M. A., Kang, G. H., Widschwendter, M., Weener, D., Buchanan, D., Koh, H., Simms, L., *et al.* (2006). CpG island methylator phenotype underlies sporadic microsatellite instability and is tightly associated with BRAF mutation in colorectal cancer. *Nat. Genet.* **38**, 787–793.

Wijnen, J., Vasen, H., Khan, P. M., Menko, F. H., van der Klift, H., van Leeuwen, C., van den Broek, M., van Leeuwen-Cornelisse, I., Nagengast, F., Meijers-Heijboer, A., *et al.* (1995). Seven new mutations in hMSH2, an HNPCC gene, identified by denaturing gradient-gel electrophoresis. *Am. J. Hum. Genet.* **56**, 1060–1066.

Wijnen, J. T., Vasen, H. F., Khan, P. M., Zwinderman, A. H., van der Klift, H., Mulder, A., Tops, C., Moller, P., and Fodde, R. (1998). Clinical findings with implications for genetic testing in families with clustering of colorectal cancer. *N. Engl. J. Med.* **339**, 511–518.

Willenbucher, R. F., Aust, D. E., Chang, C. G., Zelman, S. J., Ferrell, L. D., Moore, D. H., 2nd, and Waldman, F. M. (1999). Genomic instability is an early event during the progression pathway of ulcerative-colitis-related neoplasia. *Am. J. Pathol.* **154**, 1825–1830.

Woerner, S. M., Gebert, J., Yuan, Y. P., Sutter, C., Ridder, R., Bork, P., and von Knebel Doeberitz, M. (2001). Systematic identification of genes with coding microsatellites mutated in DNA mismatch repair-deficient cancer cells. *Int. J. Cancer* **93**, 12–19.

Woerner, S. M., Benner, A., Sutter, C., Schiller, M., Yuan, Y. P., Keller, G., Bork, P., Doeberitz, M. K., and Gebert, J. F. (2003). Pathogenesis of DNA repair-deficient cancers: a statistical meta-analysis of putative Real Common Target genes. *Oncogene* **22**, 2226–2235.

Wolf, B., Gruber, S., Henglmueller, S., Kappel, S., Bergmann, M., Wrba, F., and Karner-Hanusch, J. (2006). Efficiency of the revised Bethesda guidelines (2003) for the detection of mutations in mismatch repair genes in Austrian HNPCC patients. *Int. J. Cancer* **118**, 1465–1470.

Wong, Y. F., Cheung, T. H., Lo, K. W., Yim, S. F., Chan, L. K., Buhard, O., Duval, A., Chung, T. K., and Hamelin, R. (2006). Detection of microsatellite instability in endometrial cancer: advantages of a panel of five mononucleotide repeats over the National Cancer Institute panel of markers. *Carcinogenesis* **27**, 951–955.

Wu, Y., Berends, M. J., Post, J. G., Mensink, R. G., Verlind, E., Van Der Sluis, T., Kempinga, C., Sijmons, R. H., van der Zee, A. G., Hollema, H., Kleibeuker, J. H., Buys, C. H., *et al.* (2001a). Germline mutations of EXO1 gene in patients with hereditary nonpolyposis colorectal cancer (HNPCC) and atypical HNPCC forms. *Gastroenterology* **120**, 1580–1587.

Wu, Y., Berends, M. J., Sijmons, R. H., Mensink, R. G., Verlind, E., Kooi, K. A., van der Sluis, T., Kempinga, C., van dDer Zee, A. G., Hollema, H., Buys, C. H., Kleibeuker, J. H., *et al.* (2001b). A role for MLH3 in hereditary nonpolyposis colorectal cancer. *Nat. Genet.* **29**, 137–138.

Xicola, R. M., Llor, X., Pons, E., Castells, A., Alenda, C., Pinol, V., Andreu, M., Castellvi-Bel, S., Paya, A., Jover, R., Bessa, X., Giros, A., *et al.* (2007). Performance of different microsatellite marker panels for detection of mismatch repair-deficient colorectal tumors. *J. Natl. Cancer Inst.* **99**, 244–252.

You, J. F., Buhard, O., Ligtenberg, M. J., Kets, C. M., Niessen, R. C., Hofstra, R. M., Wagner, A., Dinjens, W. N., Colas, C., Lascols, O., Collura, A., Flejou, J. F., *et al.* (2010). Tumours with loss of MSH6 expression are MSI-H when screened with a pentaplex of five mononucleotide repeats. *Br. J. Cancer* **103**, 1840–1845.

Young, J., Simms, L. A., Biden, K. G., Wynter, C., Whitehall, V., Karamatic, R., George, J., Goldblatt, J., Walpole, I., Robin, S. A., Borten, M. M., Stitz, R., *et al.* (2001). Features of colorectal cancers with high-level microsatellite instability occurring in familial and sporadic settings: parallel pathways of tumorigenesis. *Am. J. Pathol.* **159**, 2107–2116.

Zhang, L. (2008). Immunohistochemistry versus microsatellite instability testing for screening colorectal cancer patients at risk for hereditary nonpolyposis colorectal cancer syndrome. Part II. The utility of microsatellite instability testing. *J. Mol. Diagn.* **10**, 301–307.

Zhang, J., Lindroos, A., Ollila, S., Russell, A., Marra, G., Mueller, H., Peltomaki, P., Plasilova, M., and Heinimann, K. (2006). Gene conversion is a frequent mechanism of inactivation of the wild-type allele in cancers from MLH1/MSH2 deletion carriers. *Cancer Res.* **66**, 659–664.

Zhou, X. P., Hoang, J. M., Cottu, P., Thomas, G., and Hamelin, R. (1997). Allelic profiles of mononucleotide repeat microsatellites in control individuals and in colorectal tumors with and without replication errors. *Oncogene* **15**, 1713–1718.

Zhou, X. P., Hoang, J. M., Li, Y. J., Seruca, R., Carneiro, F., Sobrinho-Simoes, M., Lothe, R. A., Gleeson, C. M., Russell, S. E., Muzeau, F., Flejou, J. F., Hoang-Xuan, K., *et al.* (1998). Determination of the replication error phenotype in human tumors without the requirement for matching normal DNA by analysis of mononucleotide repeat microsatellites. *Genes Chromosomes Cancer* **21**, 101–107.

Activation-Induced Cytidine Deaminase in Antibody Diversification and Chromosome Translocation

Anna Gazumyan,[*,†] Anne Bothmer,[*] Isaac A. Klein,[*] Michel C. Nussenzweig,[*,†] and Kevin M. McBride[‡]

[*]*Laboratory of Molecular Immunology, New York, USA*
[†]*Howard Hughes Medical Institute, The Rockefeller University, New York, USA*
[‡]*Department of Molecular Carcinogenesis, University of Texas MD Anderson Cancer Center, Smithville, Texas, USA*

I. Introduction
II. Chromosome Translocations
III. Activation-Induced Cytidine Deaminase
IV. Regulation of AID
  A. AID Is Targeted by Transcription
  B. AID Targeting by Spt5 and Stalled Pol II
V. Template Strand Targeting by Exosomes
VI. AID Genome Wide Damage
VII. CSR-Specific Cofactors
VIII. AID Influences CSR Outcome
IX. Conclusions
  Acknowledgment
  References

DNA damage, rearrangement, and mutation of the human genome are the basis of carcinogenesis and thought to be avoided at all costs. An exception is the adaptive immune system where lymphocytes utilize programmed DNA damage to effect antigen receptor diversification. Both B and T lymphocytes diversify their antigen receptors through RAG1/2 mediated recombination, but B cells undergo two additional processes—somatic hypermutation (SHM) and class-switch recombination (CSR), both initiated by activation-induced cytidine deaminase (AID). AID deaminates cytidines in DNA resulting in U:G mismatches that are processed into point mutations in SHM or double-strand breaks in CSR. Although AID activity is focused at Immunoglobulin (*Ig*) gene loci, it also targets a wide array of non-*Ig* genes including oncogenes associated with lymphomas. Here, we review the molecular basis of AID regulation, targeting, and initiation of CSR and SHM, as well as AID's role in generating chromosome translocations that contribute to lymphomagenesis. © 2012 Elsevier Inc.

I. INTRODUCTION

Adaptive immunity is an exquisitely specific immune response that vertebrates have evolved to recognize and remember specific pathogens. A key event is the somatic assembly of unique immune receptors; antibodies from Immunoglobulin (*Ig*) genes in B lymphocytes and the T-cell receptor in T lymphocytes. During development each lymphocyte is capable of generating a unique receptor specific to one of a vast array of possible antigens. For example, in humans B cells generate a repertoire of receptors that exceed trillions of unique antibodies (Abbas *et al.*, 2010). This diversity is generated by a system of programmed DNA damage that recombines and alters the coding sequence of lymphocyte immune receptors.

Initial antibody assembly occurs in the bone marrow. There, B lymphocytes express RAG1/2 to catalyze V(D)J recombination, a site-specific recombination reaction that juxtaposes variable (V), diversity (D), and joining (J) gene segments to one another (Fugmann *et al.*, 2000; Jung *et al.*, 2006). T lymphocytes assemble the T-cell receptor in a similar RAG1/2-mediated manner. However, in contrast to T cells, B cells can further diversify *Ig* genes through somatic hypermutation (SHM) and class-switch recombination (CSR) after antigen encounter (Fig. 1). SHM alters antibody affinity by introducing nucleotide changes in the antigen binding variable region of antibodies. B cells producing antibodies with improved antigen affinity are positively selected during the process of affinity maturation. CSR is a region-specific recombination reaction that replaces one antibody-constant region with another, thereby altering antibody effector function while leaving the variable region and its antigen binding specificity intact (Di Noia and Neuberger, 2007; Peled *et al.*, 2008; Rajewsky, 1996; Stavnezer *et al.*, 2008; Teng and Papavasiliou, 2007). While CSR and SHM are very different reactions, both are initiated by AID (Muramatsu *et al.*, 2000; Revy *et al.*, 2000), which introduces uracil:guanine (U:G) mismatches in transcribed DNA (Bransteitter *et al.*, 2003; Chaudhuri *et al.*, 2003; Dickerson *et al.*, 2003; Petersen-Mahrt *et al.*, 2002; Ramiro *et al.*, 2003). These U:G mismatches are converted to a mutation in the case of SHM or processed to double-stranded DNA breaks (DSBs), which serve as obligate intermediates in the recombination reaction during CSR (Di Noia and Neuberger, 2007; Stavnezer *et al.*, 2008).

II. CHROMOSOME TRANSLOCATIONS

If not properly repaired, physiological DSBs that arise during CSR may pose a threat to genome integrity. For example, they can be substrates for chromosome rearrangements such as deletions and translocations and

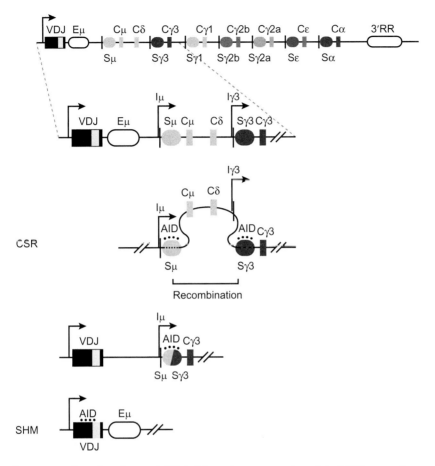

Fig. 1 AID-dependent CSR and SHM. Schematic representation of the IgH locus and rearrangement that takes place during CSR. Constant region exons depicted by solid rectangles, switch regions by solid ovals, promoters by black lines, enhancer and 3′ regulatory regions by white cylinders. Areas that accumulate DSB are denoted by discontinuous lines and mutations area denoted by black circles.

can lead to malignant transformation (Gostissa et al., 2009; Nussenzweig and Nussenzweig, 2010; Tsai and Lieber, 2010; Zhang et al., 2010). While deletions may occur by joining breaks on one chromosome in *cis*, chromosome translocations involve the joining of paired DSBs on different chromosomes. Translocations may arise in either reciprocal or nonreciprocal conformations. In the latter, formation of genetically unstable dicentric or acentric chromosomes with or without the loss of chromosome segments may result. Conversely, a reciprocal translocation involves the exchange of

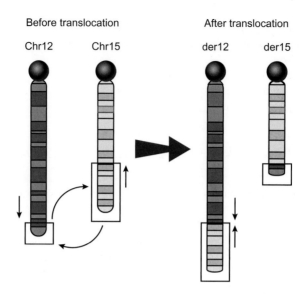

Fig. 2 Depiction of reciprocal translocations. Schematic representation of normal chromosome structures or following a reciprocal translocation such as the *c-myc/IgH*. PCR primers (arrows) can be used to specifically amplify translocations for detection in Southern assays.

telomeric and centromeric chromosome portions, resulting in the formation of two stable hybrid chromosomes with complete sequence retention (Fig. 2). These balanced events can then be stably propagated during cell division. Most hematopoietic malignancies harbor clonal reciprocal translocations (Kuppers, 2005). These translocations are frequently a consequence of genetically programmed DSB generation during lymphocyte development. And, since these events may induce oncogenic transformation through several mechanisms, they are thought to be etiologic in many cases. For example, a highly active promoter or cis-regulatory element can be juxtaposed to a protooncogene thereby deregulating its expression. An "infamous" example is the *c-myc/IgH* translocation, a hallmark of Burkitt's lymphoma, which places *IgH* regulatory elements upstream of the *c-myc* proto-oncogene. A chromosome translocation may bring together disparate coding sequences to form a chimeric fusion protein. For example, the BCR–ABL fusion, found in chronic myeloid leukemia, results in constitutive ABL kinase activity (Kuppers, 2005; Potter, 2003).

Unlike V(D)J recombination, which introduces DSBs at specific recombination signal sequences (RSSs), CSR is a regional, imprecise recombination event. Aberrant DSBs during CSR can occur in non-*Ig* loci and participate in chromosome translocations (Robbiani *et al.*, 2008). The promiscuity of this B cell-specific process is reflected in the relative prevalence of B cell lymphomas in the population; more than 90% of human lymphomas in the western world arise from B cells as opposed to T cells (Kuppers, 2005). The majority of these originate from mature B cells or postgerminal center B cell compartments where AID expression is normally induced. Therefore, AID and aberrant switching events may be substantial contributors to the molecular etiology of B cell lymphomas.

III. ACTIVATION-INDUCED CYTIDINE DEAMINASE

A seminal discovery in understanding the molecular mechanism of CSR and SHM was the identification of AID by Honjo and colleagues (Muramatsu *et al.*, 1999). The finding that AID deficiency abolished CSR and SHM in mice and humans confirmed its essential role in both processes (Muramatsu *et al.*, 2000; Revy *et al.*, 2000). Defects in the AID gene (*AICDA*), located at 12p13, are the cause of autosomal recessive hyper-IgM syndrome type 2 (HIGM2) in humans (Revy *et al.*, 2000). Initial sequence analysis of AID revealed a deaminase homologous to the mRNA editing cytidine deaminase, APOBEC1 (Muramatsu *et al.*, 1999); however, attempts to show deaminase activity on RNA have failed and no RNA substrate has been identified. In contrast, *in vitro* analysis has revealed that AID can directly deaminate single-stranded DNA (Bransteitter *et al.*, 2003; Chaudhuri *et al.*, 2003; Dickerson *et al.*, 2003; Petersen-Mahrt *et al.*, 2002; Pham *et al.*, 2003; Ramiro *et al.*, 2003; Sohail *et al.*, 2003). Indeed, the preponderance of biochemical, cell biology, and genetic evidence supports a model in which AID deaminates DNA to initiate CSR and SHM (Di Noia and Neuberger, 2007; Petersen-Mahrt *et al.*, 2002). During SHM, mutations may be generated by replication over U:G mismatches. Alternatively, lesions may be processed by uracil DNA glycosylase (UNG), or the mismatch repair proteins MSH2/MSH6 and translesion polymerases involved in error prone DNA synthesis. In CSR U:G mismatches in donor and acceptor switch regions are processed to DSBs that are joined by both the classical and alternative non-homologous end joining (NHEJ) pathways (Delker *et al.*, 2009; Di Noia and Neuberger, 2007; Maul and Gearhart, 2010; Maul *et al.*, 2011; Peled *et al.*, 2008; Stavnezer *et al.*, 2008).

IV. REGULATION OF AID

AID has significant potential for inducing genomic damage and tumorigenesis across cell types. Mice carrying a ubiquitously expressed AID transgene driven by the actin promoter develop tumors in a wide variety of tissues, but curiously not in B cells (Okazaki *et al.*, 2003; Rucci *et al.*, 2006). Development of AID-dependent B cell cancers requires additional inactivation of p53 (Ramiro et al., 2006; Robbiani et al., 2009). Normally, AID expression is cell type restricted—mainly to germinal center or activated B cells. *In vitro*, cytokines such as interleukin 4(IL-4) and transforming growth factor beta (TGF B) in combination with B cell mitogens such as CD40 ligation or LPS induce AID (Dedeoglu et al., 2004; Muramatsu et al., 1999; Zhou et al., 2003). A highly conserved regulatory region within the AID gene is thought to be responsible for this restriction (Crouch et al., 2007) by enforcing a requirement for PAX5 and E47 transcription factor binding that is balanced by inhibitors of differentiation (Id) proteins Id2 and Id3 (Gonda et al., 2003; Sayegh et al., 2003; Tran et al., 2010). A number of cytokine response elements including NFκB, STAT6, and Smad3/4 also reside there and confer response to signaling by CD40L, IL-4, and TGFβ (Tran et al., 2010).

The regulation of AID expression plays an important role in controlling its activity since AID levels are rate-limiting for both SHM and CSR. AID exhibits haploinsufficiency or hyperactivity when overexpressed by virtue of a retrovirus or transgene (McBride et al., 2008; Robbiani et al., 2009; Sernandez et al., 2008; Takizawa et al., 2008). Importantly, when overexpressed, AID not only increases the rates of physiological SHM and CSR but also generates widespread genomic instability and associated rearrangements (Robbiani et al., 2009). Therefore, AID levels are strictly controlled to maintain genomic stability and to support proper antibody development, which is achieved by multiple levels of regulation.

Once transcribed, the stability and half-life of AID mRNA is controlled by miR-155 and miR-181b binding to sites in the 3′ untranslated region (de Yebenes et al., 2008; Dorsett et al., 2008; Teng et al., 2008). Deficiency of miR-155 or mutation of the miR-155 binding site in AID mRNA results in increased AID levels and a marked increase in *c-myc/IgH* translocations. This indicates that miR-155 could act as a tumor suppressor by directly reducing AID levels (Dorsett et al., 2008). In the cytoplasm, AID protein levels are stabilized by HSP90 (heat shock protein 90) interaction, which prevents ubiquitylation and degradation. Inhibition of HSP90 results in decreased AID levels, decreased CSR rates, and fewer off target mutations (Orthwein et al., 2010).

AID, like many proteins with nuclear function, is regulated on the level of subcellular localization. Although AID is actively transported to the nucleus (Patenaude et al., 2009), its steady-state localization is predominantly cytoplasmic (Rada et al., 2002). This is accomplished by a combination of cytoplasmic retention (Patenaude et al., 2009), degradation of nuclear AID (Aoufouchi et al., 2008), and active nuclear export by virtue of a carboxyl terminal nuclear export signal (NES) recognized by the CRM1 shuttling factor (Brar et al., 2004; Ito et al., 2004; McBride et al., 2004).

Protein modification by phosphorylation is one of the most common means to alter activity. Compared to the previously mentioned mechanisms, phosphorylation has the potential to rapidly and dynamically modulate activity. In switching B cells, mass spectrometry analysis of AID has revealed phosphorylation on serine 3, serine 38, threonine 140, and tyrosine 184 (S3, S38, T140, Y184, respectively) (S3, S38, T140, Y184) (Basu *et al.*, 2005; Gazumyan *et al.*, 2011; McBride *et al.*, 2008). With the exception of Y184, these phosphorylation sites have been confirmed by AID antiphospho antibodies (Gazumyan *et al.*, 2011; McBride *et al.*, 2006, 2008). Mutation of S38 or T140 to alanine does not impact AID catalytic activity, but does interfere with CSR and SHM, indicating that they are positive regulators (Basu *et al.*, 2005; Chatterji *et al.*, 2007; Gazumyan *et al.*, 2011; McBride *et al.*, 2006; Pasqualucci *et al.*, 2006). While S38 is equally important for CSR and SHM, T140 phosphorylation preferentially affects SHM suggesting posttranslational modification may contribute to the choice between SHM and CSR (McBride *et al.*, 2008). In contrast to S38 and T140, phosphorylation of S3 acts as a negative regulator. AID carrying an S3A mutation catalyzes increased CSR and *c-myc/IgH* translocation frequencies, demonstrating that the oncogenic potential of AID can be regulated by phosphorylation (Gazumyan *et al.*, 2011).

The signaling pathways that control AID phosphorylation are unknown; however, evidence suggests PKA phosphorylates AID at S38: it is within a PKA consensus site, it is phosphorylated by PKA *in vitro*, and inhibitors of PKA decrease CSR (Basu *et al.*, 2005; Pasqualucci *et al.*, 2006). Although the overall levels of S38 phosphorylation are low (5% total), chromatin associated AID is highly phosphorylated (McBride *et al.*, 2006). Consistent with this finding, PKA is specifically recruited to switch regions (Vuong *et al.*, 2009).

There is unique control of each AID phosphorylation site, as AID-T140 can be phosphorylated by PKC, but not PKA, and protein phosphatase 2A (PP2A) preferentially controls AID-S3 phosphorylation levels (Gazumyan *et al.*, 2011; McBride *et al.*, 2008). It appears that multiple signaling pathways converge on AID to control activity through phosphorylation. The complete story of AID regulation by phosphorylation is likely to be even more complicated; little is known about the function of Y184 phosphorylation and additional phosphorylation sites at T27, S41, and S43 have been detected in AID purified from Sf9 insects cells (Pham *et al.*, 2008). Taken together, the evidence suggests that increasing AID levels or activity increases SHM and CSR efficiency at the expense of genomic integrity. Therefore, the summation of AID regulatory mechanisms may strike an important equilibrium between the rate of antibody evolution and oncogenesis.

A. AID Is Targeted by Transcription

Transcription is an absolute requirement for CSR and SHM, and their rates directly relate to the rate of transcription (Bachl *et al.*, 2001). This connection between transcription and SHM and CSR was appreciated long before the discovery of AID. Early studies of mutation patterns in B cells revealed that SHM was largely confined to the transcribed 1–2 kb region 3' to the V(D)J or switch region promoter. Promoter regions and distal downstream regions were largely devoid of mutations (Both et al., 1990; Lebecque and Gearhart, 1990; Rothenfluh et al., 1993; Steele et al., 1992; Weber et al., 1991). The finding that *Ig* transgenes undergo SHM at levels that correlated with transcription rates provided additional evidence for this connection (Bachl et al., 2001). Similar relationships between transcription and recombination of switch regions in CSR were also found (Chaudhuri and Alt, 2004; Guikema et al., 2008; Stavnezer et al., 2008). Replacement of the V(D)J region with unrelated sequences did not significantly alter SHM rates, nor did replacing the V promoter with promoters from non-*Ig* (B-globin or B29) loci (Betz et al., 1994; Tumas-Brundage and Manser, 1997). This suggests that transcription, and not specific sequence elements per se, was essential to targeting SHM. A conclusive link between SHM and transcription was shown by Storb and colleagues. They demonstrated that a normally unmutated exon underwent SHM when repositioned to a promoter proximal location, demonstrating position within a transcription unit dictated SHM (Peters and Storb, 1996). These experiments together with the spatial positioning of mutations from the promoter suggested that the active Pol II complex is playing an essential role in AID targeting *in vivo*. Based on their study Peters and Storb proposed a model linking somatic mutation to transcription. According to that model a "mutator factor", present only in mutating B cells, loads on to the transcription initiation complex assembled at the promoter and remains associated with the elongating complex (Peters and Storb, 1996).

The mechanistic role of transcription in SHM and CSR became evident after AID was discovered to be an ssDNA deaminase. *In vitro* assays with purified AID demonstrated no activity on double-stranded DNA but showed activity against single-stranded DNA and transcribed double-stranded DNA (Bransteitter et al., 2003; Chaudhuri et al., 2003; Dickerson et al., 2003; Petersen-Mahrt et al., 2002; Pham et al., 2003; Ramiro et al., 2003; Sohail et al., 2003). This suggested that transcription was creating the ssDNA substrate. The precise structure of ssDNA that is accessible to AID is not known but transcription bubbles, supercoiled DNA in the wake of the polymerase, and R loops (RNA–DNA hybrid structures that can displace the nontranscribed DNA strand) are suggested to be targets (Besmer et al., 2006; Lumsden et al., 2004; Yu et al., 2003).

Consistent with this, DNA transcribed by *Escherichia coli*, T7, and mammalian RNA Pol II all become AID targets both *in vitro* and *in vivo*. Interestingly, there is an inherent bias for AID to target the nontemplate transcribed DNA strand in *E. coli* but not in mammalian cells. This suggests the existence of a specific mechanism targeting AID to template DNA. A number of groups have suggested that RNA–DNA hybrid R loops may influence AID access (Basu *et al.*, 2011; Bransteitter *et al.*, 2003; Pham *et al.*, 2003; Ramiro *et al.*, 2003). The likely role of transcription, therefore, is to create the ssDNA substrate for AID. However, most transcribed genes do not undergo SHM in B cells (Liu *et al.*, 2008). The finding that AID physically associates with the Pol II complex in B cells suggested that there were additional factors that influence AID targeting (Nambu *et al.*, 2003; Pavri *et al.*, 2010).

B. AID Targeting by Spt5 and Stalled Pol II

Studies of Pol II, AID, and Spt5 distribution at the *Ig* locus suggested a relationship between Pol II density, elongation stalling, and AID activity. ChIP analysis of Pol II distribution demonstrated a spike in density through the core of *Ig* switch regions in activated B cells (Rajagopal *et al.*, 2009). Such density suggested Pol II accumulation due to elongation stalling. Consistent with this, nuclear run-on analysis showed a concomitant accumulation of transcripts within the transcribed switch regions (Rajagopal *et al.*, 2009). A high Pol II density within the highly repetitive switch regions seemed to be an intrinsic feature of the locus, occurring even in the absence of AID (Rajagopal *et al.*, 2009; Wang *et al.*, 2009). Impressively, Pol II density correlated with the rate and distribution of mutation through the switch region. A stalled Pol II complex would be predicted to expose ssDNA AID substrate created by transcription for prolonged periods. The observed stalling at the *Ig* switch regions provided a logical mechanism by which Pol II-associated AID would have preferential access to sites of elongation stalling. Consistent with this concept, stalling has been reported at other AID target sites including *Igk* and *c-myc* (Bentley and Groudine, 1986; Raschke *et al.*, 1999).

The study by Pavri *et al.* of the Pol II stalling factor Spt5 provided a direct mechanistic link between AID and stalled Pol II. In an unbiased loss-of-function screen, Suppressor of Ty5 homolog (Spt5) was found to be an essential factor for CSR (Pavri *et al.*, 2010). Spt5 is a Pol II interacting elongation factor that associates with stalled Pol II and ssDNA (Gilmour, 2009; Lis, 2007; Rahl *et al.*, 2010). Spt5 partners with Spt4 to form the DRB sensitivity inducing factor (DSIF) complex (Swanson *et al.*, 1991; Wada *et al.*, 1998). DSIF can interact with the negative elongation factor (NELF) complex to promote the stalling of Pol II (Yamaguchi *et al.*, 1999). This stall effect of DSIF and NELF is relieved by P-TEFb, a kinase that phosphorylates Pol II and Spt5 (Kim and Sharp, 2001; Yamada *et al.*, 2006). Co-immunoprecipitation of Spt5 interacting factors revealed an association with AID in activated B cells, while binding studies with recombinant proteins demonstrated a direct interaction (Pavri *et al.*, 2010). Since both Spt5 and AID associate with the Pol II complex (Nambu *et al.*, 2003;

Wada *et al.*, 1998), the contribution of each protein to this interaction was examined. siRNA mediated depletion of Spt5 decreased the amount of Pol II co-immunoprecipitated by AID. In contrast, Pol II depletion did not alter AID–Spt5 interaction, suggesting that Spt5 serves as a direct adaptor to recruit AID to Pol II (Pavri *et al.*, 2010).

AID induced mutations are known to be widely distributed among transcribed genes in B cells (Liu *et al.*, 2008). Genome-wide AID ChIP-Seq analysis confirmed the widespread association of AID with thousands of promoter proximal transcribed areas. Strikingly, Pol II occupancy rather than overall transcription rate correlated with AID load (Yamane *et al.*, 2011). AID was specifically enriched in Pol II dense promoter proximal regions, which are known sites of Pol II stalling (Yamane *et al.*, 2011). There was also a strong correlation between Spt5 and AID recruitment in these regions, consistent with Spt5–AID association at Pol II stalling sites (Pavri *et al.*, 2010). Importantly, a high density of AID, Spt5, and Pol II were found in the area of known switch region hypermutation (Pavri *et al.*, 2010; Yamane *et al.*, 2011). Thus, a Spt5 mediated targeting mechanism orchestrated by Pol II stalling could explain the targeting enigma of AID and the widespread promiscuity of AID mutations.

The exact mechanism by which AID–Spt5 accesses ssDNA is unknown; however, the recently solved crystal structure of *Pyrococcus furiosus* (*Pfu*) Spt 4/5 in complex with the RNA polymerase clamp domain gives some insight. RNA polymerases (RNAP) are found in all organisms and structural studies reveal universal conservation of architecture and active site mechanism throughout evolution (Cramer *et al.*, 2008; Klein *et al.*, 2011a). The only RNAP-associated factor universally conserved through all evolution is Spt5 (Cramer *et al.*, 2008; Martinez-Rucobo *et al.*, 2011). The Spt4/5-Pol II structure reveals that Spt5 is in close proximity to the upstream edge of the transcription bubble (Fig. 3) (Klein *et al.*, 2011a; Martinez-Rucobo *et al.*, 2011). Based on this model, AID interaction with Spt5 in the Pol II complex would result in close spatial proximity to ssDNA at the transcription bubble and the newly formed RNA–DNA hybrid (Fig. 3). In combination, AID interaction with Spt5 place AID near ssDNA substrates in a Pol II elongation complex, while stalling of Pol II would result in a prolonged ssDNA substrate exposure (Fig. 4).

V. TEMPLATE STRAND TARGETING BY EXOSOMES

AID targets both the template and nontemplate strand of transcribed genes *in vivo*; however, in *E. coli* or *in vitro* the nontemplate strand is preferentially targeted (Bransteitter *et al.*, 2003; Chaudhuri *et al.*, 2003;

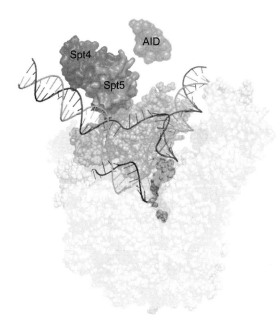

Fig. 3 Model of AID within a Spt4/5 RNA polymerase complex. Rendering based on the crystal structure of a complex of Spt4 (red)/Spt5 (purple) with RNA clamp domain (orange) (Martinez-Rucobo et al., 2011). Melted DNA is loaded into the cleft to trigger RNA (red ribbon) synthesis on template DNA (blue ribbon). Spt5 domain locks nucleic acids and contacts the nontemplate DNA (green ribbon) in the cleft, preventing collapse of the transcription bubble. The interaction with Spt5 brings AID (pink) into the proximity with the both template and nontemplate ssDNA. (See Page 3 in Color Section at the back of the book.)

Dickerson et al., 2003; Petersen-Mahrt et al., 2002; Pham et al., 2003; Ramiro et al., 2003; Sohail et al., 2003). Transcription readily exposed ssDNA on the nontemplate strand but the mechanism for template strand targeting was unclear. However, RNA–DNA hybrids such as those formed on transcribed template strands are inhibitory to AID (Bransteitter et al., 2003). The recent discovery by Alt and colleagues that the exosome complex associates with AID and provides access to the template strand *in vitro* provides an explanation (Basu et al., 2011). Components of the exosome were identified by mass spectrometry from Ramos cell fractions that enhanced deamination activity of purified AID on *in vitro* transcribed dsDNA. They were confirmed to interact *in vivo* and siRNA knockdown resulted in diminished SHM, Interestingly purified exosome resulted in the appearance of mutations on the in vitro transcribed template strand (Basu et al., 2011). The evolutionarily conserved exosome complexes are $3'$–$5'$ exoribonucleases involved in RNA processing, turnover, and

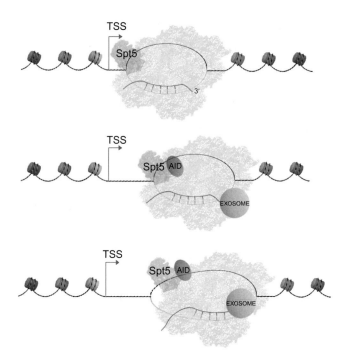

Fig. 4 A model of AID transcription mediated targeting by stalled Pol II/Spt5 elongation complex. Stalling model for AID access based on refinement of a proposed association with Pol II (Peters and Storb, 1996). (A) Elongation factors including Spt5/Spt4 (DSIF) (purple/red), and other factors (not shown) bound to RNA Pol II (gray) create the stalled state in the promoter proximal region near the transcription start site (TSS). (B) AID, via interaction with Spt5 and association with stalled polymerase, gains access to single-stranded nontemplate DNA. (C) Exosome degrades 3′ end of the nascent RNA exposed by backtracking of stalled polymerase and thus exposes the template DNA (blue) to AID. (For interpretation of the references to color in this figure legend, the reader is referred to the Web version of this chapter.)

degradation of unstable stalled transcripts (Houseley *et al.*, 2006). The exosome is only known to engage free 3′ ends of RNA that would be buried and inaccessible in an elongating Pol II complex. However, stalled polymerases may represent a special situation since they are known to undergo backtracking resulting in an exposed 3′ RNA substrate that exosomes may access (Adelman *et al.*, 2005). Furthermore, there is a known association of Spt5 and exosome components (Andrulis *et al.*, 2002). Therefore, AID associated Spt5/Pol II stalled complexes may be targets for exosome action to remove nascent RNA transcripts and expose the template strand (Pavri *et al.*, 2010). The stalling model of AID targeting would therefore provide a mechanism for not only AID access but for exposure of the template strand via exosome processing (Fig. 4).

VI. AID GENOME WIDE DAMAGE

Following the discovery that *Ig* variable regions undergo high levels of hypermutation (McKean *et al.*, 1984; Weigert *et al.*, 1970) analysis of the surrounding promoter and constant regions revealed few mutations (Both *et al.*, 1990; Lebecque and Gearhart, 1990; Rothenfluh *et al.*, 1993; Steele *et al.*, 1992; Weber *et al.*, 1991). It was therefore concluded that hypermutation was confined to the *Ig* variable region. This assumption was extended to other areas of the genome as it seemed logical that a dangerous process like hypermutation would be confined. Over the next decades, reports accumulated that a few non-*Ig* genes did undergo SHM in germinal center B cells (*Bcl-6*, *IgA*, *IgB*, *FAS*, *c-myc*, *miR-142*) (Gordon *et al.*, 2003; Muschen *et al.*, 2000; Pasqualucci *et al.*, 1998, 2001; Pavri *et al.*, 2010; Robbiani *et al.*, 2009; Shen *et al.*, 1998; Yamane *et al.*, 2011). Furthermore, it was demonstrated that a retroviral indicator gene with no *Ig* elements underwent hypermutation at multiple locations (Wang *et al.*, 2004), challenging the notion that hypermutation was confined to *Ig* loci.

A mutational analysis of ~120 highly transcribed genes in germinal center B cells showed that more than 25% accumulated mutations in an AID-dependent manner (Liu *et al.*, 2008). However, most of these genes mutated at rates much lower than the *Ig* variable region. Strikingly, almost 50% of assayed genes accumulated mutations in UNG/MSH2 double knock out B cells, also in an AID-dependent manner (Liu *et al.*, 2008). This suggested that low-level AID targeting was widespread in the genome and that high-fidelity repair served to suppress AID induced mutations (Liu and Schatz, 2009). The extent of potential AID targets genome-wide was revealed by AID-ChIP-Seq analysis that showed AID loading on more than 5000 genes (Yamane *et al.*, 2011).

In addition to mutations, AID lesions can also be processed into DSBs that serve as substrates for CSR and chromosome translocations. Recurrent translocations are frequently found in lymphomas and leukemias and, reciprocal chromosomal translocations between oncogenes and the *IgH* regulatory elements are hallmark cytogenetic abnormalities in many mature B cell lymphomas. The first evidence that AID induced DSBs act as translocation substrates came from studies examining *c-myc/IgH* translocations, a hallmark of Burkitt's lymphoma. AID was shown to be essential for the appearance of *c-myc/IgH* chromosome translocations *in vivo* (Ramiro *et al.*, 2004). Further characterization demonstrated that AID was creating the DSB necessary for *c-myc/IgH* translocations at both the *c-myc* and *IgH* loci (Robbiani *et al.*, 2008). Consistent with widespread AID hypermutation, it was soon shown that AID is capable of inducing DSB at several non-Ig loci, thereby catalyzing their rearrangement and triggering oncogenesis (Robbiani et al., 2009). Like SHM, AID induced translocations are suppressed by internal surveillance and protection pathways. Stress-induced checkpoint control proteins (ATM, Nbs1, p19-Arf, and p53) suppress translocations, the loss of any results in greatly enhanced *c-myc/IgH* translocation frequencies (Ramiro et al., 2006).

Until recently, the genome-wide incidence of AID induced DSBs had not been measured. Only specific loci, such as *c-myc*, have been conclusively shown to be targeted by AID for DSBs (Robbiani et al., 2008, 2009). While ChIP-Seq studies revealed thousands of potential AID targets (Staszewski et al., 2011; Yamane et al., 2011), and selective sequencing of specific loci suggested widespread AID activity (Liu et al., 2008), the unbiased measurement of AID DSB activity had yet to be achieved. The recently developed technique of genome-wide translocation capture sequencing facilitated the unbiased discovery of AID targets, thereby revealing the loci that suffer AID-mediated DSBs and confirming the current model of AID targeting (Chiarle et al., 2011; Klein et al., 2011b). The two groups that simultaneously developed this method engineered I-SceI meganuclease sites at the *IgH* or the *c-myc* locus. Defined DSBs were induced by retroviral introduction of the I-SceI nuclease in activated B cells in the presence or absence of AID. After short-term culture, junctions between the I-SceI induced DSB and other endogenous DSB sites were identified by PCR amplification and deep sequencing (Oliveira et al., 2011). The capture strategy allowed unbiased amplification of DNA surrounding the I-SceI site but selected against intact I-SceI sites. Thereby events that disrupted the I-SceI site, such as translocations, would be identified. The power of this methodology is the capture of single cell events in short-term culture. This allows for the unbiased analysis of rare translocation partners, providing insight into the mechanisms of genomic rearrangement at the earliest stages of oncogenic transformation. By defining recurrent AID-dependent rearrangement hotspots these studies revealed DSB targets of AID (Chiarle et al., 2011; Klein et al., 2011b).

Using this method, nearly 200,000 independent junctions were identified. Whole genome rearrangement maps reveal that the pluralities of rearrangement partners are local, within several kb of the engineered I-SceI break site at both the *IgH* and *c-myc* locus. Farther afield, double-strand breaks still showed a strong preference toward intrachromosomal rearrangements, suggesting the presence of chromosome territories (Chiarle et al., 2011; Klein et al., 2011b). This suggests that both chromosome positioning and proximal DNA damage response play a role in DSB partner ligation. The tendency toward local translocation partners could have the functional affect of abating more deleterious rearrangements to distant parts of the genome (Chiarle et al., 2011; Klein et al., 2011b).

When interchromosomal translocations did occur they did so across the genome, most frequently to genic regions near active transcription start sites. The enrichment of rearrangements in transcription start sites, which are sites of ssDNA exposure, implicates ssDNA as a significant source of genomic instability even in the absence of AID. Inducing expression of AID revealed hundreds of AID-dependent rearrangement hotspots that frequently rearranged to both the *c-myc* and *IgH* locus. When the regions bearing AID-dependent rearrangements were examined for shared genomic features several interesting trends emerged. Consistent with a ssDNA substrate, AID-dependent hotspots were clustered in the transcription start sites of transcribed genes. However, transcription itself was neither rate-limiting nor sufficient for AID DSB targeting; as translocation of thousands of highly transcribed genes was not detected (Chiarle *et al.*, 2011; Klein *et al.*, 2011b). In contrast, stalled Pol II and the presence of Spt5 correlated well with AID translocation hotspots (Klein *et al.*, 2011b; Pavri *et al.*, 2010; Yamane *et al.*, 2011). Furthermore, analysis of mutation rates at AID hotspots showed a strong correlation suggesting that the rate of SHM and DSB creation at non-*Ig* genes are proportional (Klein *et al.*, 2011b). Hence, the genome wide mapping of AID induced DSB is consistent with an AID–Spt5–Pol II targeting complex at stalled Pol II sites (Fig. 4). Finally, many loci targeted for DSBs by AID are found to participate in clonal translocations in mature B cell lymphomas. This indicates that AID is an important source of genomic instability in these tumors to generate chromosomal abnormalities that drive lymphomagenesis.

VII. CSR-SPECIFIC COFACTORS

The model of AID targeting by Pol II (Fig. 4) explains several key observations of SHM and CSR, including the transcription requirement and promoter proximal occupancy. However, the spread of mutations and DSBs over several kb suggests that AID action is coincident along the elongating Pol II. Therefore, together with targeting, the stability of this interaction may be important to maintain AID association across the switch region. Multiple mechanisms may impact this including AID phosphorylation and interactions with partner proteins. For example, the splicing factor Ptbp2 and scaffold proteins of the 14-3-3 family are critical for CSR and necessary for AID switch region association (Nowak *et al.*, 2011; Xu *et al.*, 2010). AID phosphorylation regulates AID activity without affecting catalytic function, suggesting it may also impact targeting or stability with Pol II. AID phosphorylation recruits the single-stranded DNA binding protein RPA to the switch regions where it augments CSR (Basu *et al.*, 2005; Vuong *et al.*, 2009; Yamane *et al.*, 2011). Hence, AID phosphorylation could stabilize partner interactions and therefore contribute to prolonged AID presence at the switch regions. Future studies of the basis of AID interaction with partners and regulation via posttranslational modification may provide a more complete view.

VIII. AID INFLUENCES CSR OUTCOME

The carboxyl terminus of AID, the same domain that encodes an NES, is required for CSR but not SHM. Interestingly, this domain is a nexus for multiple functions. Deletion of this region results in (i) increased nuclear AID, (ii) decreased protein stability, (iii) increased off target DNA damage with loss of CSR and hyper-IGM syndrome in humans (Barreto et al., 2003; Geisberger et al., 2009; McBride et al., 2004; Shinkura et al., 2004; Ta et al., 2003). That AID lacking the C-terminal region is able to target and mutate the switch regions but does not support CSR (Barreto et al., 2003; Ta et al., 2003) suggests that AID interacts with class-switch specific cofactors (Doi et al., 2009). Interestingly, mutant AID lacking the C-terminal region acts as a dominant negative molecule in heterozygous individuals resulting in severely compromised CSR efficiency. Examination of the switch region junctions in the few cells that accomplished CSR revealed a bias toward junctions bearing microhomology (Kracker et al., 2010). This suggested that domains of AID including the C-terminal region impacted the repair of created lesions, possibly by interacting with CSR specific factors. Consistent with this, the C-terminus of AID is involved in cooperative binding of AID, UNG, and mismatch repair proteins MSH2–MSH6 to the switch regions (Ranjit et al., 2011). Other AID interacting factors including KAP1 and Spt6 have been found to be critical for CSR but not SHM (Jeevan-Raj et al., 2011; Okazaki et al., 2011). Altogether this suggests that AID itself plays a role in dictating the downstream processing of the U:G mismatches.

IX. CONCLUSIONS

DNA damage, DSB and chromosome translocations are events fundamental to the genesis of cancer. However, the adaptive immune system induces these genetic insults as a by-product of diversifying immune receptors. Even though both T and B cells undergo RAG1/2 V(D)J recombination, lymphomas are overwhelmingly of B cell origin. The DSB and mutations created by AID during SHM and CSR *Ig* diversification are major contributors to this prevalence. Recent studies have revealed a surprisingly wide array of AID target genes, including oncogenes frequently involved in lymphoid cancer (Chiarle *et al.*, 2011; Klein *et al.*, 2011b; Yamane *et al.*, 2011). AID targeting by Spt5 to stalled Pol II gives major mechanistic insight into the basis of genome wide damage. However, AID action is highly focused at the *Ig* loci compared to elsewhere in the genome. This differential of activity could involve preferential targeting, Pol II stalling, and the stabilization of the AID complex to the *Ig* region. AID posttranslational modifications and interacting factors are likely to influence targeting and future studies addressing these should lead to a more comprehensive mechanism of AID targeting. During CSR, AID induced U:G mismatches in the switch regions triggers DSBs and DNA recombination. However, U:G mismatches also occur through spontaneously or radiation induced deamination of cytidines. These occur not infrequently in the genome but are normally faithfully repaired (Krokan *et al.*, 2002). Whether a mechanism governs preferential processing of AID lesions to DSB and mutations is still unclear. However, AID lacking the C-terminal target switch regions does not support CSR, suggesting that specific domains of AID influence the resolution of lesions. Overall, the wide targeting of AID poses a significant threat to genome stability. Therefore, mechanisms that regulate AID targeting, activity, and U:G mismatch processing represent important means to suppress pro-oncogenic damage.

ACKNOWLEDGMENT

We would like to thank Alex Scheinker for his assistance in rendering Figs. 3 and 4.

REFERENCES

Abbas, A. K., Lichtman, A. H., and Pillai, S. (2010). Cellular and Molecular Immunology. 6th edn. Saunders Elsevier, Philadelphia, PA.

Adelman, K., Marr, M. T., Werner, J., Saunders, A., Ni, Z., Andrulis, E. D., and Lis, J. T. (2005). Efficient release from promoter-proximal stall sites requires transcript cleavage factor TFIIS. *Mol. Cell* **17**, 103–112.

Andrulis, E. D., Werner, J., Nazarian, A., Erdjument-Bromage, H., Tempst, P., and Lis, J. T. (2002). The RNA processing exosome is linked to elongating RNA polymerase II in Drosophila. *Nature* **420**, 837–841.

Aoufouchi, S., Faili, A., Zober, C., D'Orlando, O., Weller, S., Weill, J. C., and Reynaud, C. A. (2008). Proteasomal degradation restricts the nuclear lifespan of AID. *J. Exp. Med.* **205**, 1357–1368.

Bachl, J., Carlson, C., Gray-Schopfer, V., Dessing, M., and Olsson, C. (2001). Increased transcription levels induce higher mutation rates in a hypermutating cell line. *J. Immunol.* **166**, 5051–5057.

Barreto, V., Reina-San-Martin, B., Ramiro, A. R., McBride, K. M., and Nussenzweig, M. C. (2003). C-terminal deletion of AID uncouples class switch recombination from somatic hypermutation and gene conversion. *Mol. Cell* **12**, 501–508.

Basu, U., Chaudhuri, J., Alpert, C., Dutt, S., Ranganath, S., Li, G., Schrum, J. P., Manis, J. P., and Alt, F. W. (2005). The AID antibody diversification enzyme is regulated by protein kinase A phosphorylation. *Nature* **438**, 508–511.

Basu, U., Meng, F. L., Keim, C., Grinstein, V., Pefanis, E., Eccleston, J., Zhang, T., Myers, D., Wasserman, C. R., Wesemann, D. R., Januszyk, K., Gregory, R. I., *et al.* (2011). The RNA exosome targets the AID cytidine deaminase to both strands of transcribed duplex DNA substrates. *Cell* **144**, 353–363.

Bentley, D. L., and Groudine, M. (1986). A block to elongation is largely responsible for decreased transcription of c-myc in differentiated HL60 cells. *Nature* **321**, 702–706.

Besmer, E., Market, E., and Papavasiliou, F. N. (2006). The transcription elongation complex directs activation-induced cytidine deaminase-mediated DNA deamination. *Mol. Cell. Biol.* **26**, 4378–4385.

Betz, A. G., Milstein, C., Gonzalez-Fernandez, A., Pannell, R., Larson, T., and Neuberger, M. S. (1994). Elements regulating somatic hypermutation of an immunoglobulin kappa gene: Critical role for the intron enhancer/matrix attachment region. *Cell* **77**, 239–248.

Both, G. W., Taylor, L., Pollard, J. W., and Steele, E. J. (1990). Distribution of mutations around rearranged heavy-chain antibody variable-region genes. *Mol. Cell. Biol.* **10**, 5187–5196.

Bransteitter, R., Pham, P., Scharff, M. D., and Goodman, M. F. (2003). Activation-induced cytidine deaminase deaminates deoxycytidine on single-stranded DNA but requires the action of RNase. *Proc. Natl. Acad. Sci. USA* **100**, 4102–4107.

Brar, S. S., Watson, M., and Diaz, M. (2004). Activation-induced cytosine deaminase (AID) is actively exported out of the nucleus but retained by the induction of DNA breaks. *J. Biol. Chem.* **279**, 26395–26401.

Chatterji, M., Unniraman, S., McBride, K. M., and Schatz, D. G. (2007). Role of activation-induced deaminase protein kinase A phosphorylation sites in Ig gene conversion and somatic hypermutation. *J. Immunol.* **179**, 5274–5280.

Chaudhuri, J., and Alt, F. W. (2004). Class-switch recombination: Interplay of transcription, DNA deamination and DNA repair. *Nat. Rev. Immunol.* **4**, 541–552.

Chaudhuri, J., Tian, M., Khuong, C., Chua, K., Pinaud, E., and Alt, F. W. (2003). Transcription-targeted DNA deamination by the AID antibody diversification enzyme. *Nature* **422**, 726–730.

Chiarle, R., Zhang, Y., Frock, R. L., Lewis, S. M., Molinie, B., Ho, Y. J., Myers, D. R., Choi, V. W., Compagno, M., Malkin, D. J., Neuberg, D., Monti, S., *et al.* (2011). Genome-wide translocation sequencing reveals mechanisms of chromosome breaks and rearrangements in B cells. *Cell* **147**, 107–119.

Cramer, P., Armache, K. J., Baumli, S., Benkert, S., Brueckner, F., Buchen, C., Damsma, G. E., Dengl, S., Geiger, S. R., Jasiak, A. J., Jawhari, A., Jennebach, S., *et al.* (2008). Structure of eukaryotic RNA polymerases. *Annu. Rev. Biophys.* **37**, 337–352.

Crouch, E. E., Li, Z., Takizawa, M., Fichtner-Feigl, S., Gourzi, P., Montano, C., Feigenbaum, L., Wilson, P., Janz, S., Papavasiliou, F. N., and Casellas, R. (2007). Regulation of AID expression in the immune response. *J. Exp. Med.* **204**, 1145–1156.

de Yebenes, V. G., Belver, L., Pisano, D. G., Gonzalez, S., Villasante, A., Croce, C., He, L., and Ramiro, A. R. (2008). miR-181b negatively regulates activation-induced cytidine deaminase in B cells. *J. Exp. Med.* **205**, 2199–2206.

Dedeoglu, F., Horwitz, B., Chaudhuri, J., Alt, F. W., and Geha, R. S. (2004). Induction of activation-induced cytidine deaminase gene expression by IL-4 and CD40 ligation is dependent on STAT6 and NFkappaB. *Int. Immunol.* **16**, 395–404.

Delker, R. K., Fugmann, S. D., and Papavasiliou, F. N. (2009). A coming-of-age story: Activation-induced cytidine deaminase turns 10. *Nat. Immunol.* **10**, 1147–1153.

Di Noia, J., and Neuberger, M. (2007). Molecular mechanisms of antibody somatic hypermutation. *Annu. Rev. Biochem.* **76**, 1–22.

Dickerson, S. K., Market, E., Besmer, E., and Papavasiliou, F. N. (2003). AID mediates hypermutation by deaminating single stranded DNA. *J. Exp. Med.* **197**, 1291–1296.

Doi, T., Kato, L., Ito, S., Shinkura, R., Wei, M., Nagaoka, H., Wang, J., and Honjo, T. (2009). The C-terminal region of activation-induced cytidine deaminase is responsible for a recombination function other than DNA cleavage in class switch recombination. *Proc. Natl. Acad. Sci. USA* **106**, 2758–2763.

Dorsett, Y., McBride, K. M., Jankovic, M., Gazumyan, A., Thai, T. H., Robbiani, D. F., Di Virgilio, M., San-Martin, B. R., Heidkamp, G., Schwickert, T. A., Eisenreich, T., Rajewsky, K., et al. (2008). MicroRNA-155 suppresses activation-induced cytidine deaminase-mediated Myc-Igh translocation. *Immunity* **28**, 630–638.

Fugmann, S. D., Lee, A. I., Shockett, P. E., Villey, I. J., and Schatz, D. G. (2000). The RAG proteins and V(D)J recombination: Complexes, ends, and transposition. *Annu. Rev. Immunol.* **18**, 495–527.

Gazumyan, A., Timachova, K., Yuen, G., Siden, E., Di Virgilio, M., Woo, F. M., Chait, B. T., Reina San-Martin, B., Nussenzweig, M. C., and McBride, K. M. (2011). Amino-terminal phosphorylation of activation-induced cytidine deaminase suppresses c-myc/IgH translocation. *Mol. Cell. Biol.* **31**, 442–449.

Geisberger, R., Rada, C., and Neuberger, M. S. (2009). The stability of AID and its function in class-switching are critically sensitive to the identity of its nuclear-export sequence. *Proc. Natl. Acad. Sci. USA* **106**, 6736–6741.

Gilmour, D. S. (2009). Promoter proximal pausing on genes in metazoans. *Chromosoma* **118**, 1–10.

Gonda, H., Sugai, M., Nambu, Y., Katakai, T., Agata, Y., Mori, K. J., Yokota, Y., and Shimizu, A. (2003). The balance between Pax5 and Id2 activities is the key to AID gene expression. *J. Exp. Med.* **198**, 1427–1437.

Gordon, M. S., Kanegai, C. M., Doerr, J. R., and Wall, R. (2003). Somatic hypermutation of the B cell receptor genes B29 (Igbeta, CD79b) and mb1 (Igalpha, CD79a). *Proc. Natl. Acad. Sci. USA* **100**, 4126–4131.

Gostissa, M., Yan, C. T., Bianco, J. M., Cogne, M., Pinaud, E., and Alt, F. W. (2009). Long-range oncogenic activation of Igh-c-myc translocations by the Igh 3′ regulatory region. *Nature* **462**, 803–807.

Guikema, J. E., Schrader, C. E., Leus, N. G., Ucher, A., Linehan, E. K., Werling, U., Edelmann, W., and Stavnezer, J. (2008). Reassessment of the role of Mut S homolog 5 in Ig class switch recombination shows lack of involvement in cis- and trans-switching. *J. Immunol.* **181**, 8450–8459.

Houseley, J., LaCava, J., and Tollervey, D. (2006). RNA-quality control by the exosome. *Nat. Rev. Mol. Cell Biol.* **7**, 529–539.

Ito, S., Nagaoka, H., Shinkura, R., Begum, N., Muramatsu, M., Nakata, M., and Honjo, T. (2004). Activation-induced cytidine deaminase shuttles between nucleus and cytoplasm like apolipoprotein B mRNA editing catalytic polypeptide 1. *Proc. Natl. Acad. Sci. USA* **101**, 1975–1980.

Jeevan-Raj, B. P., Robert, I., Heyer, V., Page, A., Wang, J. H., Cammas, F., Alt, F. W., Losson, R., and Reina-San-Martin, B. (2011). Epigenetic tethering of AID to the donor switch region during immunoglobulin class switch recombination. *J. Exp. Med.* **208**, 1649–1660.

Jung, D., Giallourakis, C., Mostoslavsky, R., and Alt, F. W. (2006). Mechanism and control of V(D)J recombination at the immunoglobulin heavy chain locus. *Annu. Rev. Immunol.* **24**, 541–570.

Kim, J. B., and Sharp, P. A. (2001). Positive transcription elongation factor B phosphorylates hSPT5 and RNA polymerase II carboxyl-terminal domain independently of cyclin-dependent kinase-activating kinase. *J. Biol. Chem.* **276**, 12317–12323.

Klein, B. J., Bose, D., Baker, K. J., Yusoff, Z. M., Zhang, X., and Murakami, K. S. (2011a). RNA polymerase and transcription elongation factor Spt4/5 complex structure. *Proc. Natl. Acad. Sci. USA* **108**, 546–550.

Klein, I. A., Resch, W., Jankovic, M., Oliveira, T., Yamane, A., Nakahashi, H., Di Virgilio, M., Bothmer, A., Nussenzweig, A., Robbiani, D. F., Casellas, R., and Nussenzweig, M. C. (2011b). Translocation-capture sequencing reveals the extent and nature of chromosomal rearrangements in B lymphocytes. *Cell* **147**, 95–106.

Krokan, H. E., Drablos, F., and Slupphaug, G. (2002). Uracil in DNA–occurrence, consequences and repair. *Oncogene* **21**, 8935–8948.

Kracker, S., Imai, K., Gardes, P., Ochs, H. D., Fischer, A., and Durandy, A. H. (2010). Impaired induction of DNA lesions during immunoglobulin class-switch recombination in humans influences end-joining repair. *Proc. Natl. Acad. Sci. USA* **107**, 22225–22230.

Kuppers, R. (2005). Mechanisms of B-cell lymphoma pathogenesis. *Nat. Rev. Cancer* **5**, 251–262.

Lebecque, S. G., and Gearhart, P. J. (1990). Boundaries of somatic mutation in rearranged immunoglobulin genes: 5′ boundary is near the promoter, and 3′ boundary is approximately 1 kb from V(D)J gene. *J. Exp. Med.* **172**, 1717–1727.

Lis, J. T. (2007). Imaging Drosophila gene activation and polymerase pausing in vivo. *Nature* **450**, 198–202.

Liu, M., and Schatz, D. G. (2009). Balancing AID and DNA repair during somatic hypermutation. *Trends Immunol.* **30**, 173–181.

Liu, M., Duke, J. L., Richter, D. J., Vinuesa, C. G., Goodnow, C. C., Kleinstein, S. H., and Schatz, D. G. (2008). Two levels of protection for the B cell genome during somatic hypermutation. *Nature* **451**, 841–845.

Lumsden, J. M., McCarty, T., Petiniot, L. K., Shen, R., Barlow, C., Wynn, T. A., Morse, H. C., 3rd, Gearhart, P. J., Wynshaw-Boris, A., Max, E. E., and Hodes, R. J. (2004). Immunoglobulin class switch recombination is impaired in Atm-deficient mice. *J. Exp. Med.* **200**, 1111–1121.

Martinez-Rucobo, F. W., Sainsbury, S., Cheung, A. C., and Cramer, P. (2011). Architecture of the RNA polymerase-Spt4/5 complex and basis of universal transcription processivity. *EMBO J.* **30**, 1302–1310.

Maul, R. W., and Gearhart, P. J. (2010). AID and somatic hypermutation. *Adv. Immunol.* **105**, 159–191.

Maul, R. W., Saribasak, H., Martomo, S. A., McClure, R. L., Yang, W., Vaisman, A., Gramlich, H. S., Schatz, D. G., Woodgate, R., Wilson, D. M., 3rd, and Gearhart, P. J. (2011). Uracil residues dependent on the deaminase AID in immunoglobulin gene variable and switch regions. *Nat. Immunol.* **12**, 70–76.

McBride, K. M., Barreto, V., Ramiro, A. R., Stavropoulos, P., and Nussenzweig, M. C. (2004). Somatic hypermutation is limited by CRM1-dependent nuclear export of activation-induced deaminase. *J. Exp. Med.* **199**, 1235–1244.

McBride, K. M., Gazumyan, A., Woo, E. M., Barreto, V. M., Robbiani, D. F., Chait, B. T., and Nussenzweig, M. C. (2006). Regulation of hypermutation by activation-induced cytidine deaminase phosphorylation. *Proc. Natl. Acad. Sci. USA* **103**, 8798–8803.

McBride, K. M., Gazumyan, A., Woo, E. M., Schwickert, T. A., Chait, B. T., and Nussenzweig, M. C. (2008). Regulation of class switch recombination and somatic mutation by AID phosphorylation. *J. Exp. Med.* **205**, 2585–2594.

McKean, D., Huppi, K., Bell, M., Staudt, L., Gerhard, W., and Weigert, M. (1984). Generation of antibody diversity in the immune response of BALB/c mice to influenza virus hemagglutinin. *Proc. Natl. Acad. Sci. USA* **81**, 3180–3184.

Muramatsu, M., Sankaranand, V. S., Anant, S., Sugai, M., Kinoshita, K., Davidson, N. O., and Honjo, T. (1999). Specific expression of activation-induced cytidine deaminase (AID), a novel member of the RNA-editing deaminase family in germinal center B cells. *J. Biol. Chem.* **274**, 18470–18476.

Muramatsu, M., Kinoshita, K., Fagarasan, S., Yamada, S., Shinkai, Y., and Honjo, T. (2000). Class switch recombination and hypermutation require activation-induced cytidine deaminase (AID), a potential RNA editing enzyme. *Cell* **102**, 553–563.

Muschen, M., Re, D., Jungnickel, B., Diehl, V., Rajewsky, K., and Kuppers, R. (2000). Somatic mutation of the CD95 gene in human B cells as a side-effect of the germinal center reaction. *J. Exp. Med.* **192**, 1833–1840.

Nambu, Y., Sugai, M., Gonda, H., Lee, C. G., Katakai, T., Agata, Y., Yokota, Y., and Shimizu, A. (2003). Transcription-coupled events associating with immunoglobulin switch region chromatin. *Science* **302**, 2137–2140.

Nowak, U., Matthews, A. J., Zheng, S., and Chaudhuri, J. (2011). The splicing regulator PTBP2 interacts with the cytidine deaminase AID and promotes binding of AID to switch region DNA. *Nat. Immunol.* **12**, 160–166.

Nussenzweig, A., and Nussenzweig, M. C. (2010). Origin of chromosomal translocations in lymphoid cancer. *Cell* **141**, 27–38.

Okazaki, I. M., Hiai, H., Kakazu, N., Yamada, S., Muramatsu, M., Kinoshita, K., and Honjo, T. (2003). Constitutive expression of AID leads to tumorigenesis. *J. Exp. Med.* **197**, 1173–1181.

Okazaki, I. M., Okawa, K., Kobayashi, M., Yoshikawa, K., Kawamoto, S., Nagaoka, H., Shinkura, R., Kitawaki, Y., Taniguchi, H., Natsume, T., Iemura, S., and Honjo, T. (2011). Histone chaperone Spt6 is required for class switch recombination but not somatic hypermutation. *Proc. Natl. Acad. Sci. USA* **108**, 7920–7925.

Oliveira, T., Resch, W., Jankovic, M., Casellas, R., Nussenzweig, M. C., and Klein, I. A. (2011). Translocation capture sequencing: A method for high throughput mapping of chromosomal rearrangements. *J. Immunol. Methods*.

Orthwein, A., Patenaude, A. M., Affar el, B., Lamarre, A., Young, J. C., and Di Noia, J. M. (2010). Regulation of activation-induced deaminase stability and antibody gene diversification by Hsp90. *J. Exp. Med.* **207**, 2751–2765.

Pasqualucci, L., Migliazza, A., Fracchiolla, N., William, C., Neri, A., Baldini, L., Chaganti, R. S., Klein, U., Kuppers, R., Rajewsky, K., and Dalla-Favera, R. (1998). BCL-6 mutations in normal germinal center B cells: Evidence of somatic hypermutation acting outside Ig loci. *Proc. Natl. Acad. Sci. USA* **95**, 11816–11821.

Pasqualucci, L., Neumeister, P., Goossens, T., Nanjangud, G., Chaganti, R. S., Kuppers, R., and Dalla-Favera, R. (2001). Hypermutation of multiple proto-oncogenes in B-cell diffuse large-cell lymphomas. *Nature* **412**, 341–346.

Pasqualucci, L., Kitaura, Y., Gu, H., and Dalla-Favera, R. (2006). PKA-mediated phosphorylation regulates the function of activation-induced deaminase (AID) in B cells. *Proc. Natl. Acad. Sci. USA* **103**, 395–400.

Patenaude, A. M., Orthwein, A., Hu, Y., Campo, V. A., Kavli, B., Buschiazzo, A., and Di Noia, J. M. (2009). Active nuclear import and cytoplasmic retention of activation-induced deaminase. *Nat. Struct. Mol. Biol.* **16**, 517–527.

Pavri, R., Gazumyan, A., Jankovic, M., Di Virgilio, M., Klein, I., Ansarah-Sobrinho, C., Resch, W., Yamane, A., San-Martin, B. R., Barreto, V., Nieland, T. J., Root, D. E., *et al.* (2010). Activation-induced cytidine deaminase targets DNA at sites of RNA polymerase II stalling by interaction with Spt5. *Cell* **143**, 122–133.

Peled, J. U., Kuang, F. L., Iglesias-Ussel, M. D., Roa, S., Kalis, S. L., Goodman, M. F., and Scharff, M. D. (2008). The biochemistry of somatic hypermutation. *Annu. Rev. Immunol.* **26**, 481–511.

Peters, A., and Storb, U. (1996). Somatic hypermutation of immunoglobulin genes is linked to transcription initiation. *Immunity* **4**, 57–65.

Petersen-Mahrt, S. K., Harris, R. S., and Neuberger, M. S. (2002). AID mutates E. coli suggesting a DNA deamination mechanism for antibody diversification. *Nature* **418**, 99–103.

Pham, P., Bransteitter, R., Petruska, J., and Goodman, M. F. (2003). Processive AID-catalysed cytosine deamination on single-stranded DNA simulates somatic hypermutation. *Nature* **424**, 103–107.

Pham, P., Smolka, M. B., Calabrese, P., Landolph, A., Zhang, K., Zhou, H., and Goodman, M. F. (2008). Impact of phosphorylation and phosphorylation-null mutants on the activity and deamination specificity of activation-induced cytidine deaminase. *J. Biol. Chem.* **283**, 17428–17439.

Potter, M. (2003). Neoplastic development in plasma cells. *Immunol. Rev.* **194**, 177–195.

Rada, C., Jarvis, J. M., and Milstein, C. (2002). AID-GFP chimeric protein increases hypermutation of Ig genes with no evidence of nuclear localization. *Proc. Natl. Acad. Sci. USA* **99**, 7003–7008.

Rahl, P. B., Lin, C. Y., Seila, A. C., Flynn, R. A., McCuine, S., Burge, C. B., Sharp, P. A., and Young, R. A. (2010). c-Myc regulates transcriptional pause release. *Cell* **141**, 432–445.

Rajagopal, D., Maul, R. W., Ghosh, A., Chakraborty, T., Khamlichi, A. A., Sen, R., and Gearhart, P. J. (2009). Immunoglobulin switch mu sequence causes RNA polymerase II accumulation and reduces dA hypermutation. *J. Exp. Med.* **206**, 1237–1244.

Rajewsky, K. (1996). Clonal selection and learning in the antibody system. *Nature* **381**, 751–758.

Ramiro, A. R., Stavropoulos, P., Jankovic, M., and Nussenzweig, M. C. (2003). Transcription enhances AID-mediated cytidine deamination by exposing single-stranded DNA on the nontemplate strand. *Nat. Immunol.* **4**, 452–456.

Ramiro, A. R., Jankovic, M., Eisenreich, T., Difilippantonio, S., Chen-Kiang, S., Muramatsu, M., Honjo, T., Nussenzweig, A., and Nussenzweig, M. C. (2004). AID is required for c-myc/IgH chromosome translocations in vivo. *Cell* **118**, 431–438.

Ramiro, A. R., Jankovic, M., Callen, E., Difilippantonio, S., Chen, H. T., McBride, K. M., Eisenreich, T. R., Chen, J., Dickins, R. A., Lowe, S. W., Nussenzweig, A., and Nussenzweig, M. C. (2006). Role of genomic instability and p53 in AID-induced c-myc-Igh translocations. *Nature* **440**, 105–109.

Ranjit, S., Khair, L., Linehan, E. K., Ucher, A. J., Chakrabarti, M., Schrader, C. E., and Stavnezer, J. (2011). AID binds cooperatively with UNG and Msh2-Msh6 to Ig switch regions dependent upon the AID C terminus. *J. Immunol.* **187**, 2464–2475.

Raschke, E. E., Albert, T., and Eick, D. (1999). Transcriptional regulation of the Ig kappa gene by promoter-proximal pausing of RNA polymerase II. *J. Immunol.* **163**, 4375–4382.

Revy, P., Muto, T., Levy, Y., Geissmann, F., Plebani, A., Sanal, O., Catalan, N., Forveille, M., Dufourcq-Labelouse, R., Gennery, A., Tezcan, I., Ersoy, F., *et al.* (2000). Activation-induced cytidine deaminase (AID) deficiency causes the autosomal recessive form of the Hyper-IgM syndrome (HIGM2). *Cell* **102**, 565–575.

Robbiani, D. F., Bothmer, A., Callen, E., Reina-San-Martin, B., Dorsett, Y., Difilippantonio, S., Bolland, D. J., Chen, H. T., Corcoran, A. E., Nussenzweig, A., and Nussenzweig, M. C. (2008). AID is required for the chromosomal breaks in c-myc that lead to c-myc/IgH translocations. *Cell* **135**, 1028–1038.

Robbiani, D. F., Bunting, S., Feldhahn, N., Bothmer, A., Camps, J., Deroubaix, S., McBride, K. M., Klein, I. A., Stone, G., Eisenreich, T. R., Ried, T., Nussenzweig, A., *et al.* (2009). AID produces DNA double-strand breaks in non-Ig genes and mature B cell lymphomas with reciprocal chromosome translocations. *Mol. Cell* **36**, 631–641.

Rothenfluh, H. S., Taylor, L., Bothwell, A. L., Both, G. W., and Steele, E. J. (1993). Somatic hypermutation in 5′ flanking regions of heavy chain antibody variable regions. *Eur. J. Immunol.* **23**, 2152–2159.

Rucci, F., Cattaneo, L., Marrella, V., Sacco, M. G., Sobacchi, C., Lucchini, F., Nicola, S., Della Bella, S., Villa, M. L., Imberti, L., Gentili, F., Montagna, C., et al. (2006). Tissue-specific sensitivity to AID expression in transgenic mouse models. *Gene* **377**, 150–158.

Sayegh, C. E., Quong, M. W., Agata, Y., and Murre, C. (2003). E-proteins directly regulate expression of activation-induced deaminase in mature B cells. *Nat. Immunol.* **4**, 586–593.

Sernandez, I. V., de Yebenes, V. G., Dorsett, Y., and Ramiro, A. R. (2008). Haploinsufficiency of activation-induced deaminase for antibody diversification and chromosome translocations both in vitro and in vivo. *PLoS One* **3**, e3927.

Shen, H. M., Peters, A., Baron, B., Zhu, X., and Storb, U. (1998). Mutation of BCL-6 gene in normal B cells by the process of somatic hypermutation of Ig genes. *Science* **280**, 1750–1752.

Shinkura, R., Ito, S., Begum, N. A., Nagaoka, H., Muramatsu, M., Kinoshita, K., Sakakibara, Y., Hijikata, H., and Honjo, T. (2004). Separate domains of AID are required for somatic hypermutation and class-switch recombination. *Nat. Immunol.* **5**, 707–712.

Sohail, A., Klapacz, J., Samaranayake, M., Ullah, A., and Bhagwat, A. S. (2003). Human activation-induced cytidine deaminase causes transcription-dependent, strand-biased C to U deaminations. *Nucleic Acids Res.* **31**, 2990–2994.

Staszewski, O., Baker, R. E., Ucher, A. J., Martier, R., Stavnezer, J., and Guikema, J. E. (2011). Activation-induced cytidine deaminase induces reproducible DNA breaks at many non-Ig Loci in activated B cells. *Mol. Cell.* **41**, 232–242.

Stavnezer, J., Guikema, J. E., and Schrader, C. E. (2008). Mechanism and regulation of class switch recombination. *Annu. Rev. Immunol.* **26**, 261–292.

Steele, E. J., Rothenfluh, H. S., and Both, G. W. (1992). Defining the nucleic acid substrate for somatic hypermutation. *Immunol. Cell Biol.* **70**(Pt 2), 129–144.

Swanson, M. S., Malone, E. A., and Winston, F. (1991). SPT5, an essential gene important for normal transcription in Saccharomyces cerevisiae, encodes an acidic nuclear protein with a carboxy-terminal repeat. *Mol. Cell. Biol.* **11**, 4286.

Ta, V. T., Nagaoka, H., Catalan, N., Durandy, A., Fischer, A., Imai, K., Nonoyama, S., Tashiro, J., Ikegawa, M., Ito, S., Kinoshita, K., Muramatsu, M., et al. (2003). AID mutant analyses indicate requirement for class-switch-specific cofactors. *Nat. Immunol.* **4**, 843–848.

Takizawa, M., Tolarova, H., Li, Z., Dubois, W., Lim, S., Callen, E., Franco, S., Mosaico, M., Feigenbaum, L., Alt, F. W., Nussenzweig, A., Potter, M., et al. (2008). AID expression levels determine the extent of cMyc oncogenic translocations and the incidence of B cell tumor development. *J. Exp. Med.* **205**, 1949–1957.

Teng, G., and Papavasiliou, F. N. (2007). Immunoglobulin somatic hypermutation. *Annu. Rev. Genet.* **41**, 107–120.

Teng, G., Hakimpour, P., Landgraf, P., Rice, A., Tuschl, T., Casellas, R., and Papavasiliou, F. N. (2008). MicroRNA-155 is a negative regulator of activation-induced cytidine deaminase. *Immunity* **28**, 621–629.

Tran, T. H., Nakata, M., Suzuki, K., Begum, N. A., Shinkura, R., Fagarasan, S., Honjo, T., and Nagaoka, H. (2010). B cell-specific and stimulation-responsive enhancers derepress Aicda by overcoming the effects of silencers. *Nat. Immunol.* **11**, 148–154.

Tsai, A. G., and Lieber, M. R. (2010). Mechanisms of chromosomal rearrangement in the human genome. *BMC Genomics* **11**(Suppl. 1), S1.

Tumas-Brundage, K., and Manser, T. (1997). The transcriptional promoter regulates hypermutation of the antibody heavy chain locus. *J. Exp. Med.* **185**, 239–250.

Vuong, B. Q., Lee, M., Kabir, S., Irimia, C., Macchiarulo, S., McKnight, G. S., and Chaudhuri, J. (2009). Specific recruitment of protein kinase A to the immunoglobulin locus regulates class-switch recombination. *Nat. Immunol.* **10**, 420–426.

Wada, T., Takagi, T., Yamaguchi, Y., Ferdous, A., Imai, T., Hirose, S., Sugimoto, S., Yano, K., Hartzog, G. A., Winston, F., Buratowski, S., and Handa, H. (1998). DSIF, a novel transcription elongation factor that regulates RNA polymerase II processivity, is composed of human Spt4 and Spt5 homologs. *Genes Dev.* **12**, 343–356.

Wang, C. L., Harper, R. A., and Wabl, M. (2004). Genome-wide somatic hypermutation. *Proc. Natl. Acad. Sci. USA* **101**, 7352–7356.

Wang, L., Wuerffel, R., Feldman, S., Khamlichi, A. A., and Kenter, A. L. (2009). S region sequence, RNA polymerase II, and histone modifications create chromatin accessibility during class switch recombination. *J. Exp. Med.* **206**, 1817–1830.

Weber, J. S., Berry, J., Manser, T., and Claflin, J. L. (1991). Position of the rearranged V kappa and its 5′ flanking sequences determines the location of somatic mutations in the J kappa locus. *J. Immunol.* **146**, 3652–3655.

Weigert, M. G., Cesari, I. M., Yonkovich, S. J., and Cohn, M. (1970). Variability in the lambda light chain sequences of mouse antibody. *Nature* **228**, 1045–1047.

Xu, Z., Fulop, Z., Wu, G., Pone, E. J., Zhang, J., Mai, T., Thomas, L. M., Al-Qahtani, A., White, C. A., Park, S. R., Steinacker, P., Li, Z., et al. (2010). 14-3-3 adaptor proteins recruit AID to 5′-AGCT-3′-rich switch regions for class switch recombination. *Nat. Struct. Mol. Biol.* **17**, 1124–1135.

Yamada, T., Yamaguchi, Y., Inukai, N., Okamoto, S., Mura, T., and Handa, H. (2006). P-TEFb-mediated phosphorylation of hSpt5 C-terminal repeats is critical for processive transcription elongation. *Mol. Cell* **21**, 227–237.

Yamaguchi, Y., Takagi, T., Wada, T., Yano, K., Furuya, A., Sugimoto, S., Hasegawa, J., and Handa, H. (1999). NELF, a multisubunit complex containing RD, cooperates with DSIF to repress RNA polymerase II elongation. *Cell* **97**, 41–51.

Yamane, A., Resch, W., Kuo, N., Kuchen, S., Li, Z., Sun, H. W., Robbiani, D. F., McBride, K., Nussenzweig, M. C., and Casellas, R. (2011). Deep-sequencing identification of the genomic targets of the cytidine deaminase AID and its cofactor RPA in B lymphocytes. *Nat. Immunol.* **12**, 62–69.

Yu, K., Chedin, F., Hsieh, C. L., Wilson, T. E., and Lieber, M. R. (2003). R-loops at immunoglobulin class switch regions in the chromosomes of stimulated B cells. *Nat. Immunol.* **4**, 442–451.

Zhang, Y., Gostissa, M., Hildebrand, D. G., Becker, M. S., Boboila, C., Chiarle, R., Lewis, S., and Alt, F. W. (2010). The role of mechanistic factors in promoting chromosomal translocations found in lymphoid and other cancers. *Adv. Immunol.* **106**, 93–133.

Zhou, C., Saxon, A., and Zhang, K. (2003). Human activation-induced cytidine deaminase is induced by IL-4 and negatively regulated by CD45: Implication of CD45 as a Janus kinase phosphatase in antibody diversification. *J. Immunol.* **170**, 1887–1893.

Opportunities and Challenges in Tumor Angiogenesis Research: Back and Forth Between Bench and Bed

Li Qin,*,† Jennifer L. Bromberg-White,‡ and Chao-Nan Qian*,‡,1

*State Key Laboratory on Oncology in South China,
Sun Yat-sen University Cancer Center,
Guangzhou, Guangdong, PR China
†Division of Pharmacoproteomics, Institute of Pharmacy and Pharmacology,
University of South China, Hengyang, Hunan, PR China
‡Laboratory of Cancer and Developmental Cell Biology, Van Andel Research
Institute, Grand Rapids, Michigan, USA

I. Introduction
II. Heterogeneity of Tumor Vasculature
III. Vessel Co-Option for Tumor Vascular Establishment
IV. Microvessels Formed by Tumor Cells and Tumor Stem-Like Cells
V. Comprehensive Angiogenesis-Related Signaling Pathways
   A. VEGF/VEGFR Signaling Pathways
   B. Ang/Tie Pathway
   C. Hypoxia Pathway and HIF-1
   D. Paradoxical Effects of NO Signaling in Tumor Angiogenesis
   E. mTOR Pathway
   F. FGFs/FGFRs Pathway
   G. Delta-Like Ligand 4 (DLL4)/Notch Pathway Interacts with the VEGF Pathway
   H. Other Molecules Regulating Tumor Angiogenesis
   I. PDGF/PDGF Receptor Pathways Regulate Endothelial Cells and Pericytes
   J. Basement Membrane in Angiogenesis
   K. Inflammation and Tumor Angiogenesis
VI. Differential Analyses of Tumor Vasculature
   A. Controversy in the Traditional Evaluation System of Tumor Microvessel Density
   B. Dissecting Tumor Vasculature is Necessary to Reveal True Prognostic Values
VII. Revascularization Within Tumors Following Withdrawal of Antiangiogenic Treatment
VIII. Vascular Normalization or Ineffective Targeting of Tumor Vasculature?
IX. Highlights in Antiangiogenic Therapy
   A. Targeting the VEGF Pathway
   B. Targeting the mTOR Pathway
   C. Targeting the Hypoxia Pathway
X. Need for a Novel Clinical Evaluation System for Antiangiogenic Therapies

[1] Corresponding author: qianchn@sysucc.org.cn

XI. Mechanisms of Resistance in Antiangiogenic Therapy
   A. Activation of Hypoxia and c-Met Pathways and Beyond
   B. Activation of Other Angiogenic Factors
   C. Vessel Co-Option Instead of Sprouting Angiogenesis
XII. Prospects for Future Development
   Acknowledgments
   References

Angiogenesis is essential for tumor growth and metastasis. Many signaling pathways are involved in regulating tumor angiogenesis, with the vascular endothelial growth factor pathway being of particular interest. The recognition of the heterogeneity in tumor vasculature has led to better predictions of prognosis through differential analyses of the vasculature. However, the clinical benefits from antiangiogenic therapy are limited, because many antiangiogenic agents cannot provide long-term survival benefits, suggesting the development of drug resistance. Activation of the hypoxia and c-Met pathways, as well as other proangiogenic factors, has been shown to be responsible for such resistance. Vessel co-option could be another important mechanism. For future development, research to improve the efficacy of antiangiogenic therapy includes (a) using tumor-derived endothelial cells for drug screening; (b) developing the drugs focusing on specific tumor types; (c) developing a better preclinical model for drug study; (d) developing more accurate biomarkers for patient selection; (e) targeting the c-Met pathway or other pathways; and (f) optimizing the dose and schedule of antiangiogenic therapy. In summary, the future of antiangiogenic therapy for cancer patients depends on our efforts to develop the right drugs, select the right patients, and optimize the treatment conditions. © 2012 Elsevier Inc.

I. INTRODUCTION

De novo development of blood vessels without preexisting vessels is called "vasculogenesis," which usually involves in the recruitment and differentiation of endothelial progenitor cells (Ding *et al.*, 2008). In contrast, the sprouting of new blood vessels from existing ones is "angiogenesis." Without an angiogenic process inside a solid tumor, the tumor volume cannot exceed 2–3cm in diameter (Hanahan and Folkman, 1996), which becomes a limiting event (Hanahan and Weinberg, 2000). The targeting of angiogenesis, as proposed by Folkman (1971), has been validated as an effective therapeutic approach to human malignancies in the kidney, liver, lung, colorectum, and pancreas (pancreatic neuroendocrine tumors; Bracarda *et al.*, 2011; Guan *et al.*, 2011; Kudo, 2011; Oudard *et al.*, 2011; Raymond *et al.*, 2011; Van Meter and Kim, 2011). Currently, seven antiangiogenic agents for treating cancer patients have been approved by the U.S. Food and Drug Administration, either as single agents or in combination with cytotoxic chemotherapeutics (Samant and Shevde, 2011). However, the clinical benefits from antiangiogenic therapy are limited: many clinical trials have produced only short-term tumor stabilization rather than long-term survival benefits, suggesting the development of drug resistance to the therapy (Bergers and Hanahan, 2008; D'Agostino, 2011; Kerbel *et al.*, 2001; Miller *et al.*, 2005).

In recent years, the complexity and abnormality of tumor vasculature, as well as the multiple paths to establishing that vasculature, have been recognized (Baluk *et al.*, 2005; Jain, 2005; Qian *et al.*, 2009; Yao *et al.*, 2007). Translating this emerging knowledge into clinical applications is an important issue, and the current strategy—of drug development aimed at inhibiting tumor angiogenesis by using nontumor endothelial cells for screening and validation—is being challenged. In this review, we discuss the heterogeneity of tumor vasculature, the establishment of such vasculature, angiogenesis signaling, treatment failure using antiangiogenic agents, and perspectives on future developments.

II. HETEROGENEITY OF TUMOR VASCULATURE

In normal vasculature, the cell types forming blood vessels include endothelial cells lining the inner surface of the vessel and pericytes on the abluminal side of the endothelial cells in the pericytic microvasculature. Endothelial cells can be classified into tip cells, stalk cells, and phalanx cells (De Smet *et al.*, 2009). Endothelial tip cells lead the way in a branching vessel, stalk cells elongate the sprout, and phalanx cells ensure quiescence and perfusion of the newly formed branch. Pericytes are solitary, smooth muscle-like mural cells and are involved in vascular stabilization through contact with endothelial cells along the length of the vessels and also through paracrine signaling (Hirschi and D'Amore, 1996). The recruitment of pericytes is essential for the formation and stabilization of mature blood vessels, and such vessels are usually covered with pericytes. Between the endothelial cells and pericytes is a basal membrane. This membrane in normal vasculature is a uniform thin layer enriched in type IV collagen and controls blood vessel penetration. High-throughput technologies have revealed that normal endothelial cells from various tissues and organs have distinct genomic and proteomic expression patterns (Zamora *et al.*, 2007; Zhang *et al.*, 2008), implying a heterogeneity of endothelial cells even in normal tissues.

The distinct characteristics of tumor vasculature versus normal vasculature have been noticed on multiple levels, including cytogenetics, protein expression, cellular structure, cellular function, and vasculature organization.

In cytogenetic level, tumor-derived endothelial cells harbor cytogenetic abnormalities (Hida *et al.*, 2004). At the protein level, these cells express surface molecules that are absent or barely detectable in normal endothelial cells, including the receptors to recognize proangiogenic growth factors and circulating leukocytes (Bussolati *et al.*, 2003; Ghosh *et al.*, 2008; Hida and Klagsbrun, 2005). For example, neural-cell adhesion molecule, which can drive the maturation of endothelial cells to form capillary structure, is expressed in renal tumor-derived endothelial cells but not in normal endothelial cells (Bussolati *et al.*, 2006).

Abnormalities in cellular structure as well as vasculature organization in tumor vasculature have been recognized (Baluk *et al.*, 2005; Jain, 2005; Nagy *et al.*, 2010). The loss of a hierarchical arrangement of arterioles, capillaries, and venules in tumors results in a lower efficiency of blood flow, although higher vascular density has been found in many solid tumors (Fig. 1). The appearance of numerous fenestration structures on endothelial cells, together with discontinuous basement membrane coverage, allows higher penetration of large molecules (e.g., albumin) into the interstitial space in a solid tumor. These structural abnormalities result in disturbed blood flow, hypoxia, hyperpermeability, and elevated interstitial pressure in many solid tumors. The elevated interstitial pressure and disturbed blood flow will impair the delivery of anticancer drugs (as well as oxygen) to the tumor site (Iwahana *et al.*, 1998; Jung *et al.*, 2000; Kerbel *et al.*, 2001).

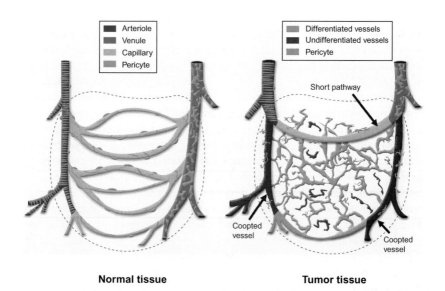

Fig. 1 Abnormalities of the tumor vasculature. In comparison to normal vasculature, tumor vasculature has several abnormalities, including loss of hierarchy, appearance of short pathway, chaotic network of tortuous microvessels, a high vascular turnover (rapid formation of new vessels and regression of the established ones), as well as high vascular density consisting of undifferentiated (without pericyte coverage), differentiated immature (without pericyte coverage), and differentiated mature (with pericyte coverage) vessels. (See Page 4 in Color Section at the back of the book.)

Categorized by endothelial markers, tumor endothelial cells can be classified into undifferentiated (CD31+, CD34−) and differentiated (CD34+), forming undifferentiated and differentiated blood vessels, respectively (Qian *et al.*, 2009; Yao *et al.*, 2007). The cellular origins and the interchangeability of these two types of endothelial cells remain unclear. Importantly, a high density of differentiated microvessels in clear cell renal cell carcinoma (CCRCC) correlates with a better prognosis while a high density of undifferentiated microvessels correlates with poor prognosis (Yao *et al.*, 2007). These vessel types have an inverse correlation: high density of one type of microvessel is accompanied by lower density of the other (Yao *et al.*, 2007), suggesting that an increased number of undifferentiated vessels might be induced by a reduction of differentiated vessels, which could result in reduction of functional vessels as well as blood supply.

Several distinct processes contribute to the formation of tumor vasculature, including sprouting angiogenesis, intussusceptive angiogenesis, vasculogenesis, vasculogenic mimicry (VM), and vessel co-option of the existing vasculature (Chen *et al.*, 2010; Hillen and Griffioen, 2007). Sprouting is the best studied mechanism of expanding the network of vessels in growing tumor and involves filopodia and endothelial stalk cells. A balance between proangiogenic and antiangiogenic growth factors and cytokines tightly controls this process. Vessel co-option has been poorly studied, but it could be crucial in our understanding of the development of tumor vasculature and will be discussed below.

Most stromal cells in solid tumors are fibroblasts, which are suggested to be an important source of growth factors and cytokines for recruiting endothelial cells. Pericytes are loosely associated with tumor vessels (Baluk *et al.*, 2005; Jain, 2005); undifferentiated tumor vessels are free of pericyte coverage (Yao *et al.*, 2007), allowing a constantly dynamic state of proliferation, remodeling, sprout formation, or regression within the tumor endothelium. It has been reported that the tumor vasculature resistant to antiangiogenic therapy has more pericyte coverage (Helfrich *et al.*, 2010). Pericyte coverage is thus thought to be responsible for different fates of tumor progression (Helfrich and Schadendorf, 2011).

The diversity of tumor vasculature can also be seen in terms of dependence on proangiogenic factors among different tumor types. For example, overexpression of vascular endothelial growth factor (VEGF) and its receptors have been found in lung cancer (Merrick *et al.*, 2005), while the expression levels of proangiogenic factors—including the VEGF family, platelet-derived growth factor (PDGF) family, angiopoietin (Ang) family, and hepatocyte growth factor (HGF)—are decreased in breast cancer (Boneberg *et al.*, 2009).

Recognizing the diversity of tumor vasculature is important in terms of predicting the patient's prognosis and in developing novel strategies for targeting tumor angiogenesis.

III. VESSEL CO-OPTION FOR TUMOR VASCULAR ESTABLISHMENT

It has been known for more than two decades that tumor vasculature may not need to develop from *de novo* vessels; tumors can "hijack" the existing blood vessels of the normal tissue surrounding tumor nests and integrate them into the tumor vasculature (Thompson *et al.*, 1987). This approach is called vessel co-option, and as a consequence, a tumor can survive and grow without the induction of sprouting angiogenesis (Pezzella *et al.*, 1997). It is also believed that vessel co-option can be the initial step in establishing tumor vasculature, earlier than sprouting angiogenesis (Holash *et al.*, 1999). In melanoma, the peritumoral vascular plexus is continuously integrated into the tumor vasculature as the tumor expands (Dome *et al.*, 2002). Angiopoietin-2 is implicated in this important process (Holash *et al.*, 1999). These integrated vessels have two important characteristics: proliferation of endothelial cells only occurs in vessel dilatation, and the integrated vessels are covered by pericytes (Dome *et al.*, 2002). It is speculated that the co-opted vessels would not be affected by angiogenesis inhibition (Leenders *et al.*, 2002), presenting another hurdle for antiangiogenic therapy.

Sentinel lymph node metastasis is the earliest and most common route of spread for many epithelial carcinomas and for melanoma. The rapid growth of the secondary tumors inside sentinel or regional lymph nodes often becomes manifest as the major sign in cancer patients (Giridharan *et al.*, 2003; Qian *et al.*, 2007). Such rapid growth cannot be simply explained by colony selection of cancer cells with a higher proliferation rate (Qian *et al.*, 2007). Accelerating systemic dissemination to other distant organs usually follow this first step of metastasis to the sentinel lymph nodes (Meier *et al.*, 2002). Therefore, the rapid growth of cancer cells inside the lymph node is a key element determining the survival time of cancer patients.

We have shown that the microenvironment inside the sentinel lymph node can be manipulated by the primary tumor to best fit the survival and growth of the later metastatic tumors (Qian *et al.*, 2006). The key component being remodeled by the primary tumor is the high endothelial venule (HEV). HEVs are special vessels only existing in lymph node and lymphoid tissues (but not in the spleen). The normal function of HEVs is not for carrying blood, but for recruiting lymphocytes for the generation of immune responses (Sackstein, 1993; von Andrian and M'Rini, 1998; Yeh *et al.*, 2001). The endothelial cells of HEVs express a specific ligand called peripheral lymph node addressin (PNAd), which can be recognized by anti-MECA79 antibody. PNAd can bind the homing receptor L-selectin expressed in the cellular membrane of lymphocytes. When the naïve lymphocytes move to the HEV and recognize PNAd, they stop there, and a significant fraction of the lymphocytes then leave the circulation and move into the lymphoid tissue. In cancer, however, the HEVs in the sentinel lymph node can be dramatically remodeled to become blood flow carriers even before the arrival of metastatic cancer cells (Qian *et al.*, 2006, 2007). After the cancer cells arrive, the remodeled HEVs can be further integrated into the tumor vasculature, losing PNAd and becoming the "mother vessels" of the metastatic tumors, shifting their role from immune response to support of the tumor (Fig. 2). Preliminary investigation of the underlying molecular mechanisms shows that

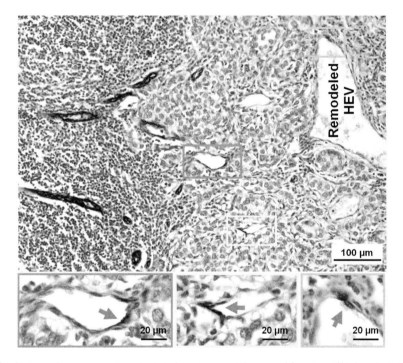

Fig. 2 Vascular co-option in metastatic breast cancer. The normal function of high endothelial venules (HEVs) in lymphoid tissue is immune response, not blood flow. In cancer, HEVs can be remodeled and integrated into the tumor vasculature, with their lumens being dilated, losing their specific marker PNAd (peripheral lymph node addressin), and carrying blood to support metastatic tumor growth. In this figure, human metastatic breast cancer tissue (right side) in an axillary lymph node is invading toward the remnant lymphoid tissue (left side). HEVs stained by antibody against PNAd are integrated into the tumor vasculature. The lower panels show enlarged images of the boxed vessels. Arrows indicate the partial expression of PNAd in the vessel wall, suggesting their HEV origin. Note that the dilated lumen can be very large in the remodeled HEV. Figure reproduced with permission from Qian *et al.*, 2006, Supplementary Fig. 2. (See Page 5 in Color Section at the back of the book.)

VEGF-D and VEGF-R2 interaction can drive the remodeling of HEVs under a cancerous scenario, while BMP-4 acts as an inhibitor of this VEGF-D-driven remodeling process (Farnsworth et al., 2011).

IV. MICROVESSELS FORMED BY TUMOR CELLS AND TUMOR STEM-LIKE CELLS

In VM, which was first found in melanoma, tumor cells can metamorphose into vascular structures that either carry blood flow or connect to the host's vasculature (Maniotis et al., 1999). Similar structures were later recognized in other malignancies including glioma, breast, lung, gastric, prostate, ovarian, and colorectal cancers, and they are associated with more aggressive tumor behavior as well as with poorer patient outcome (Baeten et al., 2009; Hendrix et al., 2003; Li et al., 2010a; Liu et al., 2002, 2011a; Passalidou et al., 2002; Sharma et al., 2002; Shirakawa et al., 2002; Sood et al., 2002). An important characteristic of VM formed by tumor cells is that they do not express endothelial markers (factor VIII-related antigen, Ulex, CD31, CD34, or VEGF-R2; Maniotis et al., 1999), but do express VE-cadherin, and EphA2 (Hess et al., 2006); they also stain positive in culture with periodic acid-Schiff (Lin et al., 2005). Accumulating evidence suggests that cancer stem-like cells can differentiate into VM structures in melanoma (Girouard and Murphy, 2011). However, it is unclear what percentage of the tumor blood supply is provided by VM and how often VM occurs in solid tumors (McDonald and Foss, 2000). In a study involving 331 cases of breast cancer, VM was found only in 7.9% (Shirakawa et al., 2002).

CD31+ endothelial cells isolated from melanoma and renal cell carcinoma possess chromosomal aberrations similar to the CD31− counterpart malignant cells (Akino et al., 2009; Hida et al., 2004), suggesting that at least a portion of the CD31+ cells in tumor vasculature have the same origin as the cancer cells. This idea was recently strengthened by the finding that 20–90% of CD31+ endothelial cells in glioblastoma carry the same genomic alterations as the tumor cells (Ricci-Vitiani et al., 2010). Other studies have shown that the glioblastoma cells harboring stem-like characteristics (CD144+) are responsible for the formation of the tumor vasculature with a cancer cell origin (Ricci-Vitiani et al., 2010; Wang et al., 2010).

The discovery that tumor cells can form vasculature might have profound impact on therapeutic strategies for targeting tumor angiogenesis. It is reasonable to assume that the cancer cells forming microvessels are more resistant than endothelial cells to antiangiogenic agents, because almost all such agents are selected based on their activities against normal endothelial cells.

V. COMPREHENSIVE ANGIOGENESIS-RELATED SIGNALING PATHWAYS

A. VEGF/VEGFR Signaling Pathways

Tumor angiogenesis is driven by multiple growth factors that activate their corresponding receptor tyrosine kinases, with VEGF–VEGF receptor signaling being the most characterized. The VEGF gene family includes four members in mice and humans (VEGF-A, -B, -C, and -D) and a related protein, placental growth factor (PlGF; Ferrara et al., 2003; Tammela et al., 2005). They are homodimeric, secreted glycoproteins whose precise regulation is essential for correct specification, assembly, and patterning of the vasculature. The best characterized of the VEGF family members is VEGF-A (commonly referred to as VEGF), and its gene consists of eight exons separated by seven introns and spans approximately 14 kb (Houck et al., 1991). VEGF is expressed either as proangiogenic or antiangiogenic factors depending on splice site choice in its terminal exon. Proximal splice site selection in exon 8 generates proangiogenic isoforms including $VEGF_{121}$, $VEGF_{165}$, $VEGF_{189}$, and $VEGF_{206}$. Distal splice site selection in exon 8 generates antiangiogenic isoforms such as $VEGF_{165}b$ (Nowak et al., 2010). Of the proangiogenic isoforms, $VEGF_{165}$ is predominant and is commonly overexpressed in a variety of human solid tumors.

The VEGF signaling depends on the binding of VEGF to three structurally similar type III receptor tyrosine kinases, designated VEGF-R1 (encoded by the *FLT1* gene in humans), VEGF-R2 (encoded by the *Flk-1* gene in mice or the *KDR* gene in humans), and VEGF-R3 (encoded by the *FLT4* gene in humans). The neuropilins (NP1 and NP2, also known as NRP1 and NRP2) act as coreceptors for the VEGFRs, increasing the binding affinity of VEGF for the VEGFR tyrosine kinase receptors (Bielenberg and Klagsbrun, 2007). VEGF-R2 is the dominant proangiogenic receptor on the surface of endothelial cells (Millauer et al., 1993). The underlying mechanism of VEGF-R1-induced angiogenesis may involve activation by PlGF, either directly or through transphosphorylation of VEGF-R2 by phosphorylated VEGF-R1 (Autiero et al., 2003). NP-1 and NP-2, which were initially characterized for their role in axon guidance, have been subsequently shown to modulate the VEGF pathway (Klagsbrun et al., 2002). NP-1 binds the heparin-binding isoform of VEGF-A and PlGF, resulting in more effective presentation of these ligands to their signaling receptors (Neufeld et al., 2002). Dual targeting of the vasculature with antibodies against VEGF and NP-1 is more effective than single-agent targeting (Batchelor et al., 2007), further confirming the cooperation between VEGF and NP1.

Signals from VEGF pathways control many biological effects, the best studied being effects on endothelial cell migration, proliferation, and cell survival, and the expression of downstream genes in endothelial cells and tumor cells (Thurston and Kitajewski, 2008).

Recently, the clinical application of targeted drugs blocking VEGF signaling has generated some patient survival benefit in the treatment of several malignancies. Equally important, some unexpected adverse effects from this novel treatment have revealed more clinically significant biological effects of VEGF signaling. The adverse effects have been noticed mainly in bevacizumab therapy. Hypertension is the most common effect, which reflects the important role of VEGF in controlling nitric oxide (NO) production. NO is a vasodilator keeping human blood pressure within normal bounds. Inhibition of VEGF signaling downregulates the expression of endothelial nitric oxide synthase (eNOS; Hood et al., 1998), resulting in a decrease of NO production in the wall of arterioles and other resistance vessels. Lack of NO results in vasoconstriction and an increase of peripheral resistance, and eventually an increase in blood pressure (Kamba and McDonald, 2007).

Proteinuria is another common adverse effect in anti-VEGF therapy. Endothelial cells, basement membranes, and podocytes form a filtration barrier in the renal glomerulus. Numerous fenestrations in the endothelial cells are sustained by VEGF signaling for normal filtration function. Blocking VEGF signaling reduces the number of fenestration structures in the glomerular capillaries (Kamba et al., 2006), disturbing the interaction between podocytes and glomerular endothelial cells and eventually disrupting the filtration barrier; the result is proteinuria (Kamba et al., 2006).

The maintenance of fenestration structures in the endothelial cells of other endocrine organs is also controlled by VEGF signaling. Blocking VEGF signaling will therefore impair the functioning of the thyroid gland and the pancreatic islets (Kamba and McDonald, 2007).

Other adverse events in blocking VEGF signaling include gastrointestinal perforation, arterial thrombosis, surgical and wound-healing complications, hemorrhage, heart failure, hematologic toxicities, reversible posterior leukoencephalopathy, lacrimation disorder, and exfoliative dermatitis (Kamba and McDonald, 2007; Randall and Monk, 2010, 2011; Schutz et al., 2011; Verma and Swain, 2011). Most of these effects are manageable (Kamba and McDonald, 2007).

B. Ang/Tie Pathway

The Ang ligands and their Tie receptors have been identified as the second vascular tissue-specific receptor Tyr kinase system. The Ang/Tie system controls endothelial cell survival and vascular maturation. Activating phosphorylation of Tie2 mediates survival signals of endothelial cells and regulates the recruitment of pericytes (Thomas and Augustin, 2009). The Ang family includes four ligands: angiopoietin-1 (Ang-1), Ang-2, and the mouse/human interspecies orthologs Ang-3/Ang-4 (Davis et al., 1996; Maisonpierre et al., 1997; Valenzuela et al., 1999), and these ligands bind to two tyrosine kinase receptors, Tie1 and Tie2 (Thomas and Augustin, 2009). Ang-1 and Ang-2 are specific ligands of Tie2 with similar binding affinities, but Ang-2 is released faster from the endothelial cell surface after binding to Tie2 (Bogdanovic et al., 2006). Ang-1 acts in a paracrine agonistic manner, inducing Tie2 phosphorylation and subsequent vessel stabilization (Huang et al., 2009). In contrast, Ang-2, also produced by endothelial cells, acts as an autocrine antagonist or partial agonist of Ang-1-mediated Tie2 activation. Ang-2 thereby primes the vascular endothelium for exogenous cytokines, and it induces vascular destabilization at higher concentrations. Ang-2 is strongly expressed in the vasculature of many tumors, suggesting that Ang-2 may act synergistically with other cytokines such as VEGF to promote tumor-associated angiogenesis and tumor progression (Hu and Cheng, 2009).

The intracellular portion is responsible for signal transduction; it contains a juxtamembrane domain, a tyrosine kinase domain, and a C-terminal regulatory tail (Bardelli et al., 1994). In the juxtamembrane domain, phosphorylation of Ser 985 can downregulate the tyrosine kinase activity (Gandino et al., 1994), while phosphorylation of Tyr 1003 can induce polyubiquitination and degradation of c-Met (Peschard et al., 2004). In the tyrosine kinase domain, phosphorylation of two tyrosine residues can regulate the kinase activity of c-Met (Longati et al., 1994). In the C-terminal regulatory tail, which is the multisubstrate docking site, phosphorylation of two tyrosine residues, Tyr 1349 and Tyr 1356, can recruit downstream adapter molecules and trigger the signal transduction of c-Met (Maina et al., 1996; Ponzetto et al., 1994). Dimerization of c-Met upon binding of HGF can induce changes of the kinetic and thermodynamic properties of c-Met and result in sustained c-Met autophosphorylation, effector recruitment, and subsequent signaling (Sheth et al., 2008).

HGF is secreted by mesenchymal cells, while c-Met is expressed on several cell types including epithelial cells, vascular endothelial cells, lymphatic endothelial cells, neural cells, hepatocytes, hematopoietic cells, and pericytes (You and McDonald, 2008). Activating phosphorylation of c-Met can be induced by HGF through an autocrine and/or paracrine loop in many tumors (Onimaru et al., 2008; Otsuka et al., 1998; Qian et al., 2002; Vadnais et al., 2002; Yi and Tsao, 2000; Yi et al., 1998). The upregulation of both HGF and c-Met or the activation of c-Met correlate with poorer prognosis in many human malignancies (Garcia et al., 2007; Miller et al., 2006; Peghini et al., 2002; Qian et al., 2002; Stellrecht et al., 2007). In addition, HGF can induce the expression of VEGF and its receptor in endothelial cells, indirectly mediating angiogenesis (Wojta et al., 1999).

C. Hypoxia Pathway and HIF-1

Tumor cells prefer to grow using glycolysis rather than oxidative phosphorylation (Fogg et al., 2011). This metabolic approach is suitable to hypoxic conditions, making cancer cells more resistant to hypoxia. Hypoxia has been recognized as a primary physiological regulator of the angiogenic switch (Giordano and Johnson, 2001; North et al., 2005).

Tumor hypoxia is well characterized in the squamous cell carcinoma of uterine cervix, with a mean pO_2 of 16mmHg in the tumor versus 40mmHg in the normal cervix (Vaupel et al., 2001). Based on causative factors, tumor hypoxia can be categorized into three types: perfusion-related (acute) hypoxia caused by inadequate blood flow in tumor tissues, diffusion-related (chronic) hypoxia caused by an increase in diffusion distances with tumor expansion, and anemic hypoxia caused by reduced oxygen transport capacity of the blood subsequent to tumor-associated or therapy-induced anemia (Vaupel and Harrison, 2004).

Generally, hypoxia can influence the biological behavior of tumor cells in two contrasting ways: one is growth impairment or cell death through growth inhibition, apoptosis, or necrosis; the other is promotion of more aggressive behavior and increased resistance to chemotherapy or radiotherapy (Hockel and Vaupel, 2001; Vaupel and Harrison, 2004). Among the numerous promoters that help cancer cells to escape their hostile hypoxic environment is hypoxia-inducible factor (HIF), a pivotal transcription factor able to induce angiogenesis.

HIF is a family of α/β-heterodimeric DNA-binding complexes that binds hypoxia response elements at target genes under hypoxic conditions (Ratcliffe, 2007). HIF consists of two subunits: the constitutively expressed subunit HIF-1β and the oxygen-regulated subunit HIF-1α (or its paralogs HIF-2α and HIF-3α; Ke and Costa, 2006). The active subunit HIF-1α confers oxygen responsiveness and is hydroxylated by oxygen-sensitive prolyl hydroxylases under normoxic conditions, followed by targeting of the von Hippel–Lindau protein and degradation through the proteasome (Semenza, 2001). Under hypoxic conditions (intracellular oxygen levels lower then 6%), HIF-1α accumulates inside the nucleus and heterodimerizes with HIF-1β, forming the HIF-1 complex, which binds to hypoxia-responsive elements (HRE) of target genes such as the proangiogenic VEGF, thereby regulating their transcription (Pfander et al., 2006).

Deletion of HIF-1α in endothelial cells disrupts an autocrine loop necessary for hypoxic induction of VEGF-R2 by VEGF signaling through VEGF-R1 (Tang et al., 2004). Disruption of a VEGF/VEGF-R1 autocrine loop in neuroblastoma cells decreased hypoxic cell survival *in vitro* and reduced angiogenesis *in vivo* (Das *et al.*, 2005), confirming the role of the hypoxic regulation of VEGF by HIF-1 for tumor angiogenesis.

Although most research on hypoxia has concentrated on the HIF-1 pathway, there are other transcription factors activated by hypoxia, including the cyclic AMP-response-element-binding protein (Taylor *et al.*, 2000) and nuclear factor-κB (NF-κB; Schmedtje *et al.*, 1997). These factors act independently of the HIF-1 pathway. Blockade of NF-κB activity can downregulate VEGF expression *in vitro* and *in vivo*, subsequently inhibiting tumor angiogenesis (Huang *et al.*, 2001). JunB is a hypoxia-responsive transcription factor activated by NF-κB, and it has been reported to be able to induce VEGF expression via direct activation of the VEGF promoter, independent of HIF activity (Schmidt *et al.*, 2007).

D. Paradoxical Effects of NO Signaling in Tumor Angiogenesis

NO, formed by NO synthase (NOS) using the substrate L-arginine, plays a critical role in angiogenesis as a mediator of signaling from VEGF and other angiogenic factors (Dimmeler and Zeiher, 1999; Roberts *et al.*, 2007). Among the three NOS isoforms expressed by mammalian cells, nNOS (predominantly in neuronal cells), and eNOS (predominantly in endothelial cells) are constitutively expressed, while iNOS is an inducible NOS often expressed in tumor cells under the stimulation of inflammatory cytokines, endotoxin, hypoxia, or oxidative stress (Fukumura *et al.*, 2006). HRE sites have been identified within the NOS promoter, and HIF can directly regulate the hypoxia-induced activity of NOS (Kimura *et al.*, 2000).

Low concentrations of NO (1–10 nmol/L or less) produced by eNOS in response to angiogenic factors stimulate endothelial cell proliferation and migration, whereas higher concentrations (>300 nmol/L) produced by iNOS inhibit angiogenesis by causing cytostasis and cell death (Roberts *et al.*, 2007).

The mechanisms of the proangiogenic effects of NO have been identified. NO-induced cGMP can activate Ras and then activates the Ras–Raf–MEK–ERK pathway, which enhances the DNA binding of activator protein 1 (AP1) for promoting the proliferation and migration of endothelial cells (Jones *et al.*, 2004; Oliveira *et al.*, 2003; Ridnour *et al.*, 2005). NO-induced cGMP can also activate, cGMP-dependent protein kinase, which can further increase the production of matrix metalloproteinase 13 (MMP13) by ERK and by AKT activation through phosphoinositide 3-kinase. MMP13 facilitates the migration of endothelial cells, and activated AKT promotes their motility (Kawasaki *et al.*, 2003; Lopez-Rivera *et al.*, 2005; Zaragoza *et al.*, 2002).

Moreover, NO produced by eNOS mediates recruitment of pericytes for the remodeling and maturation of blood vessels (Yu et al., 2005), and also induces vascular hyperpermeability (Bucci et al., 2005).

E. mTOR Pathway

Mammalian target of rapamycin (mTOR) is a 289-kDa intracellular serine/threonine kinase, a member of the phosphoinositide kinase-related kinase family, which plays an integral role in coordinating cellular growth in response to growth factors, nutrients, and microenvironmental changes (Bjornsti and Houghton, 2004; Brown et al., 1994; Polak and Hall, 2009; Wullschleger et al., 2006). mTOR exists in two structurally and functionally distinct complexes, mTORC1 and mTORC2. mTORC1 controls cell growth by regulating translation, ribosome biogenesis, autophagy, and metabolism, and these activities are sensitive to the inhibition by rapamycin (Guertin and Sabatini, 2007). mTORC2 phosphorylates Akt, SGK1, and PKC to control multiple functions including cell survival and cytoskeletal organization, and these activities are relatively resistant to rapamycin (Dowling et al., 2010; Sarbassov et al., 2006). mTOR signaling can increase the expression of HIF-1, resulting in a concomitant increase in the production and secretion of VEGF-A (Hudson et al., 2002). mTORC1 can also be activated by growth factors including VEGFR, platelet-derived growth factor receptor (PDGFR), epidermal growth factor receptor, and insulin growth factor receptor, as well as by nutrients through the phosphatidylinositol 3-kinase (PI3K) pathway (Le Tourneau et al., 2008). Some tumors with activated mTOR signaling are highly vascularized (Inoki et al., 2005). It is therefore believed that targeting mTOR pathway might be an effective way of inhibiting tumor angiogenesis.

F. FGFs/FGFRs Pathway

Fibroblast growth factors (FGFs) form a family with 22 members. These growth factors can act in concert with heparin or heparan sulfate proteoglycan (HSPG) to bind and activate four different isoforms of receptor tyrosine kinase, namely FGFR1, FGFR2, FGFR3, and FGFR4, via receptor dimerization (Eswarakumar et al., 2005; Grose and Dickson, 2005). The activated fibroblast growth factor receptors (FGFRs) can trigger tyrosine phosphorylation and thus the activation of several signaling molecules, including FGFR substrate 2 (FRS2; Eswarakumar et al., 2005; Ong et al., 2000). Phosphorylated FRS2 can further activate the Grb2/Sos1 complex and subsequent activate the MAPK pathway (Eswarakumar et al., 2005; Kouhara et al., 1997).

Activated FRS2 can also trigger PI3K for promoting cellular motility and survival (Cross *et al.*, 2000; Eswarakumar *et al.*, 2005).

FGF2 has been known for years to be related to tumor angiogenesis (Czubayko *et al.*, 1997; Kandel *et al.*, 1991). A recent study showed that in endothelial cells, FGF2 and FGF8b can activate STAT5 through the activation of tyrosine kinases Src and Janus kinase 2 and then induce endothelial cell migration, invasion, and tube formation (Yang *et al.*, 2009).

G. Delta-Like Ligand 4 (DLL4)/Notch Pathway Interacts with the VEGF Pathway

Notch receptor proteins form a family of four transmembrane proteins designated as Notch1, Notch2, Notch3, and Notch4 (Katoh, 2007). Notch receptors are activated by type I transmembrane ligands, referred collectively as DSL (Delta, Serrate, and Lag 2 proteins; Mumm and Kopan, 2000). DLL1, DLL3, DLL4, JAG1, and JAG2, which all have a disulfide-rich DSL domain, are canonical Notch ligands with higher binding affinity, while DNER, F3/Contactin, NB-3, DLK1, and DLK2, without DSL domains, are noncanonical ligands with lower binding affinity (Katoh, 2007; Sanchez-Solana *et al.*, 2011). Triggered by the Notch receptor–ligand interactions between adjacent cells, the metalloproteinase ADAM (a disintegrin and metalloproteinase) is activated for cleaving the Notch extracellular region and then inducing γ-secretase-mediated generation of the Notch intracellular domain (NICD). The NICD then translocates to the nucleus and interacts with the CSL (RBPJ) transcription factor to upregulate the expression of downstream targets, including the Hes and Hey transcription factors, which in turn, modulate the expression of many other differentiation and proliferation genes (Iso *et al.*, 2003; Mumm and Kopan, 2000).

Notch1 and Notch4 and three canonical Notch ligands, Jagged1, DLL1, and DLL4, are predominantly expressed in vascular endothelial cells for the induction of arterial-cell fate during development and for the selection of endothelial tip and stalk cells during sprouting angiogenesis (Kume, 2009).

DLL4 overexpression is found in the tumor vasculature of breast, renal, and bladder cancers (Jubb *et al.*, 2010; Patel *et al.*, 2005, 2006). It is highly expressed in the tumor cells as well as in tumor endothelial cells in ovarian and pancreatic cancers, and it correlates with poor prognosis (Chen *et al.*, 2011a; Hu *et al.*, 2011).

The interactions between VEGF and Notch signaling have been noticed in both normal angiogenesis and tumor angiogenesis. VEGF-A increases DLL4 expression in both normal and tumorous endothelial cells, especially in response to hypoxia (Hu *et al.*, 2011; Thurston and Kitajewski, 2008). Moreover, hypoxia can directly upregulate DLL4 expression, possibly through HIF-1α and hypoxia response elements in the DLL4 promoter (Diez et al., 2007; Patel et al., 2005). Increased DLL4/Notch signaling can increase the expression of VEGF-R1 and its soluble form (sFlt1) in endothelial cells, but importantly, it downregulates VEGF-R2 expression (Taylor et al., 2002). Reciprocally, blocking DLL4 function can increase the expression of VEGF-R2 (Suchting et al., 2007). Immobilized DLL4 can also downregulate VEGF-R2 expression in endothelial cells through methylation of the VEGF-R2 promoter (Hu et al., 2011). This interaction is crucial for sprouting angiogenesis in normal tissues, in which VEGF-A induces the formation and extension of filopodia as well as the expression of DLL4 protein in the tip cells. The DLL4 binds to the Notch receptors in the adjacent stalk cells for activating Notch signaling, which turns into downregulation of VEGF-R2 expression and consequently prohibits the tip-cell behavior in the stalk cells (Roca and Adams, 2007). In this local cell differentiation, the tip cells respond robustly to VEGF for vasculature extension, whereas the stalk cells respond in a more muted manner to continue to support the tip cells and to differentiate into their final tubular structure (Thurston and Kitajewski, 2008). However, this well-organized angiogenic process has not been confirmed in the angiogenesis of tumors.

H. Other Molecules Regulating Tumor Angiogenesis

Angiogenin was initially isolated from tumor cells (Fett et al., 1985). Angiogenin can activate vessel endothelial cells and trigger important biological processes including proliferation, migration, invasion, and the formation of tubular structure through its four molecular functions, that is, ribonuclease activity, basement membrane degradation, signaling transduction, and nuclear translocation (Gao and Xu, 2008). However, the direct roles of angiogenin in promoting cancer proliferation and tumorigenesis have also been revealed (Ibaragi et al., 2009). Consistent with the findings in animal studies, increased levels of angiogenin have also been found in solid tumors, which correlates with poor prognosis (Shimoyama and Kaminishi, 2000).

Netrins are laminin-like secreted glycoproteins known to be axonal guidance molecules, and they recently have been suggested to be involved in vasculature development (Eichmann et al., 2005). Conflicting reports show that Netrin-1 promotes proliferation, migration, and tube formation of human umbilical vein endothelial cells (HUVECs; Xie et al., 2011), but ectopic expression of Netrin-1 by transplanted tumor cells delays tumor angiogenesis (Larrivee et al., 2007). Moreover, Netrin-4 can inhibit the migration and tubular formation of HUVECs (Nacht et al., 2009). Thus, the exact roles of Netrins in regulating angiogenesis are unclear (Castets and Mehlen, 2010).

Robo4 is another endothelial-specific member of the family of axonal guidance molecules (Huminiecki and Bicknell, 2000). Robo4 expression is upregulated in tumor-derived endothelial cells but not in those of the normal vasculature. A soluble robo4 molecular inhibits angiogenesis *in vitro* and *in vivo* (Suchting *et al.*, 2005).

I. PDGF/PDGF Receptor Pathways Regulate Endothelial Cells and Pericytes

The PDGF family consists of the cystine knot class of growth factors including four structurally related members, namely PDGF-A, PDGF-B, PDGF-C, and PDGF-D. These are linked by disulfide bonds to form homodimers and heterodimers (PDGF-AA, PDGF-AB, PDGF-BB, PDGF-CC, and PDGF-DD; Cao *et al.*, 2002, 2008a; Liu *et al.*, 2011b; Nissen *et al.*, 2007); the monomeric forms are inactive. PDGFs bind to the protein tyrosine kinase receptors PDGF receptor-α and -β. Upon binding, the receptor isoforms dimerize, resulting in three receptor combinations (PDGFR-$\alpha\alpha$, PDGFR-$\beta\beta$, and PDGFR-$\alpha\beta$; Seifert *et al.*, 1989). PDGF-BB, -AA, and -AB are produced by endothelial cells, pericytes, and fibroblasts in angiogenic and remodeling tissues (Fredriksson *et al.*, 2004). PDGFR-β is exclusively expressed in pericytes (Winkler *et al.*, 2010).

The PDGF family possesses potent proangiogenic activity (Cao *et al.*, 2002; Ridgway *et al.*, 2006). PDGF-BB and FGF2 synergistically promote tumor angiogenesis by reciprocally increasing their endothelial cell and mural cell responses (Nissen *et al.*, 2007). In the embryonic neural tube, PDGF-B binds to the PDGFR-β receptor located on the pericyte membrane, resulting in dimerization of PDGFR-β and subsequent initiation of several signaling pathways, ultimately promoting proliferation, migration, and the recruitment of pericytes to the vascular wall of newly formed blood vessels (Heldin *et al.*, 1998; Hellstrom *et al.*, 1999; Lindahl *et al.*, 1997).

J. Basement Membrane in Angiogenesis

In the vascular basement membrane, type IV collagen and laminin form two-independent, three-dimensional network, respectively (Yurchenco and Cheng, 1993). Type IV collagen confers structural stability to the basement membrane, and the different isoforms of laminins are involved in signaling to various tissues (Hallmann *et al.*, 2005). HSPGs crosslink the collagen type IV and laminin networks and bind soluble factors (Sasaki *et al.*, 2002; Timpl *et al.*, 2003). Other minor molecules in basement membrane include collagen types VIII, XV, and XVIII, thrombospondins I and II, and osteonectin.

The basement membrane is a critical structure for angiogenesis, providing structural support to endothelial cells and pericytes (Benton et al., 2009; Davis and Senger, 2005; Hallmann et al., 2005; Iozzo et al., 2009). During angiogenesis, the basement membrane is degraded by proteinases, with membrane-type MMPs being particularly important. As angiogenesis proceeds, the basement membrane provides essential functions that support key signaling events involved in regulating endothelial cell migration, invasion, proliferation, and survival (Davis and Senger, 2005).

The effects of the basement membrane on vessel walls differ, largely depending on the vessel type and the developmental state of the vessel (Primo et al., 2010). In tumor angiogenesis, the basement membrane incorporates with VEGF and FGF2 in promoting angiogenesis (Berthod et al., 2006), with the participations of α6 integrin (Primo et al., 2010). Antiangiogenic therapy through targeting VEGF signaling can induce regression of the vascular network, but while endothelial cells disappear, the basement membrane stays. After withdrawal of antiangiogenic therapy, endothelial cells can regrow along with the remnant basement membrane as a scaffold, resulting in drug resistance (Kamba et al., 2006). On the other hand, the fragments of large molecules in the extracellular matrix and basement membrane could possess inhibitory properties for tumor angiogenesis. For instance, the angiogenesis inhibitor endostatin is a 20-kDa fragment from the C-terminal of the collagen XVIII K1-chain (O'Reilly et al., 1997). More investigations are needed to elaborate the multiple functions of basement membrane in the interaction between endothelial cells and pericytes, as well as in drug resistance to antiangiogenic therapy.

K. Inflammation and Tumor Angiogenesis

Among the various inflammatory cell types in and around tumors, tumor-associated macrophages (TAMs) have been widely studied and have been associated with tumor angiogenesis (Dirkx et al., 2006). A high density of TAMs in solid tumors correlates with poor patient prognosis (Ding et al., 2009; Fujiwara et al., 2011; Lee et al., 2008; Lin et al., 2011; Zhu et al., 2008). Multiple proangiogenic factors can be released by TAMs, including interleukin (IL)-1β, IL-6, IL-8, transforming growth factor α (TNF-α), FGF2, VEGF, EGF, and PDGF, as well as antiangiogenic factors including TGF-β, TSP1 (thrombospondin 1), and IFNγ (interferon γ).

TAMs form phenotypically and functionally distinct subsets (Squadrito and De Palma, 2011). Among the better characterized TAM subsets are the proangiogenic (TIE2$^+$) and the angiostatic/inflammatory (CD11c$^+$) macrophages, which coexist in tumors. Such antagonizing TAM subsets form distinct niches in the tumor microenvironment and regulate tumor growth (Squadrito and De Palma, 2011).

Among inflammatory cytokines, IL-8 is the one most studied in terms of promoting angiogenesis (Koch *et al.*, 1992; Waugh and Wilson, 2008). IL-8 binds to two G protein-coupled receptors (CXCR1 and CXCR2) on the cellular membrane (Schraufstatter *et al.*, 2001), triggering multiple signaling pathways that influencing gene expression through activation of numerous transcription factors, as well as modulating the translation of multiple genes, and resulting in regulation of the cytoskeleton (Waugh and Wilson, 2008). Increased expression of IL-8 and/or its receptors has been characterized in cancer cells, endothelial cells, infiltrating neutrophils, and TAMs (Waugh and Wilson, 2008). IL-8 promotes the migration of tumor-derived endothelial cells more than of normal endothelial cells (Charalambous *et al.*, 2005). IL-8 also promotes the resistance of CCRCC against antiangiogenic agent sunitinib (Huang *et al.*, 2010a).

Another important inflammatory factor regulating tumor angiogenesis is TNF-α. Released by macrophages, TNF α can stimulate chemotaxis of endothelial cells and promotes angiogenesis (Leibovich *et al.*, 1987). A recent study showed that low doses of TNF-α promoted tumor angiogenesis by activating the VEGF and HIF-1α signaling pathways (Jing *et al.*, 2011).

VI. DIFFERENTIAL ANALYSES OF TUMOR VASCULATURE

A. Controversy in the Traditional Evaluation System of Tumor Microvessel Density

The idea of measuring tumor microvasculature density (MVD) to assess the angiogenic status of a tumor was proposed in 1972 (Brem *et al.*, 1972). Only since the 1980s, with the introduction of specific endothelial markers such as factor-VII, CD31, CD34, and CD105, has the quantification of tumor vasculature been widely conducted (Nico *et al.*, 2008).

In most of the cancer types studied, higher tumor MVD correlates with more aggressive tumors and shorter patient survival. These tumor types include nasopharyngeal, other squamous head and neck, lung, gastric, colorectal, liver, pancreatic, bladder, ovarian, endometrial and breast cancers, and neuroblastoma (Cantu De Leon *et al.*, 2003; Cernea *et al.*, 2004; Chandrachud *et al.*, 1997; Couvelard *et al.*, 2005; Dickinson *et al.*, 1994; El-Assal *et al.*, 1998; Fernandez-Aguilar *et al.*, 2006; Foss *et al.*, 1996; Gasparini *et al.*, 1994; Horak *et al.*, 1992; Igarashi *et al.*, 1998; Kusamura *et al.*, 2003; Lackner *et al.*, 2004; Lindmark *et al.*, 1996; Lissbrant *et al.*, 1997; Qian *et al.*, 1997; Ribatti and Ponzoni, 2005; Shimizu *et al.*, 2000; Weidner *et al.*, 1992; Zhao *et al.*, 2005). However, other studies failed to show any significant correlation between MVD and tumor progression or patient prognosis (Bossi et al., 1995; Busam et al., 1995; Hillen et al., 2006; Marioni et al., 2005; Mooteri et al., 1996). The most confusing reports have been of a highly vascularized solid tumor, CCRCC (Qian et al., 2009). Among 21 published studied, 10 of them report a favorable correlation between higher MDV and patient prognosis, 6 report an unfavorable prognostic value with higher MVD, and 5 failed to find a significant prognostic role for MVD (Qian et al., 2009).

Although the discrepancies in the evaluation of MVD could be explained by several technical and experimental drawbacks in study design, the diversity of tumor vasculature could also be a crucial reason. Accumulating evidence shows that many important biological implications of tumor vasculature cannot be revealed if we consider all of the microvessels inside a tumor to be uniform and to have the same biological function (Qian et al., 2009). In fact, all tumor vessels are not equal in their ability to carry blood for support of tumor cells (Nico et al., 2008).

B. Dissecting Tumor Vasculature is Necessary to Reveal True Prognostic Values

There are at least two approaches to dissecting tumor vasculature so that we can identify the biological meanings of its different parts. The first approach is through morphologically classification, and the second is through molecular classification.

The vasculature of hepatocellular carcinoma (HCC) is suitable for morphological classification (Chen et al., 2011b; Ding et al., 2011). The HCC vasculature can be classified into two major types of vessel based on their morphology, namely capillary-like microvessels and the sinusoid-like vasculature (Fig. 3). HCC patients with sinusoid-like vasculature have a shorter survival time, although the MVD of a tumor with sinusoid-like vasculature is significantly less than that of a tumor with capillary-like microvessels (Chen et al., 2011b). More importantly, when tumor endothelium covers the whole tumor cell cluster, the tumor cells inside this endothelial envelope have a higher proliferation rate but a lower apoptotic rate. An increased micrometastasis rate is also significantly correlated with the presence of endothelium-coated tumor clusters. As a consequence, HCC patients having more endothelium-coated tumor clusters suffer from early relapse and shorter overall survival (Ding et al., 2011).

Molecular classification of tumor vasculature is based on the differential expression of endothelial markers and pericyte markers (Qian et al., 2009; Yao et al., 2007). In CCRCC, for example, the tumor microvessels can by classified into differentiated vessels (CD31+, CD34+) and undifferentiated

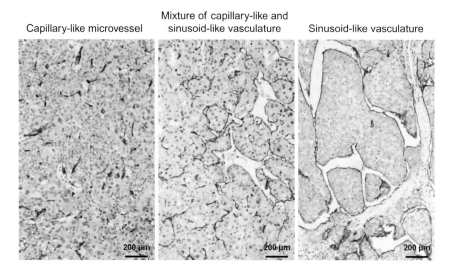

Fig. 3 Morphological classification of tumor vasculature in hepatocellular carcinoma. Three human hepatocellular carcinomas were stained with anti-CD34 antibody to show their vasculature. The vasculature can be capillary-like microvessels, sinusoid-like vasculature, or a mixture. The prognosis of patients harboring these three vasculature types differ, capillary-like microvessels having the best prognosis and sinusoid-like tumor vasculature the worst. Figure courtesy of Dr. Zhi-Yuan Chen, Sun Yat-sen University Cancer Center. (See Page 6 in Color Section at the back of the book.)

vessels (CD31+, CD34−; Qian et al., 2009; Yao et al., 2007). Inside the differentiated vessels, two subtypes of vessels can be further distinguished: differentiated mature vessels (CD34+, α-SMA+) and differentiated immature vessels (CD34+, α-SMA−). Figure 4 shows the coexistence of undifferentiated, differentiated immature, and differentiated mature vessels in a high-grade CCRCC. More importantly, a higher MVD of differentiated vessels correlates with better prognosis, while higher MVD of undifferentiated vessels correlates with poor prognosis. There is an inverse correlation between these two types of vessel: higher differentiated MVD is usually accompanies by lower undifferentiated MVD, and vice versa (Yao et al., 2007).

In urothelial carcinoma of the bladder, higher pericyte coverage of the tumor microvessels correlates with shorter progression-free survival (O'Keeffe et al., 2008). It has also been reported that tumor microvessels covered by pericytes are less sensitive to antivascular therapy (Gee et al., 2003). It is therefore necessary to separately quantify the mature microvessels supported by pericytes for predicting patient's prognosis or drug response.

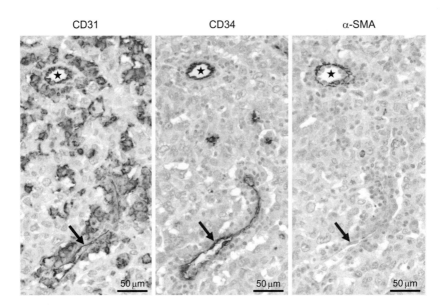

Fig. 4 Molecular classification of tumor vasculature in clear cell renal cell carcinoma. One high-grade human clear cell renal cell carcinoma was continuously sectioned for staining with CD31, CD34, and α-SMA. CD31 and CD34 are endothelial cell markers and α-SMA is a pericyte marker. A star denotes a mature differentiated vessel recognized by CD31, CD34, and α-SMA staining. This vessel is classified as a differentiated vessel because it expresses CD34; this vessel is also classified as a mature vessel because it is supported by pericytes. Arrows indicate a differentiated immature vessel, expressing CD34 and CD31 but lacking pericyte support. Numerous vessels expressing CD31 but not CD34 are classified as undifferentiated; these vessels always lack pericyte support. (See Page 7 in Color Section at the back of the book.)

In summary, differential analyses on tumor vasculature will bring us more useful information for better understanding the biological meaning of the complicated tumor vasculature.

VII. REVASCULARIZATION WITHIN TUMORS FOLLOWING WITHDRAWAL OF ANTIANGIOGENIC TREATMENT

While the field of tumor angiogenesis has focused on the regression of vasculature via VEGF targeting, little is known about regrowth of tumor vessels after cessation of such therapies. Initial observations using mouse xenograft models showed that interruption of antiangiogenic treatment led to rapid regrowth of tumors, and that multiple rounds of treatment resulted in continual regression of the tumor (Boehm *et al.*, 1997). Since then, both preclinical and clinical studies suggest that the withdrawal of antiangiogenic therapy induces rapid regrowth of tumor vessels and restart of tumor growth. However, the mechanism by how this occurs is still unclear.

Tumor regrowth, or rebound, following cessation of antiangiogenic treatment, specifically with regard to bevacizumab therapy, has been demonstrated clinically in a variety of cancer types, including high-grade gliomas (Zuniga *et al.*, 2010) and metastatic colorectal cancer (Cacheux *et al.*, 2008). However, a recent retrospective analysis of randomized phase III trials of several cancer types (including metastatic renal cell cancer, metastatic pancreatic cancer, metastatic breast cancer, and metastatic colorectal cancer) indicates that there is no difference in the rates of mortality or disease progression in patients treated with bevacizumab who stopped treatment prematurely compared to placebo treatment (Miles *et al.*, 2011). As this analysis does not support the idea that bevacizumab discontinuation is associated with accelerated disease progression, it is as yet unclear whether tumor vasculature rebound following cessation of treatment equates to disease progression.

Some answers as to how tumors respond to the cessation of antiangiogenic treatment will come from direct analysis of the tumor vasculature and of the regrowth of regressed vessels resulting from such treatments. Recent preclinical analyses have uncovered two potential mechanisms for vascular regrowth. Analyzing mammary carcinoma allografts, Hlushchuk *et al.* demonstrated that after stopping dosage of a VEGFR tyrosine kinase inhibitor (PTK787/ZK22284), tumor vasculature expanded via intussusception and not by angiogenic sprouting, that is, the new vessels originated from preserved SMA-positive vessels within the tumor cortex (Hlushchuk *et al.*, 2008). Other data, analyzing vessel regrowth in spontaneous RIP-Tag2 tumors and Lewis lung carcinoma xenografts, suggest that residual basement membrane sleeves, or ghosts, which have been shown to persist following endothelial cell degeneration (Inai *et al.*, 2004), provide a scaffold for tumor vascular regrowth (Mancuso *et al.*, 2006). Further, not only might the residual basement membrane provide tracks for new vessels to follow, but it appears to also serve as a storage site for proangiogenic factors such as VEGF (Mancuso *et al.*, 2006).

As these experiments were performed in mice under conditions that allowed for rapid vessel regrowth, it remains unclear whether prolonged treatment or multiple rounds of treatment, as is common in human patients, would alter vascular regrowth following cessation of antiangiogenic therapy. Further analysis of vascular regrowth in clinical samples following disruption of antiangiogenic therapy will provide a clearer picture of the mechanism and the importance of tumor vascular regrowth in terms of patient response and progression.

VIII. VASCULAR NORMALIZATION OR INEFFECTIVE TARGETING OF TUMOR VASCULATURE?

When Rakesh K. Jain first introduced the concept of vascular normalization, he clearly described an antiangiogenic strategy (Jain, 2001). In this vascular normalization concept, antiangiogenic therapy will eliminate the immature, ineffective vessels in the structurally and functionally abnormal tumor vasculature, but leave those mature vessels which are resistant to antiangiogenic therapy because of their pericyte coverage. Those remaining vessels look relatively normal in morphology and are more effective in blood perfusion, making drug delivery possible; yet, they are inadequate for tumor growth, resulting in growth inhibition and forcing the tumor to remain dormant. He claimed this consequence is the "ultimate goal of antiangiogenic/antivascular therapy" (Jain, 2001). As accumulating evidence partially supports this hypothesis, Jain refined it by adding the necessity of a balance between pro- and antiangiogenic factors in vascular normalization (Jain, 2005). The additional promising aspect of this theory is that by normalizing the tumor vasculature, hypoxia could be partially corrected, and the cancer cells would presumably be restored to sensitivity toward traditional radiation and chemotherapy (Peng and Chen, 2010). However, in a very recent study on an orthotopic glioma mouse model, targeting immature vessels with bevacizumab at low and medium doses can induce vascular normalization, but cannot stop tumor growth nor influence tumor cell viability (von Baumgarten et al., 2011), challenging the ultimate goal of vascular normalization theory.

Vascular normalization theory is based on observations in animal models, but tumor vasculature could be different in humans. A study in human CCRCC showed that, in addition to mature and immature differentiated vessels, there are undifferentiated vessels that have a significant correlation with poor prognosis, lack of pericyte coverage, and most of them are without a lumen, suggesting ineffective blood transport (Fig. 3; Yao et al., 2007). In a very recent study, differentiated and undifferentiated microvessels were also found in nonsmall cell lung cancer (Zhao et al., 2012). Moreover, in this study, the increased undifferentiated MVD significantly correlated with better response to combined chemotherapy integrating the VEGFR-targeted agent bevacizumab with other cytotoxic agents (Zhao et al., 2012). It is speculated that the undifferentiated vessels and the differentiated immature vessels in human tumors are more sensitive to current antiangiogenic agents than the mature differentiated vessels supported by pericytes. It is unclear whether selective targeting of immature tumor vessels can translate into effective tumor control or even survival benefit. If such selective targeting of the tumor vasculature cannot result in better tumor control or survival benefit, then more effective antiangiogenic approaches must be generated to simultaneously destroy the mature vessels in the tumor vasculature.

IX. HIGHLIGHTS IN ANTIANGIOGENIC THERAPY

A. Targeting the VEGF Pathway

Targeting the VEGF pathway can be achieved using neutralizing antibodies to VEGF or VEGFRs; soluble VEGF receptors or receptor hybrids; tyrosine kinase inhibitors (TKIs) with selectivity for VEGFRs (Casanovas et al., 2005; Schneider and Sledge, 2007); or multiple receptor tyrosine kinase inhibitors targeting multiple growth factor receptors, including VEGFRs (Alvarez et al., 2010; Gordon, 2011). Figure 5 highlights these agents which are in clinical trials or approved by the FDA of the United States. Antibody targeting of VEGF-R3, which is expressed in lymphatic vessels and tumor vasculature,

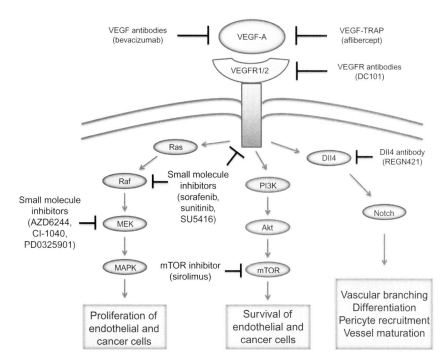

Fig. 5 Agents targeting VEGF signaling in clinical trials or approved by FDA in the United States. (For color version of this figure, the reader is referred to the Web version of this chapter.)

can enhance the antiangiogenic effect from blocking VEGF-R2 signaling (Tammela et al., 2008). However, cancer cells express VEGF and other angiogenic factors variably in a wide range, and this heterogeneity significantly influences treatment response (Hasina et al., 2008).

1. MONOCLONAL ANTIBODIES TARGETING VEGF

The first antiangiogenic agent approved by the FDA was bevacizumab (Avastin, Genentech), a humanized anti-VEGF monoclonal antibody. Administration of bevacizumab, in combination with cytotoxic chemotherapy or immunotherapy, conferred benefits to patients with metastatic colorectal cancer, nonsquamous, nonsmall cell lung carcinoma, renal cell carcinoma, and glioblastoma multiforme (Chamberlain, 2011; Hurwitz et al., 2004; Miller et al., 2007; Sandler et al., 2006; Summers et al., 2011). The limited survival benefit of bevacizumab in the treatment of metastatic breast cancer recently resulted in withdrawal of FDA approval for the use of bevacizumab in treating metastatic breast cancer (Jones and Ellis, 2011).

Three mechanisms have been proposed for the treatment effects of bevacizumab. First is an antiangiogenic mechanism based on its effects on mouse xenograft models (Duda et al., 2007; Olsson et al., 2006). Second is the inhibition of circulating endothelial cells and endothelial progenitor cells colonizing the tumor vasculature (Batchelor et al., 2007; Willett et al., 2006). Third is the effect on tumor vascular normalization (Willett et al., 2006). However, a recent study reported that tumor vascular normalization induced by bevacizumab does translate into inhibition of glioma growth in an orthotopic animal model (von Baumgarten et al., 2011). Moreover, in a recent preclinical study, bevacizumab was found to accelerate the growth of CCRCC by selection of cancer cells with increased growth capacities (Grepin et al., 2011).

2. MONOCLONAL ANTIBODIES TARGETING VEGFR

Sorafenib and sunitinib are small-molecule inhibitors targeting VEGFRs and other kinases, including Raf, platelet-derived growth factor receptor (PDGFR-B), FGFR, and the FLT3, KIT, and FMS (also known as CSF1R) receptors (Karaman et al., 2008). These two drugs have been approved by FDA for treatment of renal or hepatocellular cancer, both being highly vascularized cancers (Bergers and Hanahan, 2008; Ellis and Hicklin, 2008; Escudier et al., 2007). Based on preclinical studies, TKIs that target both VEGFRs on endothelial cells and PDGFRs on pericytes should have greater efficacy, a so-called "dual attack" on the tumor vasculature (Ellis and Hicklin, 2008).

Aflibercept (VEGF-Trap) is a recombinant chimeric soluble receptor containing structural elements from VEGF-R1 and VEGF-R2, and therefore it can scavenge VEGF-A and PlGF (Holash et al., 2002). Aflibercept has promising anticancer effects in preclinical studies (Teng *et al.*, 2010). However, in a phase II clinical trial, it provided only minimal efficacy on recurrent malignant glioma when administrated as a single agent (de Groot *et al.*, 2011).

B. Targeting the mTOR Pathway

The mTORC1 inhibitor temsirolimus can inhibit the expression and activation of HIF-1α and subsequently inhibit the production of VEGF in breast cancer cells (Del Bufalo *et al.*, 2006). Dual inhibition of mTORC1 and mTORC2 signaling reduced VEGF production in tumors as well as vessel sprouting in the tumor vasculature (Falcon *et al.*, 2011).

In the highly vascularized solid tumor, CCRCC, a phase III randomized clinical trial showed that the mTORC1 inhibitor everolimus can generate progression-free survival in patients (Beaumont *et al.*, 2011). Everolimus was later approved for clinical use as a novel targeted therapeutic after sorafenib and sunitinib in treating CCRCC (Coppin, 2010).

C. Targeting the Hypoxia Pathway

Myo-inositol trispyrophosphate (ITPP) is a membrane-permeant allosteric effector of hemoglobin that can enhance the oxygen release capacity of red blood cells (RBCs) and thus reduce the hypoxic effects in the microenvironment (Duarte *et al.*, 2010). *In vitro* experiments have shown that the ITPP-loaded RBCs, which therefore have low O_2 affinity, can dramatically reduce the production of HIF-1α and VEGF-A by cocultured endothelial cells and prevent the formation of a vascular network by the latter (Kieda *et al.*, 2006). *In vivo* animal experiments using a rat hepatocyte carcinoma model has further shown that ITPP has a high potency to inhibit tumor growth, allow long-term survival, and even cure most tumors (Aprahamian *et al.*, 2011). Obviously, clinical trials are warranted to confirm the effects of ITPP as well as to validate this novel therapeutic approach.

X. NEED FOR A NOVEL CLINICAL EVALUATION SYSTEM FOR ANTIANGIOGENIC THERAPIES

The current evaluation system for the treatment response of solid tumors in most clinical trials is called Response Evaluation Criteria in Solid Tumors (RECIST; Therasse *et al.*, 2000). This system relies mainly on computerized tomography and magnetic resonance imaging, by which the largest diameter of target lesions is measured and the sum of the longest diameter (LD) for all target lesions is used as baseline sum LD. Disappearance of all target lesions after the treatment is defined as complete response. At least a 30% decrease in the sum of the LD of target lesions, taking as a reference the baseline sum LD, is defined as partial response (PR). Stable disease is defined as neither sufficient shrinkage to qualify as PR nor sufficient increase to qualify as progressive disease (PD). At least a 20% increment in the sum of the LD of target lesions is defined as PD.

The RECIST system has been successful in evaluating treatment responses induced by traditional cytotoxic agents, in which cancer cells are the primary treatment targets and shrinkage of tumor volume is the immediate effect. However, in antiangiogenic therapy, in which the primary drug targets are not necessarily cancer cells but endothelial cells, the immediate treatment effects could be very different. For example, stable disease induced by antiangiogenic therapy is frequently seen with attenuation of tumor tissue, necrosis, and cavitation without an alteration in tumor volume (Desar et al., 2009; George et al., 2011). Thus, the efficacy of antiangiogenic therapy could be underestimated by the RECIST system, as shown in the clinical trials of sorafenib and bevacizumab against metastatic CCRCC, in which significant increase in progression-free survival was achieved but a significant objective response rate was not found (Escudier et al., 2007; Yang et al., 2003).

It is therefore reasonable to consider novel evaluation systems for antiangiogenic therapy, in which functional and molecular criteria, as well as traditional imaging criteria, can be integrated for more precise evaluation (Desar et al., 2009). The use of dynamic contrast-enhanced perfusion computed tomography has been recently validated to be effective in evaluating tumor blood flow and permeability of tumor vasculature during antiangiogenic therapy, with specific criteria of tumor blood flow, permeability surface area product, and fractional intravascular blood volume (Lim et al., 2011). However, a more accurate evaluation system for antiangiogenic therapy should be established.

XI. MECHANISMS OF RESISTANCE IN ANTIANGIOGENIC THERAPY

To date, antiangiogenic therapy alone has limited anticancer effects in human, and dishearteningly, cannot provide long-term survival benefit (Gaitskell et al., 2011; Kane et al., 2006; Karamouzis and Moschos, 2009; Kreisl et al., 2009, 2011; Williamson et al., 2010). Combination of antiangiogenic therapy with other conventional therapies is the current mainstream in clinical trials aiming to improve patient survival. The possible mechanisms for tumor relapse, invasion, and metastasis after antiangiogenic therapy have been reviewed elsewhere (Ebos and Kerbel, 2011). Here, we discuss three major mechanisms contributing to the failure of antiangiogenic therapy.

A. Activation of Hypoxia and c-Met Pathways and Beyond

In animal models of pancreatic neuroendocrine carcinoma and glioblastoma, angiogenesis inhibitors targeting the VEGF pathway not only can inhibit primary tumor growth, but also can trigger tumor progression to more aggressive phenotype (enhanced invasiveness) as well as increased metastases to lymph node and distant organs (Paez-Ribes *et al.*, 2009). Sunitinib can also accelerate metastatic tumor growth and decrease overall survival in mice hosting human breast cancer cells (Ebos *et al.*, 2009). The mechanism underlying these deteriorating treatment outcomes has been speculated to be the consequence of activating hypoxia signaling by blocking of the VEGF pathway (Loges *et al.*, 2009). Tumor hypoxia can activate multiple signaling pathways in the tumor cells, and the c-Met pathway is one that has the potent biological effect of stimulating the motility of unkilled cancer cells, resulting in more invasiveness and more metastasis (Ide *et al.*, 2007; Loges *et al.*, 2009). The speculation of c-Met pathway activation to be the cause of treatment failure is recently supported by a study that combination of blocking VEGF and c-Met signaling using the small molecules XL880 and XL18, which inhibit multiple tyrosine kinase receptors including VEGFR and c-Met, can reduce the invasiveness and metastasis rates, as well as more dramatically inhibit the primary tumor growth (You *et al.*, 2011).

In a recent clinical trial treating metastatic nasopharyngeal carcinoma using pazopanib (which is a multitargeted small-molecule inhibitor of VEGFR-1, -2, and -3, PDGF-a, PDGF-b, and c-kit tyrosine kinases), a greater reduction in the permeability of the tumor vasculature was associated with shorter progression-free survival (Lim *et al.*, 2011). This detrimental outcome is immediately explained by the mechanism in the abovementioned animal models, where targeting the VEGF pathway harshens hypoxic conditions within the tumor and subsequently triggers c-Met and other pathways, resulting in a more aggressive phenotype (Lim *et al.*, 2011). However, the human scenario could be more complicated than that of animal models. Tumor vasculature heterogeneity can correlate with tumor aggressiveness. In CCRCC, we found that high undifferentiated MVD correlates with high tumor grade and worse prognosis (Yao *et al.*, 2007). Therefore, another possibility is that the more aggressive nasopharyngeal tumors possess more undifferentiated or immature vessels, which could be more vulnerable to VEGF inhibitors. As long as the treatments do not eradicate all tumor cells, when the tumors regrow, the more aggressive ones manifest earlier. Therefore, more in-depth translational investigations are needed to better understand the mechanism of resistance to antiangiogenic therapy.

B. Activation of Other Angiogenic Factors

Alternative proangiogenic factors may also contribute to the development of resistance to anti-VEGF therapies (Bergers and Hanahan, 2008). Tumor vessels can be (or become) less sensitive to antiangiogenic agents, and sustained tumor angiogenesis can occur through VEGF-independent mechanisms (Ellis and Hicklin, 2008; Shojaei and Ferrara, 2008). More recently, sunitinib and the anti-VEGF2 antibody DC101 were reported to promote metastasis even though they inhibited the growth of primary tumors and increased overall survival in some cases (Ebos *et al.*, 2009; Paez-Ribes *et al.*, 2009). Increased invasiveness might result from enhanced expression of cytokines induced by the treatment or from hypoxia-driven effects, including transcriptional activation of the HGF receptor c-Met (Brahimi-Horn *et al.*, 2007; Knudsen and Vande Woude, 2008). Sunitinib could disrupt vascular integrity via pericyte detachment mediated by the inhibition of PDGFR-B and thus facilitate intravasation and extravasation of tumor cells.

Upregulation of VEGF and PlGF has been reported in patients treated with antiangiogenic agents targeting the VEGF pathway (Batchelor *et al.*, 2007; Polyzos, 2008). This observation suggests that PlGF may participate in tumor evasion of some antiangiogenic therapies (Crawford and Ferrara, 2009). FGFs, including the prototype members FGF1 and FGF2, are reported to promote angiogenesis independent of VEGF (Cao *et al.*, 2008b). Expression of FGFs correlated with restoration of the tumor vasculature that initially regressed in response to anti-VEGFR therapy. FGF2 has been found to be increased in the blood of patients relapsed from treatment using VEGFR inhibitors (Batchelor *et al.*, 2007). This research suggests that alternative proangiogenic factors may be responsible for the acquired resistance of patients.

Resistance also may arise from incomplete suppression of the VEGF pathway in endothelial cells. Some studies have shown that inhibiting Dll4 via antibody or Dll4-Fc inhibits tumor growth by deregulating angiogenesis, resulting in more (but nonfunctional) vessels (Noguera-Troise *et al.*, 2007; Ridgway *et al.*, 2006). Anti-Dll4 therapy reduced the growth of tumors moderately responsive to anti-VEGF. Cellular and molecular research has indicated that signals from the stromal compartment play an important role in refractoriness and, potentially, in acquired resistance to antiangiogenic therapy using VEGF blockers (Shojaei and Ferrara, 2007). Recent studies indicate that at least one population of bone marrow-derived cells (CD11b+

Gr1+) and tumor-associated fibroblasts can provide alternative growth factors to sustain tumor angiogenesis when VEGFR signaling is inhibited (Shojaei *et al.*, 2008).

C. Vessel Co-Option Instead of Sprouting Angiogenesis

The failure to inhibit tumor expansion after vascular normalization by blocking VEGF signaling (von Baumgarten *et al.*, 2011) suggests that the growing tumor does not always rely on sprouting angiogenesis. Metastatic melanoma and breast cancer are reported to establish tumor vasculature by vessel co-option (Kienast *et al.*, 2010; Qian *et al.*, 2006). The co-opted vessels are usually supported by pericytes and therefore insensitive to antiangiogenic therapy. It has been noticed that after antiangiogenic therapy using sunitinib against CCRCC, the remnant tumor tissues found in the tumor rim are supported mainly by vessels with pericyte coverage (Huang *et al.*, 2010b).

XII. PROSPECTS FOR FUTURE DEVELOPMENT

The outcome of antiangiogenic therapy against various malignancies has not been as profound as expected; most antiangiogenic agents are less effective in clinical applications than in preclinical studies (Carmeliet and Jain, 2011). There are several reasons for this.

First, most antiangiogenic drugs are selected based on their efficacy in inhibiting the growth of normal endothelial cells. The endothelial cells and angiogenesis model have been used to screen and validate antiangiogenic agents include HUVECs, human umbilical artery endothelial cells, human microvascular endothelial cells, human dermal microvascular endothelial cells, myogenic endothelial cells, zebrafish vasculature, corneal angiogenesis assay, and the chick embryo chorioallantoic membrane assay. All of these cellular models and assays for drug screening are using normal endothelial cells or normal vasculature systems. In fact, tumor vasculature and tumor-derived endothelial cells could be very different from their normal counterparts in many ways, including cytogenetics, protein expression pattern, cellular structure, cellular functions, and vasculature organization (Baluk *et al.*, 2005; Bussolati *et al.*, 2003, 2006; Ghosh *et al.*, 2008; Hida and Klagsbrun, 2005; Hida *et al.*, 2004; Jain, 2005; Nagy *et al.*, 2010). These differences might contribute to differing sensitivity to antiangiogenic drugs, environmental adaptability to hypoxic conditions, and genomic abnormalities favoring survivability. Thus, using normal endothelial cells as target cells for drug screening and effect validation could be a problem. More effective and more specific antiangiogenic agents can be expected by using next-generation drug screening based on tumor-derived endothelial cells or the tumor vasculature.

Second, the endothelial cells in different tumor vasculatures could differ. We know of differences in genetic and proteomic expression profiles among normal endothelial cells from various organs (Zamora et al., 2007; Zhang et al., 2008). The inhibition rates of antiangiogenic agents also differ among normal endothelial cell types (Li et al., 2010b). It is quite possible that the tumor-derived endothelial cells from different tumors could differ significantly in their sensitivity to antiangiogenic agents. Actually, the structural differences among different tumor vasculatures have been recognized. For instance, the sinusoid-like vasculature in aggressive hepatocyte carcinoma (Chen et al., 2011b) is not commonly seen in other types of malignancy. Therefore, our future development of antiangiogenic agents should be more focused on specific tumor types rather than broad-spectrum antiangiogenesis.

Third, most drug testing studies use xenograft models. The ability to grow xenograft tumors in immunodeficient mice does not prove that the vasculature inside those tumors is similar to that in human tumors. In fact, the complexities of the human tumor vasculature are not always found in xenograft tumors (Cao, 2011). Further, the fast-growing characteristics of xenograft tumors are in contrast to relatively slow-growing human tumors (Cao, 2011; Carmeliet and Jain, 2011). All of these disparities upon using current animal models contribute to disparate outcomes of preclinical studies and clinical treatments. Obviously, we need to develop better preclinical animal models for antiangiogenic drug development.

Fourth, only a fraction of cancer patients respond to a specific antiangiogenic drug. Therefore, we ideally want to select those patients who will benefit from the drug prior to the treatment. This is especially critical to further improve the response rate of existing antiangiogenic agents. In one preliminary study, we showed that the MVD of undifferentiated vessels in nonsmall cell lung cancer before treatment positively correlated with tumor shrinkage after combined chemotherapy of bevacizumab plus cytotoxic agents (Zhao et al., 2012). It is still unclear whether this better tumor response can further translate into survival benefits. Neither is it clear whether the undifferentiated MDV is a marker of better response to bevacizumab treatment or just a marker for combination therapy. However, this was a promising attempt to identify patients suitable for antiangiogenic therapy by differentially analyzing tumor vasculature. From a practical point of view, detecting serum biomarkers for patient selection is easy and cost-effective. Elevated serum IL-8 level is one of the promising serum markers for sunitinib treatment resistance (Bellmunt et al., 2011; Huang et al., 2010a).

Fifth, targeting one proangiogenic factor or receptor is obviously not enough, as tumor cells can evade such treatment by using other signaling pathways. Promising anticancer results from a preclinical study by blocking both the VEGF and c-Met signaling pathways strengthen the concept that targeting several key molecules is a reasonable approach to increasing the efficacy of antiangiogenic therapy (You *et al.*, 2011). At the same time, the risk of more severe adverse effects resulted from multitargeted therapy demands the development of personalized cancer therapy.

Finally, the timing and dosage of administrating antiangiogenic agents could also be crucial. Optimization of the dose and schedule of antiangiogenic and chemotherapeutic drugs is suggested for further improvement of current antiangiogenic therapy (Jain *et al.*, 2009).

In summary, the future of antiangiogenic therapy for cancer patients relies on our endeavors to develop the right drugs, select the right patients, and optimize the treatment conditions.

ACKNOWLEDGMENTS

This work was supported by Grants from the State Key Program of National Natural Science Foundation of China (Grant No. 81030043), the Youth Science Fund of the National Natural Science Foundation of China (Grant No. 81000946), the National Key Sci-Tech Special Project of China (Grant No. 2008ZX10002-019), the Outstanding Youth Project of Educational Department of Hunan Province, China (Grant No. 10B090), and the Van Andel Foundation. We thank David Nadziejka, Grand Rapids, Michigan for critical reading of the manuscript.

REFERENCES

Akino, T., Hida, K., Hida, Y., *et al.* (2009). Cytogenetic abnormalities of tumor-associated endothelial cells in human malignant tumors. *Am. J. Pathol.* **175**, 2657–2667.

Alvarez, R. H., Valero, V., and Hortobagyi, G. N. (2010). Emerging targeted therapies for breast cancer. *J. Clin. Oncol.* **28**, 3366–3379.

Aprahamian, M., Bour, G., Akladios, C. Y., *et al.* (2011). Myo-InositolTrisPyroPhosphate treatment leads to HIF-1alpha suppression and eradication of early hepatoma tumors in rats. *Chembiochem* **12**, 777–783.

Autiero, M., Waltenberger, J., Communi, D., *et al.* (2003). Role of PlGF in the intra- and intermolecular cross talk between the VEGF receptors Flt1 and Flk1. *Nat. Med.* **9**, 936–943.

Baeten, C. I., Hillen, F., Pauwels, P., de Bruine, A. P., and Baeten, C. G. (2009). Prognostic role of vasculogenic mimicry in colorectal cancer. *Dis. Colon Rectum* **52**, 2028–2035.

Baluk, P., Hashizume, H., and McDonald, D. M. (2005). Cellular abnormalities of blood vessels as targets in cancer. *Curr. Opin. Genet. Dev.* **15**, 102–111.

Bardelli, A., Ponzetto, C., and Comoglio, P. M. (1994). Identification of functional domains in the hepatocyte growth factor and its receptor by molecular engineering. *J. Biotechnol.* **37**, 109–122.

Batchelor, T. T., Sorensen, A. G., di Tomaso, E., *et al.* (2007). AZD2171, a pan-VEGF receptor tyrosine kinase inhibitor, normalizes tumor vasculature and alleviates edema in glioblastoma patients. *Cancer Cell* **11**, 83–95.

Beaumont, J. L., Butt, Z., Baladi, J., et al. (2011). Patient-reported outcomes in a phase iii study of everolimus versus placebo in patients with metastatic carcinoma of the kidney that has progressed on vascular endothelial growth factor receptor tyrosine kinase inhibitor therapy. *Oncologist* **16**, 632–640.

Bellmunt, J., Gonzalez-Larriba, J. L., Prior, C., et al. (2011). Phase II study of sunitinib as first-line treatment of urothelial cancer patients ineligible to receive cisplatin-based chemotherapy: baseline interleukin-8 and tumor contrast enhancement as potential predictive factors of activity. *Ann. Oncol.* **22**, 2646–2653.

Benton, G., George, J., Kleinman, H. K., and Arnaoutova, I. P. (2009). Advancing science and technology via 3D culture on basement membrane matrix. *J. Cell. Physiol.* **221**, 18–25.

Bergers, G., and Hanahan, D. (2008). Modes of resistance to anti-angiogenic therapy. *Nat. Rev. Cancer* **8**, 592–603.

Berthod, F., Germain, L., Tremblay, N., and Auger, F. A. (2006). Extracellular matrix deposition by fibroblasts is necessary to promote capillary-like tube formation in vitro. *J. Cell. Physiol.* **207**, 491–498.

Bielenberg, D. R., and Klagsbrun, M. (2007). Targeting endothelial and tumor cells with semaphorins. *Cancer Metastasis Rev.* **26**, 421–431.

Bjornsti, M. A., and Houghton, P. J. (2004). The TOR pathway: a target for cancer therapy. *Nat. Rev. Cancer* **4**, 335–348.

Boehm, T., Folkman, J., Browder, T., and O'Reilly, M. S. (1997). Antiangiogenic therapy of experimental cancer does not induce acquired drug resistance. *Nature* **390**, 404–407.

Bogdanovic, E., Nguyen, V. P., and Dumont, D. J. (2006). Activation of Tie2 by angiopoietin-1 and angiopoietin-2 results in their release and receptor internalization. *J. Cell Sci.* **119**, 3551–3560.

Boneberg, E. M., Legler, D. F., Hoefer, M. M., et al. (2009). Angiogenesis and lymphangiogenesis are downregulated in primary breast cancer. *Br. J. Cancer* **101**, 605–614.

Bossi, P., Viale, G., Lee, A. K., Alfano, R., Coggi, G., and Bosari, S. (1995). Angiogenesis in colorectal tumors: microvessel quantitation in adenomas and carcinomas with clinicopathological correlations. *Cancer Res.* **55**, 5049–5053.

Bracarda, S., Bellmunt, J., Melichar, B., et al. (2011). Overall survival in patients with metastatic renal cell carcinoma initially treated with bevacizumab plus interferon-alpha2a and subsequent therapy with tyrosine kinase inhibitors: a retrospective analysis of the phase III AVOREN trial. *BJU Int.* **107**, 214–219.

Brahimi-Horn, M. C., Chiche, J., and Pouysségur, J. (2007). Hypoxia and cancer. *J. Mol. Med.* **85**, 1301–1307.

Brem, S., Cotran, R., and Folkman, J. (1972). Tumor angiogenesis: a quantitative method for histologic grading. *J. Natl. Cancer Inst.* **48**, 347–356.

Brown, E. J., Albers, M. W., Shin, T. B., et al. (1994). A mammalian protein targeted by G1-arresting rapamycin-receptor complex. *Nature* **369**, 756–758.

Bucci, M., Roviezzo, F., Posadas, I., et al. (2005). Endothelial nitric oxide synthase activation is critical for vascular leakage during acute inflammation in vivo. *Proc. Natl. Acad. Sci. USA* **102**, 904–908.

Busam, K. J., Berwick, M., Blessing, K., et al. (1995). Tumor vascularity is not a prognostic factor for malignant melanoma of the skin. *Am. J. Pathol.* **147**, 1049–1056.

Bussolati, B., Deambrosis, I., Russo, S., Deregibus, M. C., and Camussi, G. (2003). Altered angiogenesis and survival in human tumor-derived endothelial cells. *FASEB J.* **17**, 1159–1161.

Bussolati, B., Grange, C., Bruno, S., et al. (2006). Neural-cell adhesion molecule (NCAM) expression by immature and tumor-derived endothelial cells favors cell organization into capillary-like structures. *Exp. Cell Res.* **312**, 913–924.

Cacheux, W., Boisserie, T., Staudacher, L., et al. (2008). Reversible tumor growth acceleration following bevacizumab interruption in metastatic colorectal cancer patients scheduled for surgery. *Ann. Oncol.* **19**, 1659–1661.

Cantu De Leon, D., Lopez-Graniel, C., Frias Mendivil, M., Chanona Vilchis, G., Gomez, C., and De La Garza Salazar, J. (2003). Significance of microvascular density (MVD) in cervical cancer recurrence. *Int. J. Gynecol. Cancer* **13**, 856–862.

Cao, Y. (2011). Antiangiogenic cancer therapy: why do mouse and human patients respond in a different way to the same drug? *Int. J. Dev. Biol.* **55**, 557–562.

Cao, R., Brakenhielm, E., Li, X., et al. (2002). Angiogenesis stimulated by PDGF-CC, a novel member in the PDGF family, involves activation of PDGFR-alphaalpha and -alphabeta receptors. *FASEB J.* **16**, 1575–1583.

Cao, Y., Cao, R., and Hedlund, E. M. (2008a). Regulation of tumor angiogenesis and metastasis by FGF and PDGF signaling pathways. *J. Mol. Med. (Berl.)* **86**, 785–789.

Cao, Y., Cao, R., and Hedlund, E. M. (2008b). R Regulation of tumor angiogenesis and metastasis by FGF and PDGF signaling pathways. *J. Mol. Med.* **86**, 785–789.

Carmeliet, P., and Jain, R. K. (2011). Molecular mechanisms and clinical applications of angiogenesis. *Nature* **473**, 298–307.

Casanovas, O., Hicklin, D. J., Bergers, G., and Hanahan, D. (2005). Drug resistance by evasion of antiangiogenic targeting of VEGF signaling in late-stage pancreatic islet tumors. *Cancer Cell* **8**, 299–309.

Castets, M., and Mehlen, P. (2010). Netrin-1 role in angiogenesis: to be or not to be a pro-angiogenic factor? *Cell Cycle* **9**, 1466–1471.

Cernea, C. R., Ferraz, A. R., de Castro, I. V., et al. (2004). Angiogenesis and skin carcinomas with skull base invasion: a case-control study. *Head Neck* **26**, 396–400.

Chamberlain, M. C. (2011). Bevacizumab for the treatment of recurrent glioblastoma. *Clin. Med. Insights Oncol.* **5**, 117–129.

Chandrachud, L. M., Pendleton, N., Chisholm, D. M., Horan, M. A., and Schor, A. M. (1997). Relationship between vascularity, age and survival in non-small-cell lung cancer. *Br. J. Cancer* **76**, 1367–1375.

Charalambous, C., Pen, L. B., Su, Y. S., Milan, J., Chen, T. C., and Hofman, F. M. (2005). Interleukin-8 differentially regulates migration of tumor-associated and normal human brain endothelial cells. *Cancer Res.* **65**, 10347–10354.

Chen, J. A., Shi, M., Li, J. Q., and Qian, C. N. (2010). Angiogenesis: multiple masks in hepatocellular carcinoma and liver regeneration. *Hepatol. Int.* **4**, 537–547.

Chen, H. T., Cai, Q. C., Zheng, J. M., et al. (2011). High Expression of Delta-Like Ligand 4 Predicts Poor Prognosis After Curative Resection for Pancreatic Cancer. *Ann. Surg. Oncol.* [Epub ahead of print].

Chen, Z. Y., Wei, W., Guo, Z. X., Lin, J. R., Shi, M., and Guo, R. P. (2011a). Morphologic classification of microvessels in hepatocellular carcinoma is associated with the prognosis after resection. *J. Gastroenterol. Hepatol.* **26**, 866–874.

Coppin, C. (2010). Everolimus: the first approved product for patients with advanced renal cell cancer after sunitinib and/or sorafenib. *Biologics* **4**, 91–101.

Couvelard, A., O'Toole, D., Turley, H., et al. (2005). Microvascular density and hypoxia-inducible factor pathway in pancreatic endocrine tumours: negative correlation of microvascular density and VEGF expression with tumour progression. *Br. J. Cancer* **92**, 94–101.

Crawford, Y., and Ferrara, N. (2009). Tumor and stromal pathways mediating refractoriness/resistance to anti-angiogenic therapies. *Trends Pharmacol. Sci.* **30**, 624–630.

Cross, M. J., Hodgkin, M. N., Roberts, S., Landgren, E., Wakelam, M. J., and Claesson-Welsh, L. (2000). Tyrosine 766 in the fibroblast growth factor receptor-1 is required for FGF-stimulation of phospholipase C, phospholipase D, phospholipase A(2), phosphoinositide 3-kinase and cytoskeletal reorganisation in porcine aortic endothelial cells. *J. Cell Sci.* **113**(Pt. 4), 643–651.

Czubayko, F., Liaudet-Coopman, E. D., Aigner, A., Tuveson, A. T., Berchem, G. J., and Wellstein, A. (1997). A secreted FGF-binding protein can serve as the angiogenic switch in human cancer. *Nat. Med.* **3**, 1137–1140.

D'Agostino, R. B., Sr. (2011). Changing end points in breast-cancer drug approval—the Avastin story. *N. Engl. J. Med.* **365**, e2.

Das, B., Yeger, H., Tsuchida, R., *et al.* (2005). A hypoxia-driven vascular endothelial growth factor/Flt1 autocrine loop interacts with hypoxia-inducible factor-1alpha through mitogen-activated protein kinase/extracellular signal-regulated kinase 1/2 pathway in neuroblastoma. *Cancer Res.* **65**, 7267–7275.

Davis, G. E., and Senger, D. R. (2005). Endothelial extracellular matrix: biosynthesis, remodeling, and functions during vascular morphogenesis and neovessel stabilization. *Circ. Res.* **97**, 1093–1107.

Davis, S., Aldrich, T. H., Jones, P. F., *et al.* (1996). Isolation of angiopoietin-1, a ligand for the TIE2 receptor, by secretion-trap expression cloning. *Cell* **87**, 1161–1169.

de Groot, J. F., Lamborn, K. R., Chang, S. M., *et al.* (2011). Phase II study of aflibercept in recurrent malignant glioma: a North American Brain Tumor Consortium study. *J. Clin. Oncol.* **29**, 2689–2695.

De Smet, F., Segura, I., De Bock, K., Hohensinner, P. J., and Carmeliet, P. (2009). Mechanisms of vessel branching: filopodia on endothelial tip cells lead the way. *Arterioscler. Thromb. Vasc. Biol.* **29**, 639–649.

Del Bufalo, D., Ciuffreda, L., Trisciuoglio, D., *et al.* (2006). Antiangiogenic potential of the Mammalian target of rapamycin inhibitor temsirolimus. *Cancer Res.* **66**, 5549–5554.

Desar, I. M., van Herpen, C. M., van Laarhoven, H. W., Barentsz, J. O., Oyen, W. J., and van der Graaf, W. T. (2009). Beyond RECIST: molecular and functional imaging techniques for evaluation of response to targeted therapy. *Cancer Treat. Rev.* **35**, 309–321.

Dickinson, A. J., Fox, S. B., Persad, R. A., Hollyer, J., Sibley, G. N., and Harris, A. L. (1994). Quantification of angiogenesis as an independent predictor of prognosis in invasive bladder carcinomas. *Br. J. Urol.* **74**, 762–766.

Diez, H., Fischer, A., Winkler, A., *et al.* (2007). Hypoxia-mediated activation of Dll4-Notch-Hey2 signaling in endothelial progenitor cells and adoption of arterial cell fate. *Exp. Cell Res.* **313**, 1–9.

Dimmeler, S., and Zeiher, A. M. (1999). Nitric oxide—an endothelial cell survival factor. *Cell Death Differ.* **6**, 964–968.

Ding, Y. T., Kumar, S., and Yu, D. C. (2008). The role of endothelial progenitor cells in tumour vasculogenesis. *Pathobiology* **75**, 265–273.

Ding, T., Xu, J., Wang, F., *et al.* (2009). High tumor-infiltrating macrophage density predicts poor prognosis in patients with primary hepatocellular carcinoma after resection. *Hum. Pathol.* **40**, 381–389.

Ding, T., Xu, J., Zhang, Y., *et al.* (2011). Endothelium-coated tumor clusters are associated with poor prognosis and micrometastasis of hepatocellular carcinoma after resection. *Cancer* **117**, 4878–4889.

Dirkx, A. E., Oude Egbrink, M. G.., Wagstaff, J., and Griffioen, A. W. (2006). Monocyte/macrophage infiltration in tumors: modulators of angiogenesis. *J. Leukoc. Biol.* **80**, 1183–1196.

Dome, B., Paku, S., Somlai, B., and Timar, J. (2002). Vascularization of cutaneous melanoma involves vessel co-option and has clinical significance. *J. Pathol.* **197**, 355–362.

Dowling, R. J., Topisirovic, I., Fonseca, B. D., and Sonenberg, N. (2010). Dissecting the role of mTOR: lessons from mTOR inhibitors. *Biochim. Biophys. Acta* **1804**, 433–439.

Duarte, C. D., Greferath, R., Nicolau, C., and Lehn, J. M. (2010). myo-Inositol trispyrophosphate: a novel allosteric effector of hemoglobin with high permeation selectivity across the red blood cell plasma membrane. *Chembiochem* **11**, 2543–2548.

Duda, D. G., Jain, R. K., and Willett, C. G. (2007). Antiangiogenics: the potential role of integrating this novel treatment modality with chemoradiation for solid cancers. *J. Clin. Oncol.* **25**, 4033–4042.

Ebos, J. M., and Kerbel, R. S. (2011). Antiangiogenic therapy: impact on invasion, disease progression, and metastasis. *Nat. Rev. Clin. Oncol.* **8**, 210–221.

Ebos, J. M., Lee, C. R., Cruz-Munoz, W., Bjarnason, G. A., Christensen, J. G., and Kerbel, R. S. (2009). Accelerated metastasis after short-term treatment with a potent inhibitor of tumor angiogenesis. *Cancer Cell* **15**, 232–239.

Eichmann, A., Le Noble, F., Autiero, M., and Carmeliet, P. (2005). Guidance of vascular and neural network formation. *Curr. Opin. Neurobiol.* **15**, 108–115.

El-Assal, O. N., Yamanoi, A., Soda, Y., et al. (1998). Clinical significance of microvessel density and vascular endothelial growth factor expression in hepatocellular carcinoma and surrounding liver: possible involvement of vascular endothelial growth factor in the angiogenesis of cirrhotic liver. *Hepatology* **27**, 1554–1562.

Ellis, L. M., and Hicklin, D. J. (2008). VEGF-targeted therapy: mechanisms of anti-tumour activity. *Nat. Rev. Cancer* **8**, 579–591.

Escudier, B., Eisen, T., Stadler, W. M., et al. (2007). Sorafenib in advanced clear-cell renal-cell carcinoma. *N. Engl. J. Med.* **356**, 125–134.

Eswarakumar, V. P., Lax, I., and Schlessinger, J. (2005). Cellular signaling by fibroblast growth factor receptors. *Cytokine Growth Factor Rev.* **16**, 139–149.

Falcon, B. L., Barr, S., Gokhale, P. C., et al. (2011). Reduced VEGF production, angiogenesis, and vascular regrowth contribute to the antitumor properties of dual mTORC1/mTORC2 inhibitors. *Cancer Res.* **71**, 1573–1583.

Farnsworth, R. H., Karnezis, T., Shayan, R., et al. (2011). A role for bone morphogenetic protein-4 in lymph node vascular remodeling and primary tumor growth. *Cancer Res.* **71**, 6547–6557.

Fernandez-Aguilar, S., Jondet, M., Simonart, T., and Noel, J. C. (2006). Microvessel and lymphatic density in tubular carcinoma of the breast: comparative study with invasive low-grade ductal carcinoma. *Breast* **15**, 782–785.

Ferrara, N., Gerber, H. P., and LeCouter, J. (2003). The biology of VEGF and its receptors. *Nat. Med.* **9**, 669–676.

Fett, J. W., Strydom, D. J., Lobb, R. R., et al. (1985). Isolation and characterization of angiogenin, an angiogenic protein from human carcinoma cells. *Biochemistry* **24**, 5480–5486.

Fogg, V. C., Lanning, N. J., and Mackeigan, J. P. (2011). Mitochondria in cancer: at the crossroads of life and death. *Chin. J. Cancer* **30**, 526–539.

Folkman, J. (1971). Tumor angiogenesis: therapeutic implications. *N. Engl. J. Med.* **285**, 1182–1186.

Foss, A. J., Alexander, R. A., Jefferies, L. W., Hungerford, J. L., Harris, A. L., and Lightman, S. (1996). Microvessel count predicts survival in uveal melanoma. *Cancer Res.* **56**, 2900–2903.

Fredriksson, L., Li, H., and Eriksson, U. (2004). The PDGF family: four gene products form five dimeric isoforms. *Cytokine Growth Factor Rev.* **15**, 197–204.

Fujiwara, T., Fukushi, J., Yamamoto, S., et al. (2011). Macrophage infiltration predicts a poor prognosis for human Ewing sarcoma. *Am. J. Pathol.* **179**, 1157–1170.

Fukumura, D., Kashiwagi, S., and Jain, R. K. (2006). The role of nitric oxide in tumour progression. *Nat. Rev. Cancer* **6**, 521–534.

Gaitskell, K., Martinek, I., Bryant, A., Kehoe, S., Nicum, S., and Morrison, J. (2011). Angiogenesis inhibitors for the treatment of ovarian cancer. *Cochrane Database Syst. Rev.* **9**, CD007930.

Gandino, L., Longati, P., Medico, E., Prat, M., and Comoglio, P. M. (1994). Phosphorylation of serine 985 negatively regulates the hepatocyte growth factor receptor kinase. *J. Biol. Chem.* **269**, 1815–1820.

Gao, X., and Xu, Z. (2008). Mechanisms of action of angiogenin. *Acta. Biochim. Biophys. Sin. (Shanghai)* **40**, 619–624.

Garcia, S., Dales, J. P., Charafe-Jauffret, E., et al. (2007). Overexpression of c-Met and of the transducers PI3K, FAK and JAK in breast carcinomas correlates with shorter survival and neoangiogenesis. *Int. J. Oncol.* **31**, 49–58.

Gasparini, G., Weidner, N., Bevilacqua, P., et al. (1994). Tumor microvessel density, p53 expression, tumor size, and peritumoral lymphatic vessel invasion are relevant prognostic markers in node-negative breast carcinoma. *J. Clin. Oncol.* **12**, 454–466.

Gee, M. S., Procopio, W. N., Makonnen, S., Feldman, M. D., Yeilding, N. M., and Lee, W. M. (2003). Tumor vessel development and maturation impose limits on the effectiveness of antivascular therapy. *Am. J. Pathol.* **162**, 183–193.

George, S., Shah, S. N., and Bukowski, R. M. (2011). Stable disease in renal cell carcinoma after using signal transduction inhibitors. *Rev. Recent Clin. Trials* **5**, 117–122.

Ghosh, K., Thodeti, C. K., Dudley, A. C., Mammoto, A., Klagsbrun, M., and Ingber, D. E. (2008). Tumor-derived endothelial cells exhibit aberrant Rho-mediated mechanosensing and abnormal angiogenesis in vitro. *Proc. Natl. Acad. Sci. USA* **105**, 11305–11310.

Giordano, F. J., and Johnson, R. S. (2001). Angiogenesis: the role of the microenvironment in flipping the switch. *Curr. Opin. Genet. Dev.* **11**, 35–40.

Giridharan, W., Hughes, J., Fenton, J. E., and Jones, A. S. (2003). Lymph node metastases in the lower neck. *Clin. Otolaryngol. Allied Sci.* **28**, 221–226.

Girouard, S. D., and Murphy, G. F. (2011). Melanoma stem cells: not rare, but well done. *Lab. Invest.* **91**, 647–664.

Gordon, M. S. (2011). Antiangiogenic therapies: is VEGF-A inhibition alone enough? *Expert Rev. Anticancer Ther.* **11**, 485–496.

Grepin, R., Guyot, M., Jacquin, M., et al. (2011). Acceleration of clear cell renal cell carcinoma growth in mice following bevacizumab/Avastin treatment: The role of CXCL cytokines. *Oncogene* **206**, 1–12.

Grose, R., and Dickson, C. (2005). Fibroblast growth factor signaling in tumorigenesis. *Cytokine Growth Factor Rev.* **16**, 179–186.

Guan, Z. Z., Xu, J. M., Luo, R. C., et al. (2011). Efficacy and safety of bevacizumab plus chemotherapy in Chinese patients with metastatic colorectal cancer: a randomized phase III ARTIST trial. *Chin. J. Cancer* **30**, 682–689.

Guertin, D. A., and Sabatini, D. M. (2007). Defining the role of mTOR in cancer. *Cancer Cell* **12**, 9–22.

Hallmann, R., Horn, N., Selg, M., Wendler, O., Pausch, F., and Sorokin, L. M. (2005). Expression and function of laminins in the embryonic and mature vasculature. *Physiol. Rev.* **85**, 979–1000.

Hanahan, D., and Folkman, J. (1996). Patterns and emerging mechanisms of the angiogenic switch during tumorigenesis. *Cell* **86**, 353–364.

Hanahan, D., and Weinberg, R. A. (2000). The hallmarks of cancer. *Cell* **100**, 57–70.

Hasina, R., Whipple, M. E., Martin, L. E., Kuo, W. P., Ohno-Machado, L., and Lingen, M. W. (2008). Angiogenic heterogeneity in head and neck squamous cell carcinoma: biological and therapeutic implications. *Lab. Invest.* **88**, 342–353.

Heldin, C. H., Ostman, A., and Ronnstrand, L. (1998). Signal transduction via platelet-derived growth factor receptors. *Biochim. Biophys. Acta* **1378**, F79–F113.

Helfrich, I., and Schadendorf, D. (2011). Blood vessel maturation, vascular phenotype and angiogenic potential in malignant melanoma: one step forward for overcoming anti-angiogenic drug resistance? *Mol. Oncol.* **5**, 137–149.

Helfrich, I., Scheffrahn, I., Bartling, S., et al. (2010). Resistance to antiangiogenic therapy is directed by vascular phenotype, vessel stabilization, and maturation in malignant melanoma. *J. Exp. Med.* **207**, 491–503.

Hellstrom, M., Kalen, M., Lindahl, P., Abramsson, A., and Betsholtz, C. (1999). Role of PDGF-B and PDGFR-beta in recruitment of vascular smooth muscle cells and pericytes during embryonic blood vessel formation in the mouse. *Development* **126**, 3047–3055.

Hendrix, M. J., Seftor, E. A., Hess, A. R., and Seftor, R. E. (2003). Vasculogenic mimicry and tumour-cell plasticity: lessons from melanoma. *Nat. Rev. Cancer* **3**, 411–421.

Hess, A. R., Seftor, E. A., Gruman, L. M., Kinch, M. S., Seftor, R. E., and Hendrix, M. J. (2006). VE-cadherin regulates EphA2 in aggressive melanoma cells through a novel signaling pathway: implications for vasculogenic mimicry. *Cancer Biol. Ther.* **5**, 228–233.

Hida, K., and Klagsbrun, M. (2005). A new perspective on tumor endothelial cells: unexpected chromosome and centrosome abnormalities. *Cancer Res.* **65**, 2507–2510.

Hida, K., Hida, Y., Amin, D. N., *et al.* (2004). Tumor-associated endothelial cells with cytogenetic abnormalities. *Cancer Res.* **64**, 8249–8255.

Hillen, F., and Griffioen, A. W. (2007). Tumour vascularization: sprouting angiogenesis and beyond. *Cancer Metastasis Rev.* **26**, 489–502.

Hillen, F., van de Winkel, A., Creytens, D., Vermeulen, A. H., and Griffioen, A. W. (2006). Proliferating endothelial cells, but not microvessel density, are a prognostic parameter in human cutaneous melanoma. *Melanoma Res.* **16**, 453–457.

Hirschi, K. K., and D'Amore, P. A. (1996). Pericytes in the microvasculature. *Cardiovasc. Res.* **32**, 687–698.

Hlushchuk, R., Riesterer, O., Baum, O., *et al.* (2008). Tumor recovery by angiogenic switch from sprouting to intussusceptive angiogenesis after treatment with PTK787/ZK222584 or ionizing radiation. *Am. J. Pathol.* **173**, 1173–1185.

Hockel, M., and Vaupel, P. (2001). Tumor hypoxia: definitions and current clinical, biologic, and molecular aspects. *J. Natl. Cancer Inst.* **93**, 266–276.

Holash, J., Maisonpierre, P. C., Compton, D., *et al.* (1999). Vessel cooption, regression, and growth in tumors mediated by angiopoietins and VEGF. *Science* **284**, 1994–1998.

Holash, J., Davis, S., Papadopoulos, N., *et al.* (2002). VEGF-Trap: a VEGF blocker with potent antitumor effects. *Proc. Natl. Acad. Sci. USA* **99**, 11393–11398.

Hood, J. D., Meininger, C. J., Ziche, M., and Granger, H. J. (1998). VEGF upregulates ecNOS message, protein, and NO production in human endothelial cells. *Am. J. Physiol.* **274**, H1054–H1058.

Horak, E. R., Leek, R., Klenk, N., *et al.* (1992). Angiogenesis, assessed by platelet/endothelial cell adhesion molecule antibodies, as indicator of node metastases and survival in breast cancer. *Lancet* **340**, 1120–1124.

Houck, K. A., Ferrara, N., Winer, J., Cachianes, G., Li, B., and Leung, D. W. (1991). The vascular endothelial growth factor family: identification of a fourth molecular species and characterization of alternative splicing of RNA. *Mol. Endocrinol.* **5**, 1806–1814.

Hu, B., and Cheng, S. Y. (2009). Angiopoietin-2: development of inhibitors for cancer therapy. *Curr. Oncol. Rep.* **11**, 111–116.

Hu, W., Lu, C., Han, H. D., *et al.* (2011). Biological roles of the Delta family Notch ligand Dll4 in tumor and endothelial cells in ovarian cancer. *Cancer Res.* **71**, 6030–6039.

Huang, S., Pettaway, C. A., Uehara, H., Bucana, C. D., and Fidler, I. J. (2001). Blockade of NF-kappaB activity in human prostate cancer cells is associated with suppression of angiogenesis, invasion, and metastasis. *Oncogene* **20**, 4188–4197.

Huang, J., Bae, J. O., Tsai, J. P., *et al.* (2009). Angiopoietin-1/Tie-2 activation contributes to vascular survival and tumor growth during VEGF blockade. *Int. J. Oncol.* **34**, 79–87.

Huang, D., Ding, Y., Zhou, M., *et al.* (2010a). Interleukin-8 mediates resistance to antiangiogenic agent sunitinib in renal cell carcinoma. *Cancer Res.* **70**, 1063–1071.

Huang, D., Ding, Y., Li, Y., *et al.* (2010b). Sunitinib acts primarily on tumor endothelium rather than tumor cells to inhibit the growth of renal cell carcinoma. *Cancer Res.* **70**, 1053–1062.

Hudson, C. C., Liu, M., Chiang, G. G., *et al.* (2002). Regulation of hypoxia-inducible factor 1alpha expression and function by the mammalian target of rapamycin. *Mol. Cell. Biol.* **22**, 7004–7014.

Huminiecki, L., and Bicknell, R. (2000). In silico cloning of novel endothelial-specific genes. *Genome Res.* **10**, 1796–1806.

Hurwitz, H., Fehrenbacher, L., Novotny, W., et al. (2004). Bevacizumab plus irinotecan, fluorouracil, and leucovorin for metastatic colorectal cancer. *N. Engl. J. Med.* **350**, 2335–2342.

Ibaragi, S., Yoshioka, N., Kishikawa, H., et al. (2009). Angiogenin-stimulated rRNA transcription is essential for initiation and survival of AKT-induced prostate intraepithelial neoplasia. *Mol. Cancer Res.* **7**, 415–424.

Ide, T., Kitajima, Y., Miyoshi, A., et al. (2007). The hypoxic environment in tumor-stromal cells accelerates pancreatic cancer progression via the activation of paracrine hepatocyte growth factor/c-Met signaling. *Ann. Surg. Oncol.* **14**, 2600–2607.

Igarashi, M., Dhar, D. K., Kubota, H., Yamamoto, A., El-Assal, O., and Nagasue, N. (1998). The prognostic significance of microvessel density and thymidine phosphorylase expression in squamous cell carcinoma of the esophagus. *Cancer* **82**, 1225–1232.

Inai, T., Mancuso, M., Hashizume, H., et al. (2004). Inhibition of vascular endothelial growth factor (VEGF) signaling in cancer causes loss of endothelial fenestrations, regression of tumor vessels, and appearance of basement membrane ghosts. *Am. J. Pathol.* **165**, 35–52.

Inoki, K., Corradetti, M. N., and Guan, K. L. (2005). Dysregulation of the TSC-mTOR pathway in human disease. *Nat. Genet.* **37**, 19–24.

Iozzo, R. V., Zoeller, J. J., and Nystrom, A. (2009). Basement membrane proteoglycans: modulators par excellence of cancer growth and angiogenesis. *Mol. Cells* **27**, 503–513.

Iso, T., Kedes, L., and Hamamori, Y. (2003). HES and HERP families: multiple effectors of the Notch signaling pathway. *J. Cell. Physiol.* **194**, 237–255.

Iwahana, M., Utoguchi, N., Mayumi, T., Goryo, M., and Okada, K. (1998). Drug resistance and P-glycoprotein expression in endothelial cells of newly formed capillaries induced by tumors. *Anticancer Res.* **18**, 2977–2980.

Jain, R. K. (2001). Normalizing tumor vasculature with anti-angiogenic therapy: a new paradigm for combination therapy. *Nat. Med.* **7**, 987–989.

Jain, R. K. (2005). Normalization of tumor vasculature: an emerging concept in antiangiogenic therapy. *Science* **307**, 58–62.

Jain, R. K., Duda, D. G., Willett, C. G., et al. (2009). Biomarkers of response and resistance to antiangiogenic therapy. *Nat. Rev. Clin. Oncol.* **6**, 327–338.

Jing, Y., Ma, N., Fan, T., et al. (2011). Tumor necrosis factor-alpha promotes tumor growth by inducing vascular endothelial growth factor. *Cancer Invest.* **29**, 485–493.

Jones, A., and Ellis, P. (2011). Potential withdrawal of bevacizumab for the treatment of breast cancer. *BMJ* **343**, d4946.

Jones, M. K., Tsugawa, K., Tarnawski, A. S., and Baatar, D. (2004). Dual actions of nitric oxide on angiogenesis: possible roles of PKC, ERK, and AP-1. *Biochem. Biophys. Res. Commun.* **318**, 520–528.

Jubb, A. M., Soilleux, E. J., Turley, H., et al. (2010). Expression of vascular notch ligand delta-like 4 and inflammatory markers in breast cancer. *Am. J. Pathol.* **176**, 2019–2028.

Jung, Y. D., Ahmad, S. A., Akagi, Y., et al. (2000). Role of the tumor microenvironment in mediating response to anti-angiogenic therapy. *Cancer Metastasis Rev.* **19**, 147–157.

Kamba, T., and McDonald, D. M. (2007). Mechanisms of adverse effects of anti-VEGF therapy for cancer. *Br. J. Cancer* **96**, 1788–1795.

Kamba, T., Tam, B. Y., Hashizume, H., et al. (2006). VEGF-dependent plasticity of fenestrated capillaries in the normal adult microvasculature. *Am. J. Physiol. Heart Circ. Physiol.* **290**, H560–H576.

Kandel, J., Bossy-Wetzel, E., Radvanyi, F., Klagsbrun, M., Folkman, J., and Hanahan, D. (1991). Neovascularization is associated with a switch to the export of bFGF in the multistep development of fibrosarcoma. *Cell* **66**, 1095–1104.

Kane, R. C., Farrell, A. T., Saber, H., *et al.* (2006). Sorafenib for the treatment of advanced renal cell carcinoma. *Clin. Cancer Res.* **12**, 7271–7278.

Karaman, M. W., Herrgard, S., Treiber, D. K., *et al.* (2008). A quantitative analysis of kinase inhibitor selectivity. *Nat. Biotechnol.* **26**, 127–132.

Karamouzis, M. V., and Moschos, S. J. (2009). The use of endostatin in the treatment of solid tumors. *Expert Opin. Biol. Ther.* **9**, 641–648.

Katoh, M. (2007). Notch signaling in gastrointestinal tract (review). *Int. J. Oncol.* **30**, 247–251.

Kawasaki, K., Smith, R. S., Jr., Hsieh, C. M., Sun, J., Chao, J., and Liao, J. K. (2003). Activation of the phosphatidylinositol 3-kinase/protein kinase Akt pathway mediates nitric oxide-induced endothelial cell migration and angiogenesis. *Mol. Cell. Biol.* **23**, 5726–5737.

Ke, Q., and Costa, M. (2006). Hypoxia-inducible factor-1 (HIF-1). *Mol. Pharmacol.* **70**, 1469–1480.

Kerbel, R. S., Yu, J., Tran, J., *et al.* (2001). Possible mechanisms of acquired resistance to antiangiogenic drugs: implications for the use of combination therapy approaches. *Cancer Metastasis Rev.* **20**, 79–86.

Kieda, C., Greferath, R., Crola da Silva, C., Fylaktakidou, K. C., Lehn, J. M., and Nicolau, C. (2006). Suppression of hypoxia-induced HIF-1alpha and of angiogenesis in endothelial cells by myo-inositol trispyrophosphate-treated erythrocytes. *Proc. Natl. Acad. Sci. USA* **103**, 15576–15581.

Kienast, Y., von Baumgarten, L., Fuhrmann, M., *et al.* (2010). Real-time imaging reveals the single steps of brain metastasis formation. *Nat. Med.* **16**, 116–122.

Kimura, H., Weisz, A., Kurashima, Y., *et al.* (2000). Hypoxia response element of the human vascular endothelial growth factor gene mediates transcriptional regulation by nitric oxide: control of hypoxia-inducible factor-1 activity by nitric oxide. *Blood* **95**, 189–197.

Klagsbrun, M., Takashima, S., and Mamluk, R. (2002). The role of neuropilin in vascular and tumor biology. *Adv. Exp. Med. Biol.* **515**, 33–48.

Knudsen, B. S., and Vande Woude, G. (2008). Showering c-MET-dependent cancers with drugs. *Curr. Opin. Genet. Dev.* **18**, 87–96.

Koch, A. E., Polverini, P. J., Kunkel, S. L., *et al.* (1992). Interleukin-8 as a macrophage-derived mediator of angiogenesis. *Science* **258**, 1798–1801.

Kouhara, H., Hadari, Y. R., Spivak-Kroizman, T., *et al.* (1997). A lipid-anchored Grb2-binding protein that links FGF-receptor activation to the Ras/MAPK signaling pathway. *Cell* **89**, 693–702.

Kreisl, T. N., Kim, L., Moore, K., *et al.* (2009). Phase II trial of single-agent bevacizumab followed by bevacizumab plus irinotecan at tumor progression in recurrent glioblastoma. *J. Clin. Oncol.* **27**, 740–745.

Kreisl, T. N., Zhang, W., Odia, Y., *et al.* (2011). A phase II trial of single-agent bevacizumab in patients with recurrent anaplastic glioma. *Neuro Oncol.* **13**, 1143–1150.

Kudo, M. (2011). Future treatment option for hepatocellular carcinoma: a focus on brivanib. *Dig. Dis.* **29**, 316–320.

Kume, T. (2009). Novel insights into the differential functions of Notch ligands in vascular formation. *J. Angiogenes. Res.* **1**, 8.

Kusamura, S., Derchain, S., Alvarenga, M., Gomes, C. P., Syrjanen, K. J., and Andrade, L. A. (2003). Expression of p53, c-erbB-2, Ki-67, and CD34 in granulosa cell tumor of the ovary. *Int. J. Gynecol. Cancer* **13**, 450–457.

Lackner, C., Jukic, Z., Tsybrovskyy, O., *et al.* (2004). Prognostic relevance of tumour-associated macrophages and von Willebrand factor-positive microvessels in colorectal cancer. *Virchows Arch.* **445**, 160–167.

Larrivee, B., Freitas, C., Trombe, M., *et al.* (2007). Activation of the UNC5B receptor by Netrin-1 inhibits sprouting angiogenesis. *Genes Dev.* **21**, 2433–2447.

Le Tourneau, C., Faivre, S., Serova, M., and Raymond, E. (2008). mTORC1 inhibitors: is temsirolimus in renal cancer telling us how they really work? *Br. J. Cancer* **99**, 1197–1203.

Lee, C. H., Espinosa, I., Vrijaldenhoven, S., *et al*. (2008). Prognostic significance of macrophage infiltration in leiomyosarcomas. *Clin. Cancer Res.* **14**, 1423–1430.

Leenders, W. P., Kusters, B., and de Waal, R. M. (2002). Vessel co-option: how tumors obtain blood supply in the absence of sprouting angiogenesis. *Endothelium* **9**, 83–87.

Leibovich, S. J., Polverini, P. J., Shepard, H. M., Wiseman, D. M., Shively, V., and Nuseir, N. (1987). Macrophage-induced angiogenesis is mediated by tumour necrosis factor-alpha. *Nature* **329**, 630–632.

Li, M., Gu, Y., Zhang, Z., *et al*. (2010). Vasculogenic mimicry: a new prognostic sign of gastric adenocarcinoma. *Pathol. Oncol. Res.* **16**, 259–266.

Li, Y., Zhang, Z. F., Chen, J., *et al*. (2010). VX680/MK-0457, a potent and selective Aurora kinase inhibitor, targets both tumor and endothelial cells in clear cell renal cell carcinoma. *Am. J. Transl. Res.* **2**, 296–308.

Lim, W. T., Ng, Q. S., Ivy, P., *et al*. (2011). A phase II study of pazopanib in Asian patients with recurrent/metastatic nasopharyngeal carcinoma. *Clin. Cancer Res.* **17**, 5481–5489.

Lin, A. Y., Maniotis, A. J., Valyi-Nagy, K., *et al*. (2005). Distinguishing fibrovascular septa from vasculogenic mimicry patterns. *Arch. Pathol. Lab. Med.* **129**, 884–892.

Lin, J. Y., Li, X. Y., Tadashi, N., and Dong, P. (2011). Clinical significance of tumor-associated macrophage infiltration in supraglottic laryngeal carcinoma. *Chin. J. Cancer* **30**, 280–286.

Lindahl, P., Johansson, B. R., Leveen, P., and Betsholtz, C. (1997). Pericyte loss and microaneurysm formation in PDGF-B-deficient mice. *Science* **277**, 242–245.

Lindmark, G., Gerdin, B., Sundberg, C., Pahlman, L., Bergstrom, R., and Glimelius, B. (1996). Prognostic significance of the microvascular count in colorectal cancer. *J. Clin. Oncol.* **14**, 461–466.

Lissbrant, I. F., Stattin, P., Damber, J. E., and Bergh, A. (1997). Vascular density is a predictor of cancer-specific survival in prostatic carcinoma. *Prostate* **33**, 38–45.

Liu, C., Huang, H., Donate, F., *et al*. (2002). Prostate-specific membrane antigen directed selective thrombotic infarction of tumors. *Cancer Res.* **62**, 5470–5475.

Liu, X. M., Zhang, Q. P., Mu, Y. G., *et al*. (2011a). Clinical significance of vasculogenic mimicry in human gliomas. *J. Neurooncol.* **105**, 173–179.

Liu, K. W., Hu, B., and Cheng, S. Y. (2011b). Platelet-derived growth factor signaling in human malignancies. *Chin. J. Cancer* **30**, 581–584.

Loges, S., Mazzone, M., Hohensinner, P., and Carmeliet, P. (2009). Silencing or fueling metastasis with VEGF inhibitors: antiangiogenesis revisited. *Cancer Cell* **15**, 167–170.

Longati, P., Bardelli, A., Ponzetto, C., Naldini, L., and Comoglio, P. M. (1994). Tyrosines1234-1235 are critical for activation of the tyrosine kinase encoded by the MET proto-oncogene (HGF receptor). *Oncogene* **9**, 49–57.

Lopez-Rivera, E., Lizarbe, T. R., Martinez-Moreno, M., *et al*. (2005). Matrix metalloproteinase 13 mediates nitric oxide activation of endothelial cell migration. *Proc. Natl. Acad. Sci. USA* **102**, 3685–3690.

Maina, F., Casagranda, F., Audero, E., *et al*. (1996). Uncoupling of Grb2 from the Met receptor in vivo reveals complex roles in muscle development. *Cell* **87**, 531–542.

Maisonpierre, P. C., Suri, C., Jones, P. F., *et al*. (1997). Angiopoietin-2, a natural antagonist for Tie2 that disrupts in vivo angiogenesis. *Science* **277**, 55–60.

Mancuso, M. R., Davis, R., Norberg, S. M., *et al*. (2006). Rapid vascular regrowth in tumors after reversal of VEGF inhibition. *J. Clin. Invest.* **116**, 2610–2621.

Maniotis, A. J., Folberg, R., Hess, A., *et al*. (1999). Vascular channel formation by human melanoma cells in vivo and in vitro: vasculogenic mimicry. *Am. J. Pathol.* **155**, 739–752.

Marioni, G., Gaio, E., Giacomelli, L., *et al*. (2005). Endoglin (CD105) expression in head and neck basaloid squamous cell carcinoma. *Acta Otolaryngol.* **125**, 307–311.

McDonald, D. M., and Foss, A. J. (2000). Endothelial cells of tumor vessels: abnormal but not absent. *Cancer Metastasis Rev.* **19**, 109–120.

Meier, F., Will, S., Ellwanger, U., et al. (2002). Metastatic pathways and time courses in the orderly progression of cutaneous melanoma. *Br. J. Dermatol.* **147**, 62–70.

Merrick, D. T., Haney, J., Petrunich, S., et al. (2005). Overexpression of vascular endothelial growth factor and its receptors in bronchial dypslasia demonstrated by quantitative RT-PCR analysis. *Lung Cancer* **48**, 31–45.

Miles, D., Harbeck, N., Escudier, B., et al. (2011). Disease course patterns after discontinuation of bevacizumab: pooled analysis of randomized phase III trials. *J. Clin. Oncol.* **29**, 83–88.

Millauer, B., Wizigmann-Voos, S., Schnurch, H., et al. (1993). High affinity VEGF binding and developmental expression suggest Flk-1 as a major regulator of vasculogenesis and angiogenesis. *Cell* **72**, 835–846.

Miller, K. D., Sweeney, C. J., and Sledge, G. W., Jr. (2005). Can tumor angiogenesis be inhibited without resistance? *EXS* **94**, 95–112.

Miller, C. T., Lin, L., Casper, A. M., et al. (2006). Genomic amplification of MET with boundaries within fragile site FRA7G and upregulation of MET pathways in esophageal adenocarcinoma. *Oncogene* **25**, 409–418.

Miller, K., Wang, M., Gralow, J., et al. (2007). Paclitaxel plus bevacizumab versus paclitaxel alone for metastatic breast cancer. *N. Engl. J. Med.* **357**, 2666–2676.

Mooteri, S., Rubin, D., Leurgans, S., Jakate, S., Drab, E., and Saclarides, T. (1996). Tumor angiogenesis in primary and metastatic colorectal cancers. *Dis. Colon Rectum* **39**, 1073–1080.

Mumm, J. S., and Kopan, R. (2000). Notch signaling: from the outside in. *Dev. Biol.* **228**, 151–165.

Nacht, M., St Martin, T. B., Byrne, A., et al. (2009). Netrin-4 regulates angiogenic responses and tumor cell growth. *Exp. Cell Res.* **315**, 784–794.

Nagy, J. A., Chang, S. H., Shih, S. C., Dvorak, A. M., and Dvorak, H. F. (2010). Heterogeneity of the tumor vasculature. *Semin. Thromb. Hemost.* **36**, 321–331.

Neufeld, G., Cohen, T., Shraga, N., Lange, T., Kessler, O., and Herzog, Y. (2002). The neuropilins: multifunctional semaphorin and VEGF receptors that modulate axon guidance and angiogenesis. *Trends Cardiovasc. Med.* **12**, 13–19.

Nico, B., Benagiano, V., Mangieri, D., Maruotti, N., Vacca, A., and Ribatti, D. (2008). Evaluation of microvascular density in tumors: pro and contra. *Histol. Histopathol.* **23**, 601–607.

Nissen, L. J., Cao, R., Hedlund, E. M., et al. (2007). Angiogenic factors FGF2 and PDGF-BB synergistically promote murine tumor neovascularization and metastasis. *J. Clin. Invest.* **117**, 2766–2777.

Noguera-Troise, I., Daly, C., Papadopoulos, N. J., et al. (2007). Blockade of Dll4 inhibits tumour growth by promoting non-productive angiogenesis. *Novartis Found. Symp.* **283**, 106–120. discussion 21–5, 238–41.

North, S., Moenner, M., and Bikfalvi, A. (2005). Recent developments in the regulation of the angiogenic switch by cellular stress factors in tumors. *Cancer Lett.* **218**, 1–14.

Nowak, D. G., Amin, E. M., Rennel, E. S., et al. (2010). Regulation of vascular endothelial growth factor (VEGF) splicing from pro-angiogenic to anti-angiogenic isoforms—a novel therapeutic strategy for angiogenesis. *J. Biol. Chem.* **285**, 5532–5540.

O'Keeffe, M. B., Devlin, A. H., Burns, A. J., et al. (2008). Investigation of pericytes, hypoxia, and vascularity in bladder tumors: association with clinical outcomes. *Oncol. Res.* **17**, 93–101.

Oliveira, C. J., Schindler, F., Ventura, A. M., et al. (2003). Nitric oxide and cGMP activate the Ras-MAP kinase pathway-stimulating protein tyrosine phosphorylation in rabbit aortic endothelial cells. *Free Radic. Biol. Med.* **35**, 381–396.

Olsson, A. K., Dimberg, A., Kreuger, J., and Claesson-Welsh, L. (2006). VEGF receptor signalling—in control of vascular function. *Nat. Rev. Mol. Cell Biol.* **7**, 359–371.

Ong, S. H., Guy, G. R., Hadari, Y. R., et al. (2000). FRS2 proteins recruit intracellular signaling pathways by binding to diverse targets on fibroblast growth factor and nerve growth factor receptors. *Mol. Cell. Biol.* **20**, 979–989.

Onimaru, Y., Tsukasaki, K., Murata, K., et al. (2008). Autocrine and/or paracrine growth of aggressive ATLL cells caused by HGF and c-Met. *Int. J. Oncol.* **33**, 697–703.

O'Reilly, M. S., Boehm, T., Shing, Y., et al. (1997). Endostatin: an endogenous inhibitor of angiogenesis and tumor growth. *Cell* **88**, 277–285.

Otsuka, T., Takayama, H., Sharp, R., et al. (1998). c-Met autocrine activation induces development of malignant melanoma and acquisition of the metastatic phenotype. *Cancer Res.* **58**, 5157–5167.

Oudard, S., Beuselinck, B., Decoene, J., and Albers, P. (2011). Sunitinib for the treatment of metastatic renal cell carcinoma. *Cancer Treat. Rev.* **37**, 178–184.

Paez-Ribes, M., Allen, E., Hudock, J., et al. (2009). Antiangiogenic therapy elicits malignant progression of tumors to increased local invasion and distant metastasis. *Cancer Cell* **15**, 220–231.

Passalidou, E., Trivella, M., Singh, N., et al. (2002). Vascular phenotype in angiogenic and non-angiogenic lung non-small cell carcinomas. *Br. J. Cancer* **86**, 244–249.

Patel, N. S., Li, J. L., Generali, D., Poulsom, R., Cranston, D. W., and Harris, A. L. (2005). Up-regulation of delta-like 4 ligand in human tumor vasculature and the role of basal expression in endothelial cell function. *Cancer Res.* **65**, 8690–8697.

Patel, N. S., Dobbie, M. S., Rochester, M., et al. (2006). Up-regulation of endothelial delta-like 4 expression correlates with vessel maturation in bladder cancer. *Clin. Cancer Res.* **12**, 4836–4844.

Peghini, P. L., Iwamoto, M., Raffeld, M., et al. (2002). Overexpression of epidermal growth factor and hepatocyte growth factor receptors in a proportion of gastrinomas correlates with aggressive growth and lower curability. *Clin. Cancer Res.* **8**, 2273–2285.

Peng, F., and Chen, M. (2010). Antiangiogenic therapy: a novel approach to overcome tumor hypoxia. *Chin. J. Cancer* **29**, 715–720.

Peschard, P., Ishiyama, N., Lin, T., Lipkowitz, S., and Park, M. (2004). A conserved DpYR motif in the juxtamembrane domain of the Met receptor family forms an atypical c-Cbl/Cbl-b tyrosine kinase binding domain binding site required for suppression of oncogenic activation. *J. Biol. Chem.* **279**, 29565–29571.

Pezzella, F., Pastorino, U., Tagliabue, E., et al. (1997). Non-small-cell lung carcinoma tumor growth without morphological evidence of neo-angiogenesis. *Am. J. Pathol.* **151**, 1417–1423.

Pfander, D., Swoboda, B., and Cramer, T. (2006). The role of HIF-1alpha in maintaining cartilage homeostasis and during the pathogenesis of osteoarthritis. *Arthritis Res. Ther.* **8**, 104.

Polak, P., and Hall, M. N. (2009). mTOR and the control of whole body metabolism. *Curr. Opin. Cell Biol.* **21**, 209–218.

Polyzos, A. (2008). Activity of SU11248, a multitargeted inhibitor of vascular endothelial growth factor receptor and platelet-derived growth factor receptor, in patients with metastatic renal cell carcinoma and various other solid tumors. *J. Steroid Biochem. Mol. Biol.* **108**, 261–266.

Ponzetto, C., Bardelli, A., Zhen, Z., et al. (1994). A multifunctional docking site mediates signaling and transformation by the hepatocyte growth factor/scatter factor receptor family. *Cell* **77**, 261–271.

Primo, L., Seano, G., Roca, C., et al. (2010). Increased expression of alpha6 integrin in endothelial cells unveils a proangiogenic role for basement membrane. *Cancer Res.* **70**, 5759–5769.

Qian, C. N., Min, H. Q., Liang, X. M., Zheng, S. S., and Lin, H. L. (1997). Primary study of neovasculature correlating with metastatic nasopharyngeal carcinoma using computer image analysis. *J. Cancer Res. Clin. Oncol.* **123**, 645–651.

Qian, C. N., Guo, X., Cao, B., et al. (2002). Met protein expression level correlates with survival in patients with late-stage nasopharyngeal carcinoma. *Cancer Res.* **62**, 589–596.

Qian, C. N., Berghuis, B., Tsarfaty, G., et al. (2006). Preparing the "soil": the primary tumor induces vasculature reorganization in the sentinel lymph node before the arrival of metastatic cancer cells. *Cancer Res.* **66**, 10365–10376.

Qian, C. N., Resau, J. H., and Teh, B. T. (2007). Prospects for vasculature reorganization in sentinel lymph nodes. *Cell Cycle* **6**, 514–517.

Qian, C. N., Huang, D., Wondergem, B., and Teh, B. T. (2009). Complexity of tumor vasculature in clear cell renal cell carcinoma. *Cancer* **115**, 2282–2289.

Randall, L. M., and Monk, B. J. (2010). Bevacizumab toxicities and their management in ovarian cancer. *Gynecol. Oncol.* **117**, 497–504.

Randall, L. M., and Monk, B. J. (2011). Bevacizumab toxicities and their management in ovarian cancer. *Gynecol. Oncol.* **117**, 497–504.

Ratcliffe, P. J. (2007). HIF-1 and HIF-2: working alone or together in hypoxia? *J. Clin. Invest.* **117**, 862–865.

Raymond, E., Dahan, L., Raoul, J. L., et al. (2011). Sunitinib malate for the treatment of pancreatic neuroendocrine tumors. *N. Engl. J. Med.* **364**, 501–513.

Ribatti, D., and Ponzoni, M. (2005). Antiangiogenic strategies in neuroblastoma. *Cancer Treat. Rev.* **31**, 27–34.

Ricci-Vitiani, L., Pallini, R., Biffoni, M., et al. (2010). Tumour vascularization via endothelial differentiation of glioblastoma stem-like cells. *Nature* **468**, 824–828.

Ridgway, J., Zhang, G., Wu, Y., et al. (2006). Inhibition of Dll4 signalling inhibits tumour growth by deregulating angiogenesis. *Nature* **444**, 1083–1087.

Ridnour, L. A., Isenberg, J. S., Espey, M. G., Thomas, D. D., Roberts, D. D., and Wink, D. A. (2005). Nitric oxide regulates angiogenesis through a functional switch involving thrombospondin-1. *Proc. Natl. Acad. Sci. USA* **102**, 13147–13152.

Roberts, D. D., Isenberg, J. S., Ridnour, L. A., and Wink, D. A. (2007). Nitric oxide and its gatekeeper thrombospondin-1 in tumor angiogenesis. *Clin. Cancer Res.* **13**, 795–798.

Roca, C., and Adams, R. H. (2007). Regulation of vascular morphogenesis by Notch signaling. *Genes Dev.* **21**, 2511–2524.

Sackstein, R. (1993). Physiologic migration of lymphocytes to lymph nodes following bone marrow transplantation: role in immune recovery. *Semin. Oncol.* **20**, 34–39.

Samant, R. S., and Shevde, L. A. (2011). Recent advances in anti-angiogenic therapy of cancer. *Oncotarget* **2**, 122–134.

Sanchez-Solana, B., Nueda, M. L., Ruvira, M. D., et al. (2011). The EGF-like proteins DLK1 and DLK2 function as inhibitory non-canonical ligands of NOTCH1 receptor that modulate each other's activities. *Biochim. Biophys. Acta* **1813**, 1153–1164.

Sandler, A., Gray, R., Perry, M. C., et al. (2006). Paclitaxel-carboplatin alone or with bevacizumab for non-small-cell lung cancer. *N. Engl. J. Med.* **355**, 2542–2550.

Sarbassov, D. D., Ali, S. M., Sengupta, S., et al. (2006). Prolonged rapamycin treatment inhibits mTORC2 assembly and Akt/PKB. *Mol. Cell* **22**, 159–168.

Sasaki, T., Hohenester, E., and Timpl, R. (2002). Structure and function of collagen-derived endostatin inhibitors of angiogenesis. *IUBMB Life* **53**, 77–84.

Schmedtje, J. F., Jr., Ji, Y. S., Liu, W. L., DuBois, R. N., and Runge, M. S. (1997). Hypoxia induces cyclooxygenase-2 via the NF-kappaB p65 transcription factor in human vascular endothelial cells. *J. Biol. Chem.* **272**, 601–608.

Schmidt, D., Textor, B., Pein, O. T., et al. (2007). Critical role for NF-kappaB-induced JunB in VEGF regulation and tumor angiogenesis. *EMBO J.* **26**, 710–719.

Schneider, B. P., and Sledge, G. W., Jr. (2007). Drug insight: VEGF as a therapeutic target for breast cancer. *Nat. Clin. Pract. Oncol.* **4**, 181–189.

Schraufstatter, I. U., Chung, J., and Burger, M. (2001). IL-8 activates endothelial cell CXCR1 and CXCR2 through Rho and Rac signaling pathways. *Am. J. Physiol. Lung Cell. Mol. Physiol.* **280**, L1094–L1103.

Schutz, F. A., Jardim, D. L., Je, Y., and Choueiri, T. K. (2011). Haematologic toxicities associated with the addition of bevacizumab in cancer patients. *Eur. J. Cancer* **47**, 1161–1174.

Seifert, R. A., Hart, C. E., Phillips, P. E., et al. (1989). Two different subunits associate to create isoform-specific platelet-derived growth factor receptors. *J. Biol. Chem.* **264**, 8771–8778.

Semenza, G. L. (2001). HIF-1, O(2), and the 3 PHDs: how animal cells signal hypoxia to the nucleus. *Cell* **107**, 1–3.

Sharma, N., Seftor, R. E., Seftor, E. A., et al. (2002). Prostatic tumor cell plasticity involves cooperative interactions of distinct phenotypic subpopulations: role in vasculogenic mimicry. *Prostate* **50**, 189–201.

Sheth, P. R., Hays, J. L., Elferink, L. A., and Watowich, S. J. (2008). Biochemical basis for the functional switch that regulates hepatocyte growth factor receptor tyrosine kinase activation. *Biochemistry* **47**, 4028–4038.

Shimizu, T., Hino, K., Tauchi, K., Ansai, Y., and Tsukada, K. (2000). Predication of axillary lymph node metastasis by intravenous digital subtraction angiography in breast cancer, its correlation with microvascular density. *Breast Cancer Res. Treat.* **61**, 261–269.

Shimoyama, S., and Kaminishi, M. (2000). Increased angiogenin expression in gastric cancer correlated with cancer progression. *J. Cancer Res. Clin. Oncol.* **126**, 468–474.

Shirakawa, K., Wakasugi, H., Heike, Y., et al. (2002). Vasculogenic mimicry and pseudo-comedo formation in breast cancer. *Int. J. Cancer* **99**, 821–828.

Shojaei, F., and Ferrara, N. (2007). Antiangiogenic therapy for cancer: an update. *Cancer J.* **13**, 345–348.

Shojaei, F., and Ferrara, N. (2008). Role of the microenvironment in tumor growth and in refractoriness/resistance to anti-angiogenic therapies. *Drug Resist. Updat.* **11**, 219–230.

Shojaei, F., Zhong, C., Wu, X., Yu, L., and Ferrara, N. (2008). Role of myeloid cells in tumor angiogenesis and growth. *Trends Cell Biol.* **18**, 372–378.

Sood, A. K., Fletcher, M. S., Zahn, C. M., et al. (2002). The clinical significance of tumor cell-lined vasculature in ovarian carcinoma: implications for anti-vasculogenic therapy. *Cancer Biol. Ther.* **1**, 661–664.

Squadrito, M. L., and De Palma, M. (2011). Macrophage regulation of tumor angiogenesis: implications for cancer therapy. *Mol. Aspects Med.* **32**, 123–145.

Stellrecht, C. M., Phillip, C. J., Cervantes-Gomez, F., and Gandhi, V. (2007). Multiple myeloma cell killing by depletion of the MET receptor tyrosine kinase. *Cancer Res.* **67**, 9913–9920.

Suchting, S., Heal, P., Tahtis, K., Stewart, L. M., and Bicknell, R. (2005). Soluble Robo4 receptor inhibits in vivo angiogenesis and endothelial cell migration. *FASEB J.* **19**, 121–123.

Suchting, S., Freitas, C., le Noble, F., et al. (2007). The Notch ligand Delta-like 4 negatively regulates endothelial tip cell formation and vessel branching. *Proc. Natl. Acad. Sci. USA* **104**, 3225–3230.

Summers, J., Cohen, M. H., Keegan, P., and Pazdur, R. (2011). FDA drug approval summary: bevacizumab plus interferon for advanced renal cell carcinoma. *Oncologist* **15**, 104–111.

Tammela, T., Enholm, B., Alitalo, K., and Paavonen, K. (2005). The biology of vascular endothelial growth factors. *Cardiovasc. Res.* **65**, 550–563.

Tammela, T., Zarkada, G., Wallgard, E., et al. (2008). Blocking VEGFR-3 suppresses angiogenic sprouting and vascular network formation. *Nature* **454**, 656–660.

Tang, N., Wang, L., Esko, J., et al. (2004). Loss of HIF-1alpha in endothelial cells disrupts a hypoxia-driven VEGF autocrine loop necessary for tumorigenesis. *Cancer Cell* **6**, 485–495.

Taylor, C. T., Furuta, G. T., Synnestvedt, K., and Colgan, S. P. (2000). Phosphorylation-dependent targeting of cAMP response element binding protein to the ubiquitin/proteasome pathway in hypoxia. *Proc. Natl. Acad. Sci. USA* **97**, 12091–12096.

Taylor, K. L., Henderson, A. M., and Hughes, C. C. (2002). Notch activation during endothelial cell network formation in vitro targets the basic HLH transcription factor HESR-1 and downregulates VEGFR-2/KDR expression. *Microvasc. Res.* **64,** 372–383.

Teng, L. S., Jin, K. T., He, K. F., Zhang, J., Wang, H. H., and Cao, J. (2010). Clinical applications of VEGF-trap (aflibercept) in cancer treatment. *J. Chin. Med. Assoc.* **73,** 449–456.

Therasse, P., Arbuck, S. G., Eisenhauer, E. A., et al. (2000). New guidelines to evaluate the response to treatment in solid tumors. European Organization for Research and Treatment of Cancer, National Cancer Institute of the United States, National Cancer Institute of Canada. *J. Natl. Cancer Inst.* **92,** 205–216.

Thomas, M., and Augustin, H. G. (2009). The role of the angiopoietins in vascular morphogenesis. *Angiogenesis* **12,** 125–137.

Thompson, W. D., Shiach, K. J., Fraser, R. A., McIntosh, L. C., and Simpson, J. G. (1987). Tumours acquire their vasculature by vessel incorporation, not vessel ingrowth. *J. Pathol.* **151,** 323–332.

Thurston, G., and Kitajewski, J. (2008). VEGF and Delta-Notch: interacting signalling pathways in tumour angiogenesis. *Br. J. Cancer* **99,** 1204–1209.

Timpl, R., Sasaki, T., Kostka, G., and Chu, M. L. (2003). Fibulins: a versatile family of extracellular matrix proteins. *Nat. Rev. Mol. Cell Biol.* **4,** 479–489.

Vadnais, J., Nault, G., Daher, Z., et al. (2002). Autocrine activation of the hepatocyte growth factor receptor/met tyrosine kinase induces tumor cell motility by regulating pseudopodial protrusion. *J. Biol. Chem.* **277,** 48342–48350.

Valenzuela, D. M., Griffiths, J. A., Rojas, J., et al. (1999). Angiopoietins 3 and 4: diverging gene counterparts in mice and humans. *Proc. Natl. Acad. Sci. USA* **96,** 1904–1909.

Van Meter, M. E., and Kim, E. S. (2011). Bevacizumab: current updates in treatment. *Curr. Opin. Oncol.* **22,** 586–591.

Vaupel, P., and Harrison, L. (2004). Tumor hypoxia: causative factors, compensatory mechanisms, and cellular response. *Oncologist* **9**(Suppl. 5), 4–9.

Vaupel, P., Kelleher, D. K., and Hockel, M. (2001). Oxygen status of malignant tumors: pathogenesis of hypoxia and significance for tumor therapy. *Semin. Oncol.* **28,** 29–35.

Verma, N., and Swain, S. M. (2011). Bevacizumab and heart failure risk in patients with breast cancer: a thorn in the side? *J. Clin. Oncol.* **29,** 603–606.

von Andrian, U. H., and M'Rini, C. (1998). In situ analysis of lymphocyte migration to lymph nodes. *Cell Adhes. Commun.* **6,** 85–96.

von Baumgarten, L. D., Brucker, D., Tirniceru, A. L., et al. (2011). Bevacizumab has differential and dose-dependent effects on glioma blood vessels and tumor cells. *Clin. Cancer Res.* **17,** 6192–6205.

Wang, R., Chadalavada, K., Wilshire, J., et al. (2010). Glioblastoma stem-like cells give rise to tumour endothelium. *Nature* **468,** 829–833.

Waugh, D. J., and Wilson, C. (2008). The interleukin-8 pathway in cancer. *Clin. Cancer Res.* **14,** 6735–6741.

Weidner, N., Folkman, J., Pozza, F., et al. (1992). Tumor angiogenesis: a new significant and independent prognostic indicator in early-stage breast carcinoma. *J. Natl. Cancer Inst.* **84,** 1875–1887.

Willett, C. G., Kozin, S. V., Duda, D. G., et al. (2006). Combined vascular endothelial growth factor-targeted therapy and radiotherapy for rectal cancer: theory and clinical practice. *Semin. Oncol.* **33,** S35–S40.

Williamson, S. K., Moon, J., Huang, C. H., et al. (2010). Phase II evaluation of sorafenib in advanced and metastatic squamous cell carcinoma of the head and neck: Southwest Oncology Group Study S0420. *J. Clin. Oncol.* **28,** 3330–3335.

Winkler, E. A., Bell, R. D., and Zlokovic, B. V. (2010). Pericyte-specific expression of PDGF beta receptor in mouse models with normal and deficient PDGF beta receptor signaling. *Mol. Neurodegener.* **5**, 32.

Wojta, J., Kaun, C., Breuss, J. M., et al. (1999). Hepatocyte growth factor increases expression of vascular endothelial growth factor and plasminogen activator inhibitor-1 in human keratinocytes and the vascular endothelial growth factor receptor flk-1 in human endothelial cells. *Lab. Invest.* **79**, 427–438.

Wullschleger, S., Loewith, R., and Hall, M. N. (2006). TOR signaling in growth and metabolism. *Cell* **124**, 471–484.

Xie, H., Zou, L., Zhu, J., and Yang, Y. (2011). Effects of netrin-1 and netrin-1 knockdown on human umbilical vein endothelial cells and angiogenesis of rat placenta. *Placenta* **32**, 546–553.

Yang, J. C., Haworth, L., Sherry, R. M., et al. (2003). A randomized trial of bevacizumab, an anti-vascular endothelial growth factor antibody, for metastatic renal cancer. *N. Engl. J. Med.* **349**, 427–434.

Yang, X., Qiao, D., Meyer, K., and Friedl, A. (2009). Signal transducers and activators of transcription mediate fibroblast growth factor-induced vascular endothelial morphogenesis. *Cancer Res.* **69**, 1668–1677.

Yao, X., Qian, C. N., Zhang, Z. F., et al. (2007). Two distinct types of blood vessels in clear cell renal cell carcinoma have contrasting prognostic implications. *Clin. Cancer Res.* **13**, 161–169.

Yeh, J. C., Hiraoka, N., Petryniak, B., et al. (2001). Novel sulfated lymphocyte homing receptors and their control by a Core1 extension beta 1,3-N-acetylglucosaminyltransferase. *Cell* **105**, 957–969.

Yi, S., and Tsao, M. S. (2000). Activation of hepatocyte growth factor-met autocrine loop enhances tumorigenicity in a human lung adenocarcinoma cell line. *Neoplasia* **2**, 226–234.

Yi, S., Chen, J. R., Viallet, J., Schwall, R. H., Nakamura, T., and Tsao, M. S. (1998). Paracrine effects of hepatocyte growth factor/scatter factor on non-small-cell lung carcinoma cell lines. *Br. J. Cancer* **77**, 2162–2170.

You, W. K., and McDonald, D. M. (2008). The hepatocyte growth factor/c-Met signaling pathway as a therapeutic target to inhibit angiogenesis. *BMB Rep.* **41**, 833–839.

You, W. K., Sennino, B., Williamson, C. W., et al. (2011). VEGF and c-Met blockade amplify angiogenesis inhibition in pancreatic islet cancer. *Cancer Res.* **71**, 4758–4768.

Yu, J., deMuinck, E. D., Zhuang, Z., et al. (2005). Endothelial nitric oxide synthase is critical for ischemic remodeling, mural cell recruitment, and blood flow reserve. *Proc. Natl. Acad. Sci. USA* **102**, 10999–11004.

Yurchenco, P. D., and Cheng, Y. S. (1993). Self-assembly and calcium-binding sites in laminin. A three-arm interaction model. *J. Biol. Chem.* **268**, 17286–17299.

Zamora, D. O., Riviere, M., Choi, D., et al. (2007). Proteomic profiling of human retinal and choroidal endothelial cells reveals molecular heterogeneity related to tissue of origin. *Mol. Vis.* **13**, 2058–2065.

Zaragoza, C., Soria, E., Lopez, E., et al. (2002). Activation of the mitogen activated protein kinase extracellular signal-regulated kinase 1 and 2 by the nitric oxide-cGMP-cGMP-dependent protein kinase axis regulates the expression of matrix metalloproteinase 13 in vascular endothelial cells. *Mol. Pharmacol.* **62**, 927–935.

Zhang, J., Burridge, K. A., and Friedman, M. H. (2008). In vivo differences between endothelial transcriptional profiles of coronary and iliac arteries revealed by microarray analysis. *Am. J. Physiol. Heart Circ. Physiol.* **295**, H1556–H1561.

Zhao, Z. S., Zhou, J. L., Yao, G. Y., Ru, G. Q., Ma, J., and Ruan, J. (2005). Correlative studies on bFGF mRNA and MMP-9 mRNA expressions with microvascular density, progression, and prognosis of gastric carcinomas. *World J. Gastroenterol.* **11**, 3227–3233.

Zhao, Y. Y., Xue, C., Jiang, W., *et al.* (2012). Predictive value of intratumoral microvascular density in patients with advanced non-small cell lung cancer receiving chemotherapy plus bevacizumab. *J. Thorac. Oncol.* **7**, 71–75.

Zhu, X. D., Zhang, J. B., Zhuang, P. Y., *et al.* (2008). High expression of macrophage colony-stimulating factor in peritumoral liver tissue is associated with poor survival after curative resection of hepatocellular carcinoma. *J. Clin. Oncol.* **26**, 2707–2716.

Zuniga, R. M., Torcuator, R., Jain, R., *et al.* (2010). Rebound tumour progression after the cessation of bevacizumab therapy in patients with recurrent high-grade glioma. *J. Neurooncol* **99**, 237–242.

Molecular Logic Underlying Chromosomal Translocations, Random or Non-Random?

Chunru Lin,[*,†] Liuqing Yang,[*,†] and Michael G. Rosenfeld[*,†]

[*]*Howard Hughes Medical Institute, University of California, San Diego, School of Medicine, La Jolla, California, USA*
[†]*Department of Medicine, Division of Endocrinology and Metabolism, University of California, San Diego, School of Medicine, La Jolla, California, USA*

I. Introduction
   A. The Types of Chromosomal Translocations
   B. Relationship Between Chromosomal Translocations and Cancer
   C. Induced Chromosomal Translocations in Cell-Based Models
II. Mechanisms of Chromosomal Translocations
   A. Generation of DSBs
   B. Transcription and Genome Instability
   C. Transposable Elements and Endonuclease
   D. Spatial Proximity
   E. Ligation
III. The Role of Epigenetics in Chromosomal Translocations
   A. DNA Methylation and Genome Instability
   B. Chromatin Modifications and Genome Instability
IV. Conclusion
   Acknowledgments
   References

Chromosomal translocations serve as essential diagnostic markers and therapeutic targets for leukemia, lymphoma, and many types of solid tumors. Understanding the mechanisms of chromosomal translocation generation has remained a central biological question for decades. Rather than representing a random event, recent studies indicate that chromosomal translocation is a non-random event in a spatially regulated, site-specific, and signal-driven manner, reflecting actions involved in transcriptional activation, epigenetic regulation, three-dimensional nuclear architecture, and DNA damage-repair. In this review, we will focus on the progression toward understanding the molecular logic underlying chromosomal translocation events and implications of new strategies for preventing chromosomal translocations. © 2012 Elsevier Inc.

I. INTRODUCTION

As a result of the unfaithful ligation of broken chromatin, the exchanged or deleted genetic materials usually associate with chimeric proteins or altered protein expression status. Various types of human cancer harbor specific types of chromosomal translocations, which play significant roles in promoting cancer progression (Kuppers and Dalla-Favera, 2001; Mitelman et al., 2007; Rowley, 2001; Shaffer and Pandolfi, 2006; Sjoblom et al., 2006). Some chromosomal translocations result in augmented or activated oncogenes, which provide growth advantage to the neoplastic cells over normal cells and are amplified under growth selection. Because of the tissue specificity of a given chromosomal translocation, targeting a particular chromosomal translocation serves as a critical diagnostic marker or therapeutic target.

Chromosomal translocation is defined as the interchange of genetic materials between two non-homologous chromosomes (Rowley, 1973). In general, the chromosomal translocations can be classified as reciprocal (non-Robertsonian) and nonreciprocal (Robertsonian) translocations. Reciprocal (non-Robertsonian) translocation is usually the result of an exchange of segments between two non-homologous chromosomes (Fig. 1A). This type of chromosomal translocation can be balanced (the exchanged genetic material is either extra or missing and fully functional) or unbalanced (the

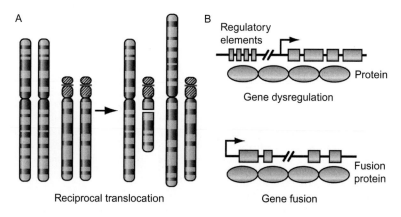

Fig. 1 Consequence of chromosomal translocations. The figure is adapted and modified from Nat. Med. 2003;9(4). (A) Reciprocal translocation results from the exchange of segments between two non-homologous chromosomes. (B) Recombination between a gene coding region and the regulatory elements of another gene leads to upregulation of the former gene. The chromosomal translocations occur between coding regions of two genes and can result in the formation of a chimeric gene, which is further translated to a chimeric protein.

exchanged chromosomes are unequal and lead to extra or missing genes). The carriers of reciprocal translocations are usually normal with no symptoms with an incidence of about 1 in 500 (Mackie Ogilvie and Scriven, 2002); however, the gametes of carriers have a higher risk of unbalanced chromosomal translocation and genetic abnormalities (Oliver-Bonet et al., 2002).

Nonreciprocal translocations (Robertsonian) occur when the short arms of two chromosomes fuse together near the centromeric region resulting in the loss of short arms and a reduction of chromosomal number (Hansmann et al., 1977). On average, every thousand newborns harbor one Robertsonian translocation. A balanced Robertsonian translocation carrier could be healthy without discovering the translocation for life span, while a newborn carrying an unbalanced Robertsonian translocation may suffer from genetic disorders. For example, a newborn harboring three copies of the long arm of chromosome 21, instead of two, in the genome translocation causes Down's syndrome (Epstein, 2006; Nelson and Gibbs, 2004).

Depending on the site, there are two types of chromosomal translocations that result in the deregulation of particular genes or formation of novel fusion genes (Fig. 1B). Recombination between a gene coding region and the transcriptionally active promoter or enhancer region of another gene leads to upregulation of the former gene. Examples of such events include *TMPRSS2-ERG* and *TMPRSS2-ETV1* translocations where *ERG* or *ETV1* are the target genes. In such cases, *ERG* or *ETV1* genes are dramatically upregulated due to their repositioning after *TMPRSS2* transcription start site, which are actively transcribed in prostate cancer (Kumar-Sinha et al., 2008; Tomlins et al., 2007, 2005). Another example is the chromatin rearrangement of the *MYC* gene with the promoter elements of *IgH* switch regions in B-cell tumors, which is a characteristic feature of diffuse large B-cell lymphoma (Jankovic et al., 2007; Kuppers and Dalla-Favera, 2001; Leder et al., 1983; Nussenzweig and Nussenzweig, 2010).

The chromosomal translocations occur between the coding regions of two genes and can result in the formation of a chimeric gene, which is further translated to a chimeric protein. The fused protein may gain constitutive kinase activity, which is permissive for cancer development or progression. The in frame fusion between ABL gene on chromosome 9 and BCR (breakpoint cluster region) gene on chromsome 22 encodes a constitutively activated tyrosine kinase, known as the chimeric protein BCR-ABL. This translocation has been documented to drive chronic myelogenous leukemia (CML) (Kuppers, 2005). Inhibitors specifically targeting BCR–ABL such as Gleevec has been applied in clinical treatment and proved to increase the survival rate of CML patients (Deininger et al., 2005).

Another type of chromosomal translocation involves recombination between coding regions of one gene with a noncoding region, contributed by the second locus, which generates truncated forms of coding proteins. One example of such a truncation is a product of *MYC* mRNA in human T cell leukemia, which is suggested to increase *MYC* mRNA stability (Aghib *et al.*, 1990). In lymphocytes, genomic rearrangements also occur in normal processes involved in V(D)J recombination and class-switch recombination (CSR), which are crucial for the antigen receptor generation (Gostissa *et al.*, 2011; Jung and Alt, 2004; Stavnezer *et al.*, 2008).

Chromosomal abnormalities taking place during the formation of reproductive cells can be random. Down syndrome, characterized by trisomy 21, usually results from errors in cell division. It has been suggested that chromosomal translocations are also random events, with any given chromosomal segment inserted into any other chromosomal regions. As an example, translocations caused by transposable element transposition are considered to be random. In cancer, a growth advantage is believed to be the selective pressure for chromosomal translocation selection, which facilitates inheritance of enhanced proliferation or antiapoptotic events facilitating passage of chromosomal translocations to the next generation. However, recent studies suggest that initiation of genomic rearrangements has cell- and tissue-type specificity and are affected by cell type, cell stage, and genomic context (Lin *et al.*, 2009; Mani and Chinnaiyan, 2010; Nussenzweig and Nussenzweig, 2010). Tissue-specific signaling pathways and cellular stress can trigger the formation of a particular type of chromosomal translocation in a combinational manner. Indeed, the fusion sites of chromosomal translocations exhibit sequence preference and site-specificity. Therefore, increasing evidence suggests that the generation of chromosomal translocation might be a tightly regulated non-random event.

A. The Types of Chromosomal Translocations

In leukemia and lymphoma, chromosomal translocations are hallmarks for diagnosis and therapeutic purposes. Many such chromosomal translocations and their associated lymphoid malignancies have been widely reviewed (Korsmeyer, 1992; Nussenzweig and Nussenzweig, 2010; Zhang *et al.*, 2010). Therefore, in this review, we will not summarize and list the types of chromosomal translocation in myeloid malignancy.

Although most cancer types originate from epithelial cells, chromosomal translocations with clinical relevance continue to be identified. In prostate cancer, deletions including PTEN, 10q; RB1, 13q; or NKX3.1, 8p are characteristic (Asatiani *et al.*, 2005; Hollander *et al.*, 2011; Li *et al.*, 1998). Amplification of androgen receptor (AR) gene (Xq11.2-q12) can also be observed in recurrent prostate cancer (Koivisto *et al.*, 1997). More recently, a group of chromosomal translocation leading to fusion of AR target gene *TMPRSS2* and *ETS* family members have been recognized in prostate cancer (Tomlins et al., 2005). Depending on the patient group and method of biopsy, the occurrence of *TMPRESS2–ERG* gene fusion occupy 27–79% of prostate cancer patients (Barwick et al., 2010; Magi-Galluzzi et al., 2011; Mehra et al., 2007; Tomlins et al., 2005; Turner and Watson, 2008; Yoshimoto et al., 2008) (Fig. 2). In addition, numerous ETS family genes can fuse to other AR target genes, including *SLC45A3* (1q32.1)-*ETV5* (3q28), *KLK2* (19q13.41)-*ETV4* (17q21), *C15orf21* (15q21.1)-*ETV1* (7p21.3), *SLC45A3* (1q32.1)-*ETV1* (7p21.3), *SLC45A3* (1q32.1)-*BRAF* (7q34), and *ESRP1* (8q22.1)-*RAF1* (3p25.1) (Maher et al., 2009; Palanisamy et al., 2010; Tomlins et al., 2007), or housekeeping gene (Attard et al., 2008).

For breast cancer, the incidence and consequence of genomic rearrangement is less well studied. Carriers of constitutive balanced translocation between the long arm of chromosome 11 and 22 t (11q:22q) are reported to exhibit increased risk of breast cancer (Lindblom et al., 1994; Wieland et al., 2006). However, the conclusions are controversial (Carter et al., 2010). A recurrent translocation between *ETV6* (TEL) and *NTRK3* (tyrosine kinase receptor for neurotrophin-3), t (12;15) was reported in secretory breast carcinoma (Tognon et al., 2002). This type of chromosomal translocation has been detected in up to 92% of secretory breast cancer samples (Schmitz et al., 2004). The fused gene encodes a chimeric protein harboring both oligomerization domain of ETV6 and tyrosine kinase domain of NTRK3 with potent transforming activity (Wai et al., 2000). Constitutive expression of this fusion gene is suggested to facilitate the initiation of breast cancer through activation of MAP kinase, PI3K, and AP-1 complex (Li et al., 2007; Tognon et al., 2002). Furthermore, an *EML4–ALK* fusion gene was identified in 6.7% of non-small-cell lung cancer cases (Soda et al., 2007).

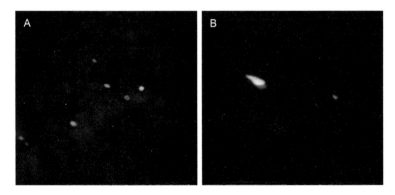

Fig. 2 Detection of chromosomal translocation by FISH technology. (A, B) K562 cells or tissue sample from prostate cancer patient were subjected to FISH technology to detect *BCR–ABL* fusion (A) and *TMPRSS2–ERG* fusion (B). The fused chromosomes exhibit overlapping signals. (See Page 7 in Color Section at the back of the book.)

Although chromosomal abnormalities in epithelial-derived tumors has been reported as early as the 1960s (Sandberg and Yamada, 1966), the discoveries of gene fusions in solid tumors was not evident. One of the reasons is that in solid tumor tissue samples, there is a deficiency of well-spread metaphases and this poor quality of chromosome separation limits the use of classic cytogenetics to identify chromosomal translocations (Varella-Garcia, 2003). With advanced cytogenetic technologies and pair-end genome-wide sequencing, it is likely that more chromosomal translocations will be identified in epithelial cancers.

B. Relationship Between Chromosomal Translocations and Cancer

As early as 1914, genome instability and chromosomal abnormalities were proposed as causal factors for cancer initiation (Rabbitts, 1994; Stern, 1950). The discovery of the Philadelphia chromosome in CML patients provides the first solid evidence to support the view that chromosome abnormalities play important roles in cancer development (Nowell, 2007). The advent of chromosome banding permits detection of many specific chromosome aberrations in various tumor types (Kearney and Horsley, 2005; Speicher and Carter, 2005). Understanding the molecular genetic consequences of chromatin abnormalities further support the role of chromosome abnormalities in tumor evolution.

Thus far, 337 gene fusions have been identified in benign and malignant neoplastic disorders (Mitelman Database of Chromosome Aberrations and Gene Fusions in Cancer, http://cgap.nci.nih.gov/Chromosomes/Mitelman). Cytogenetic characterization of these translocations has been the most important technique in tumor diagnosis and prognosis. Recurrent balanced rearrangements have been detected in most tumor types (Mitelman Database of Chromosome Aberrations and Gene Fusions in Cancer, http://cgap.nci.nih.gov/Chromosomes/Mitelman). Many of these balanced translocations are associated with specific tissue types, defined clinical features, and distinctive global gene expression profiles (Iljin *et al.*, 2006; Mani *et al.*, 2011; Pierotti *et al.*, 1992; Soda *et al.*, 2007). The information revealed from chromosomal translocation has become an important tool in the management of cancer patients, helping in establishing a correct diagnosis, selecting the appropriate treatment, and predicting outcome.

Spectral karyotyping (SKY) and fluorescence *in situ* hybridization (FISH) technologies are used to screen patient samples for translocations (Rao *et al.*, 1998; Schrock *et al.*, 1996; van der Burg *et al.*, 2004; Yoshimoto *et al.*, 2006). These methods provide sensitive identification of an individual's chromosomal abnormalities. SKY technology distinguish each chromosomes by a mixture of DNA probes with fluorescent that create distinct fluorescent signal for each chromosomes. When hybridized to metaphase chromosome, uniformly colored probes are hybridized to normal chromosomes, while multicolor probes recognize translocation chromosomes.

FISH technology is a powerful tool to detect and localize the presence of specific chromosome translocations using fluorescent probes targeting defined chromosome regions. By using a split-apart strategy, two adjacent FISH probes with overlapping signals under a fluorescence microscope exhibit separate signals when there is a chromosome break between the two FISH probe targeted sites (Tomlins et al., 2005). For example, two FISH probes targeting the ERG gene loci will break apart when TMPRSS2–ERG translocation occurs in prostate cancer tissue. In contrast, when two fragments of different chromosomes join together, the two FISH probes targeting the two chromosomes display overlapping signals indicating the chromosomal translocations.

Most recently, paired-end sequencing technology has been applied to patient tissue samples to identify genomic rearrangement and translocation on a genome-wide scale. The DNA fragments extracted from tumor samples are converted to a library and subjected to high-resolution sequencing, for example, the Illumina Genome Analyzer or SOLiD system. The sequencing readouts are mapped back to the human genome sequence for identification of genome-wide survey of gene fusions and genomic rearrangements in a given tumor sample. By applying this method, new chromosomal translocations and other types of genetic disorders (Asmann et al., 2011; Palanisamy et al., 2010; Slade et al., 2010) have been identified to reveal fusion sites.

In many cases, the rearrangement of genetic material has been linked to cancer initiation and progression. Chromosomal translocation can lead to genomic region deletions or duplications in the case of unbalanced chromosomal translocations. Deletion of chromosomal region 9p21 is detected in many types of cancers (Cairns et al., 1995), in which the cyclin-dependent kinase inhibitor 2A gene is inactivated due to deletion (Williamson et al., 1995), facilitating cancer initiation, progression, and metastasis (Liggett and Sidransky, 1998). Gene duplications or amplifications also contribute to cancer. For example, amplification of the ERBB2 gene leads to overexpression of human epidermal growth factor receptor 2 in certain advanced breast cancer types (Pollack et al., 1999), encoding a receptor tyrosine kinase involved in cell growth and differentiation.

The chimeric protein encoded from a gene fusion could produce an oncogene, for example, the Philadelphia chromosome that was initially identified in patients with CML. Molecular analyses revealed that the translocation fused the coding sequence of the BCR (breakpoint cluster region) gene on chromosome 22 with the coding sequence of the ABL gene on chromosome 9. The fusion protein encoded by BCR-ABL translocation acts as a tyrosine kinase constitutively activating signaling pathways involved in cell growth and proliferation (Ren, 2005; Wong and Witte, 2004). Understanding of this particular breakpoint and sequence information has led to definitive study of the crystal structure of BCR-ABL protein. As consequence, scientific researchers developed specific drugs that inhibit this protein's activity, which has been successfully applied for CML treatment. (Helgason *et al.*, 2011; Rosti *et al.*, 2011).

Another consequence of chromosomal translocation is to cause dysregulation, especially upregulation of transcription factors by placing the coding region of transcription factors under the control of the regulatory element of a transcriptionally active gene. The first example of this type of translocation was identified in patients with Burkitt's lymphoma, who harbor a translocation involving chromosomes 8 and 14. This particular translocation relocates the *MYC* proto-oncogene from chromosome 8 to chromosome 14 (Taub *et al.*, 1982). Under the control of the promoter of immunoglobin heavy chain gene (*IGH*), the translocation causes high levels of MYC overexpression in lymphoid cells, where the *IGH* gene promoter is normally active. When the coding region of *ETS* is fused to *TMPRSS2* regulatory elements, the upregulation of the Ets transcription factor promotes expression of genes involved in cell metabolism, cell adhesion, and cell invasion (Kumar-Sinha *et al.*, 2008).

The database of catalogued chromosomal translocations continues to grow as investigators uncover new links between translocations and cancer. Molecular mechanisms underlying chromosomal translocations continue to be investigated to uncover new therapeutic strategies for cancer treatment. The development of small molecules for kinase inhibition for controlling the dysregulated oncogenes by small RNA interference, and monoclonal antibody blockage will offer new treatment options for cancer patients.

C. Induced Chromosomal Translocations in Cell-Based Models

It is important to establish a cell culture model to study the mechanism of chromosomal translocation. Recently, chromosomal translocations have been induced from various cancer and even normal cell lines (Bastus *et al.*, 2010; Lin *et al.*, 2009; Mani *et al.*, 2009; Weinstock *et al.*, 2008). Understanding the molecular mechanisms that underlie tumor translocations requires a model that mimics *in vivo* events prior to any selective advantages of cell growth. Based on the critical roles of AR in prostate development and tumor progression, and the observation that genotoxic stress is able to rapidly induce chromosomal translocations (Caudill *et al.*, 2005; Deininger *et al.*, 1998), an androgen agonist (such as dihydrotestosterone, DHT) and genotoxic stress, either alone or in combination, might induce chromosomal translocations of *TMPRSS2–ERG* and *TMPRSS2–ETV1* in prostate cancer cell line LNCaP or normal prostate epithelial cells.

Treatment of LNCaP cells with both DHT and irradiation (IR) resulted in an enhanced appearance of both *TMPRSS2–ERG* and *TMPRSS2–ETV1* fusion transcripts in 24 h without growth selection. Other modalities that cause genotoxic stress, including Etoposide and Doxorubicin, when combined with the DHT treatment, caused similar effects with respect to inducing tumor translocations (Lin *et al.*, 2009). After combined treatment of DHT with IR (1 or 3 Gy) *TMPRSS2–ERG* fusion transcripts were detected in single cell clones (Mani *et al.*, 2009). The expression of *TMPRSS2–ERG* transcripts in positive LNCaP clones was similar to that in VCaP prostate cancer cells (Tomlins *et al.*, 2005), which endogenously harbor this particular type of gene fusion.

Because prostate cancer usually develops years after the male reproductive system has matured, long-term androgen exposure may be required for prostate cancer development. Continuous treatment of immortalized non-malignant prostate epithelial cell lines, PNT1a and PNT2, with high doses (300 and 3000 nM) of the AR ligand, DHT, lead to accumulation of *TMPRSS2–ERG* translocations after 5 months of treatment (Bastus *et al.*, 2010). The incidence of gene fusion was DHT dose-dependent. These observations suggest that androgen-induced fusions represent an event that can occur prior to prostate cancer development, rather than as a secondary genetic change after the cancer has developed.

The induced chromosomal translocations are cell-type specific, as demonstrated by the failure of induction of detectable *TMPRSS2–ERG* translocation in human B-cell lymphoma (Lin *et al.*, 2009). The effort to identify all potential chromosomal translocation isoforms when combining treatment of DHT and IR revealed that the frequency of each induced *TMPRSS2–ERG* fusion isoform in LNCaP cells (reported by Lin *et al.*, 2009) was very similar to the *TMPRSS2–ERG* fusion types identified in prostate cancer tissues (Wang *et al.*, 2006). The AR signaling pathways in prostate cancer may explain the cell-type and tissue-type specificity of chromosomal translocations. The binding of AR to target sites may facilitate local single-stranded chromatin formation, the spatial proximity of potential translocation partners, and also recruitment of the enzymatic machinery to cleavage site-specific chromatin regions, which will be discussed in detail below.

In B-cell lymphoma, random cutting of chromosomes by insertion of I-Sec sites throughout the genome resulted in chromosomal translocation during IgH class switching. Surprisingly, a large number of induced chromosomal translocation included the *MYC* locus, indicating the cell-type specificity in chromosomal translocation initiation. Furthermore, the translocation partners of the *MYC* locus also exhibit a non-random selection (Chiarle *et al.*, 2011; Klein *et al.*, 2011).

Therefore, instead of the long-prevailing view that chromosomal translocation occurs randomly and that the inherited chromosomal translocations are selected by growth pressure, the initiation of chromosomal translocation is programmed by cell state, signaling pathways, and other internal factors.

II. MECHANISMS OF CHROMOSOMAL TRANSLOCATIONS

How are chromosomal translocations induced? Theoretically, improperly ligated broken chromatin ends could be jointed to form a chromosomal translocation. The initiation of chromosomal translocations has been presumed to be random events occurring under selection of growth pressure. However, recent publications investigating numerous factors that affect the spatial proximity of chromatins, DNA double-stranded breaks (DSBs) and the translocation breakpoints, suggest the possibility that chromosomal translocation does not occur randomly. In this review, we will focus on recent insights into the mechanisms of chromosomal translocation formation.

Although spatial proximity and DSBs are both critical for chromosomal translocation to occur, whether physical association happens first or DSBs appear first remains to be defined. In the "contact-first" model, physical association of translocation partners occurs before DSB formation under the assumption that broken chromatin is limited to a regional area. In contrast, in a "breakage-first" model, translocations occur between broken chromosome ends that undergo large-scale "motion" within the nucleus to achieve proximity between translocation partners. Using the *TMPRESS2–ERG* chromosomal translocation model system, recent research supports both of these models. AR is suggested to promote the spatial proximity between *TMPRESS2* and *ERG* or *ETV* gene loci (Lin et al., 2009; Mani et al., 2009). The effect of androgen agonists also facilitates DSBs to occur near translocation sites. The broken chromatin ends undergo large regions of movement (Haffner et al., 2010), suggesting that free DSBs are able to move freely in the nucleus. Therefore, it is possible that both spatial proximity and site-specific DSBs occur at the same time, are required for chromosomal translocation formation, and are influenced by the same factors.

A. Generation of DSBs

DSBs are indispensable for initiation of chromosomal translocations. Many factors can induce DSBs, including cellular stress and genotoxic stress. Endogenous DSBs also contribute to *de novo* chromosomal translocation formation during transcription and DNA replication.

1. GENOTOXIC STRESS

Physical or chemical agents that potentially alter the nucleotide structure of DNA sequences are considered to exert as genotoxic stresses. To guard the integrity of the genome and the faithful transmission of the genetic information to the next generation, DNA damage caused by genotoxic stress needs to be repaired. Genotoxic stress can be triggered from exposure to toxic agents, including highly active molecules produced during metabolism, the sun's ultraviolet rays, ionizing radiation from environment, and chemical ingredients in food. To repair the DNA damaged by genotoxic stress, mutated or cross-linked DNA nucleotides are removed and a new, complementary sequence to the DNA template is synthesized on damaged sites (Lieber, 2010). Failure to repair genotoxic stress-induced DSBs causes mutations or chromatin abnormalities that accumulate during aging resulting in a higher risk of neoplasm formation (Medema and Macurek, 2011).

Although the prevailing view is that genotoxic stress randomly generates DNA breaks throughout the genome, recent research works suggest that the generation of DSBs is extensive, non-random, and site-specific (Chiarle et al., 2011; Klein et al., 2011). DNA fragile sites are prone to genotoxic stress-induced DNA damage. These sites, including purine and pyrimidine repetitive elements, breakpoint cluster regions, or scaffold/matrix attachment regions often show increased susceptibility to genotoxic stress (Aplan, 2006; Boehm et al., 1989). DNA containing purine and pyrimidine residues may form a type of left-handed helical structure, called Z-DNA, which is more sensitive to DNA damage and more likely to be involved in chromosomal translocations (Nambiar and Raghavan, 2011).

Genotoxic stress-activated downstream signaling pathways may contribute to site-specific DNA breaks via upregulation of genotoxic stress-induced growth arrest and DNA damage-inducible protein 45 (GADD45) and activation-induced cytidine deaminase (AID) complex (Lin et al., 2009). In prostate cancer, this complex is recruited to specific loci by the ligand-bound AR. The binding of AID to single-stranded DNA and the deaminase activity of AID contribute to sensitizing the AR binding sites. The upregulation of GADD45 and AID complexes also facilitates site-specific DNA breaks by demethylation of the promoter of Long Interspersed Element (LINE) repeats. The removal of DNA methylation-mediated transcriptional repression derepresses LINE1 transcript-encoded endonuclease upon genotoxic stress (Ohka et al., 2011). The encoded endonuclease performs preferential cleavage, leading to further DNA damage.

Another mechanism causing site-specific DNA damage is mediated by topoisomerase II (TOPII). In prostate cancer, activated AR recruits TOPII to *TMPRSS2* and *ERG* chromosomal translocation sites, triggering TOPII-mediated DNA breaks and subsequent *TMPRESS2–ERG* genomic recombination (Haffner *et al.*, 2010). In addition, TOPII ligase inhibitors, such as etoposide and teniposide that are used in the treatment for acute lymphoblastic leukemia (ALL), are also involved in treatment-related acute myeloid leukemia (t-AML) (Mistry *et al.*, 2005). Research studies suggest that treatment of leukemia patients with TOPII ligase inhibitors may induce formation of secondary translocations involving several genes, but primarily the MLL locus. The mechanisms of TOPII-induced chromosomal translocations in patients with t-AML are not clear.

2. OTHER CELLULAR STRESS

Reactive oxygen species (ROS) are reactive molecules derived from metabolic processes that play important roles in oxidative DNA damage in normal tissues (Cooke *et al.*, 2003; McCord and Fridovich, 1969). In the presence of environmental stress (e.g., ionizing or ultraviolet radiation or heat exposure), ROS levels increase dramatically resulting in significant cellular damage. ROS may damage all kinds of macromolecules including DNA, leading to the modification of nucleotides, which are mutagenic, inducing conformational changes in DNA in addition to structural alterations of the native base itself.

Increased ROS is one of the major contributors to genomic instability in DNA repair deficient genetic disorders (Burhans and Weinberger, 2007; Ragu *et al.*, 2007). Failure to scavage reactive radical molecules in yeast results in the accumulation of genetic mutations and gross chromosomal rearrangements (Huang and Kolodner, 2005). Recently, ROS has been suggested to play a role in mitochondria genomic rearrangement (Shedge *et al.*, 2010). Mechanically, in non-homologue end-joining (NHEJ) deficient mice, ROS causes endogenous DNA damage and breaks, which increase the error rate of DNA repair and the incidence of myeloid leukemia (Rassool *et al.*, 2007).

Another major cause of genome instability is replication stress (Burhans and Weinberger, 2007). Inefficient DNA replication causing slow progression, stalling, or collapse of replication forks can lead to replication stress. Because of single-stranded replicating DNA at the replication forks, single-stranded lesions within unwound DNA are easily converted to DSBs (Branzei and Foiani, 2005). The single-stranded DNA is sensitive to damage from radiation, ROS, and other factors. The DNA damage occurring within poised replication forks facilitates genomic instability and chromosomal rearrangements. To maintain the integrity of genetic material, DNA replication is monitored by cell cycle checkpoint regulators, DNA repair mechanisms and proofreading activity of DNA polymerase.

Lesions generated from replication stress have been linked to genomic instability and chromosomal rearrangements (Ichijima *et al.*, 2010; Zhang *et al.*, 2009). Recent studies suggest that replication stress-induced single-stranded DNA lesions could progress to the M phase, increasing the risk of chromosome-bridge and tetraploidy generation (Ichijima *et al.*, 2010).

B. Transcription and Genome Instability

When transcription takes place on a DNA segment that undergoes DNA replication, the two machineries collapse and as a consequence, genome stability is impaired. Therefore, the increased transcription rates are associated with higher DNA mutations and recombination incidence (Aguilera, 2002). Indeed, this phenomenon was observed more than 30 years ago based on studies using *Escherichia coli* and yeast (Beletskii and Bhagwat, 1996; Brock, 1971; Datta and Jinks-Robertson, 1995). Transcription could also increase the frequency of homologue recombination (HR) as observed in λ phage, *E. coli*, and yeast (Ikeda and Matsumoto, 1979; Vilette *et al.*, 1995; Voelkel-Meiman *et al.*, 1987). Blockage of the transcription process by using a mutant RNA polymerase I or insertion of a transcription terminator abolishes transcription-coupled hyperrecombination between DNA repeats (Huang and Keil, 1995; Voelkel-Meiman *et al.*, 1987). In T- and B-cell lineage development, transcription is coupled to generation of genetic diversity in normal biological processes called somatic hypermutation and class switching recombination (CSR) (Chiarle *et al.*, 2011; Klein *et al.*, 2011).

One of the mechanisms that potentially explains transcription-coupled mutation and recombination is based on the collision between the DNA replication and transcription machinery. The frequent collisions between DNA replication forks and transcription elongation complexes cause replication forks to collapse and produces chromatin DSBs (Marians, 2000). Based on studies using direct-repeat containing constructs, the interactions of RNA polymerase complexes and replication forks dramatically increase transcription-coupled recombination rates (Aguilera, 2005; Prado and Aguilera, 2005). Multiple pathways have been demonstrated to reduce collisions and to repair broken DNA, including reinitiation of DNA replication forks, mechanisms preventing encounters between transcription and replication processes, and termination of the RNA polymerase complexes (Srivatsan *et al.*, 2010; Tehranchi *et al.*, 2010; Trautinger *et al.*, 2005; Washburn and Gottesman, 2011).

The transient accumulation of both positively and negatively supercoiled DNA also contributes to the higher mutation and recombination rates that are coupled with transcription activity (Brill *et al.*, 1987; Wu *et al.*, 1988). Mutations of Top1, Top2, and Top3 in yeast lead to hyperrecombination incidence, which suggests that the supercoils during transcription contribute to DSBs and recombination (Christman *et al.*, 1988; Trigueros and Roca,

2001). In mammalian cells, both estrogen and androgen recruit the enzyme TOPII to the promoters of target genes (Bartek et al., 2010; Ju et al., 2006). The expression of TOPII in prostate cancer is linked to prostate specific *TMPRSS2–ERG* chromosomal translocations (Haffner et al., 2010). TOPI restrains transcription-coupled mutation and recombination by relaxing transcription-induced negative supercoiling (Masse and Drolet, 1999).

When an RNA molecule partially or completely hybridizes to one strand of a double-stranded DNA, leaving the other strand unpaired, the whole structure is an R-loop (Fig. 3A). R-loop structures are observed in more than 60% of nascent RNA transcripts in *in vitro* experiments (Kadesch and Chamberlin, 1982). Under *in vivo* conditions, it is generally considered that extensive R-loops are rare except for short RNA–DNA hybrids that appear in the transcription bubble (Gnatt et al., 2001; Kettenberger et al., 2004; Westover et al., 2004). Certain DNA-binding proteins, for example, replication protein A (RPA), facilitate the stabilization of single-stranded DNA (ssDNA) and RNA–DNA hybrids (Chaudhuri et al., 2004; Fanning et al., 2006).

The formation of R-loops during transcription is highly correlated with DNA mutation and recombination, indicating the impact of R-loops on genome integrity. Single-stranded DNA segments are more vulnerable to mutagens than double-stranded DNA (Lindahl and Nyberg, 1974). During transcription and elongation, the maintained DNA single-stranded regions (R-loop), facilitated by the negatively supercoiled DNA behind the RNA polymerase complex, are susceptible to chemical reactions or oxidants generated from internal cell metabolism. In support of this hypothesis, single-stranded DNA is more than 100-fold more susceptible to spontaneous deamination than double-stranded DNA (Frederico et al., 1990). Furthermore, the nontranscribed strand of single-stranded DNA fragment is prone to greater mutation than the transcribed strand (Fix and Glickman, 1987; Skandalis et al., 1994).

Furthermore, enzymes recruited to R-loop structures might be involved in generating the single-stranded DNA lesions, and subsequent accumulation of DSBs and mutations (Fig. 3B). AID was first identified as a B-cell specific factor critical for CSR (Longerich et al., 2006; Muramatsu et al., 2000, 1999). The deaminase activity of AID can deaminate cytidine to uracil, generating U:G mismatches, which are subsequently removed by uracil DNA glycosylase (UNG) and apurinic/apyrimidinic endonuclease (APE) (Di Noia and Neuberger, 2002; Petersen-Mahrt et al., 2002; Rada et al., 2002). The generated DNA nicks are further converted into double-stranded DNA breaks, which are required for CSR (Chaudhuri and Alt, 2004). Transcription and elongation enhance and stabilize the non-template single-stranded DNA, which are subjected to AID-mediated cytidine deamination (Ramiro et al., 2003; Robbiani et al., 2009; Sohail et al., 2003). In AR signaling, AR interacting with its ligand recruits AID to AR binding sites, where R-loops potentially exist, to introduce DNA breaks in a "site-

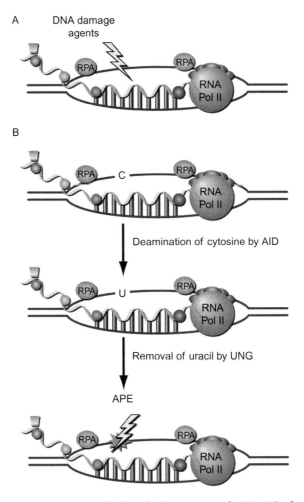

Fig. 3 Transcription activation and R-loop facilitate site-specific DSBs. The figure is adapted and modified from *Genes Dev.* 2006;20: 1838–1847. (A) The formation of an R-loop exposes a nontranscribed DNA strand susceptible to DNA damage agents. The binding of RPA stabilizes the single-stranded chromatin regions. (B) The formation of R-loop facilitates the AID-mediated cytosine deamination to generate abasic sites, which will be further removed by UNG- and APE-mediated base excision repair pathways to create a single-stranded DNA nick.

specific" manner (Lin *et al.*, 2009). Recent studies using translocation capture or genomic translocation sequencing methods (Chiarle *et al.*, 2011; Klein *et al.*, 2011) indicate that site-specific DSBs are primary located in transcriptionally active genomic regions and defined chromatin territories. The DNA damage-repair machinery may facilitate the nuclear architecture rearrangement and physical proximity of translocation partners.

Furthermore, in B-cell lymphoma, AID plays significant roles in mediating DSBs generation in transcription start sites and promotes genomic rearrangements with non-Ig genes (Chiarle et al., 2011; Klein et al., 2011).

C. Transposable Elements and Endonuclease

Transposable elements are sequences of DNA segments that have the ability to move or transpose themselves within cellular genomes. Because of their promiscuous nature, it has been hypothesized that transposable elements are linked to genomic instability and chromosomal rearrangements (Belancio et al., 2009). Recently, solid evidence indicates that transposable elements are involved in generating DNA DSBs, insertional mutagenesis, and chromosomal rearrangements (Collier and Largaespada, 2007; Lonnig and Saedler, 2002).

There are two major classes of transposable elements, retroelements and DNA transposons, which transpose through a RNA- or DNA-dependent mechanism, respectively. Long interspersed elements (LINEs) belong to retroelements with two open reading frames that encode a reverse transcriptase and an endonuclease in intact L1 elements (Feng et al., 1996; Martin and Bushman, 2001; Mathias et al., 1991) (Fig. 4). To copy itself and insert into another site of the genome, the endonuclease creates a single-stranded DNA nick on the target site; the reverse transcriptase subsequently uses the nicked DNA as a primer for reverse transcription to insert the LINE RNA sequence to the target site (Kazazian, 2004; Ostertag and Kazazian, 2001; Yoder et al., 1997). It is generally believed that the LINE-1 transposable machinery is responsible for most of the retrotransposition events in the genome. Although the feature of transposable elements to generate insertional mutation and recombination can be considered as driving forces in evolution, the deleterious aspect of genomic instability and cancer formation has raised considerable attention to the potential negative aspects of these interactions.

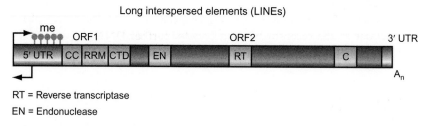

Fig. 4 Long interspersed elements (LINEs). LINE-1 elements are transcribed to RNA using a promoter located in the 5′UTR, which is highly DNA methylated to silence LINE-1 element. LINE-1 codes open reading frames, ORF1 and ORF2, which encode reverse transcriptase and endonuclease, respectively. Figure is adapted from Beck, CR et al., Annu. Rev. Genomics Hum. Genet. 2011 12:187-215. (See Page 8 in Color Section at the back of the book.)

The consequence of active transposition could be diverse. Insertion of the element into a coding gene could lead to disruption of a gene's function. The insertion events in a regulatory region of a given gene may cause gene deregulation and abnormal gene expression. Deletion or insertion of genomic information, reversion of gene function by mutation, or recombination could occur during the transposition process. In addition, a high repeat number of transposable elements distributed across the genome can trigger recombinatorial events due to the inherent mobility, abundance, and high identical repeat sequences (Belancio et al., 2010; Konkel and Batzer, 2010).

The DNA damage and repair process could activate many types of transposable elements (Handler and Gomez, 1997; Izsvak and Ivics, 2004). Gamma IR and etoposide treatment could both enhance the transcription and retrotransposition of LINE-1 and SINE-1 elements (Asakawa et al., 2004; Hagan et al., 2003; Rudin and Thompson, 2001). The promoter of LINE-1 elements has been shown to be DNA methylated. Demethylation of the LINE-1 promoter derepresses LINE-1 transcription as well as retrotransposition activity (Hata and Sakaki, 1997). Genotoxic stress reduces the activity and expression of DNA methyltransferases, which leads to global hypomethylation (Batra et al., 2004; Graziano et al., 2006; Pogribny et al., 2004). Indeed, increased LINE-1 element transcription has been observed in *DNMT1* and *DNMT3L* knockout mice (Walsh et al., 1998). On the other hand, LINE-1 elements could integrate into existing broken DNA ends in an endonuclease activity-independent manner. This type of insertion is more likely to be accompanied by large deletions (Morrish et al., 2002).

In many cancer cell lines and tissues, LINE-1 elements are hypomethylated and expressed in a cell type- and tissue type-specific manner (Ergun et al., 2004; Sunami et al., 2011; Wilhelm et al., 2010). In human cancer, hypomethylation of LINE-1 elements dispersed in the genome could facilitate chromosome breakage and recombination, mediated by a NHEJ mechanism. Genome-wide studies have shown a strong association between LINE-1 sequences and deletion endpoints (Cox et al., 2005; Florl et al., 1999). Recent research indicates that cell type-specific signaling pathways, for example, AR signaling in prostate cancer cell lines, recruit LINE-1 endonuclease to the receptor binding sites for further DNA cleavage in the presence of genotoxic stress (Lin et al., 2009).

D. Spatial Proximity

As tumor chromosomal translocations exhibit cell- and tissue-type specificities, a possible explanation of this phenomenon is that in different cell and tissue types, the nuclear interior is non-randomly positioned as a consequence of higher-order nuclear architecture. Growing evidence indicates that the formation of tumor translocations is partially affected by higher-order genomic organization in nuclear territory. Cell type-specific signaling pathways may affect or induce the spatial proximity of translocation partners to facilitate the formation of chromosomal translocations. Investigating the mechanisms of tissue-specific genome organization and dynamics of nuclear architecture will elucidate the fundamental aspects of chromosomal translocation formation.

In the nucleus of eukaryotes, each interphase chromosome occupies a discrete, spatially defined region known as chromatin territory (Cremer and Cremer, 2001; Misteli, 2009; Misteli and Soutoglou, 2009). The position and organization of human chromosomes are non-random (Fig. 5). The current view is that chromosome territories are well-organized space zones within the nucleus with complex-folded surfaces. According to gene density and gene transcription status, the position of each chromosome is assigned relative to the nuclear center. Specifically, the interior of the nucleus harbors the gene-rich chromosomes and the nuclear periphery preferentially situates gene-poor chromosomes (Misteli, 2005). Similarly, transcriptionally active genes are favored to be located on a chromatin loop, which is distant from centromeric heterochromatin. On the contrary, it is proposed that transcriptionally repressed genes are recruited to centromeric heterochromatin for silencing (Cremer and Cremer, 2001; Murmann et al., 2005; Zhao et al., 2009). For transcription activation, chromatin loops extend from the chromosome territory surface into the nuclear matrix for transcription, splicing, and DNA replication (Kosak et al., 2002; Meaburn and Misteli, 2007; Meaburn et al., 2007; Skok et al., 2001).

Cell cycle progression and cell/tissue type-specific signaling could induce reorganization of chromosome territory. The distance between individual chromosome territory and the center of the nucleus are distinct in different tissues. In quiescent cells, chromosomes are more likely arranged in radical positioning according to gene density. When human cells enter the cell cycle, gene-poor chromosomes could be repositioned to the nuclear periphery while gene-rich chromosomes locate toward the nuclear interior (Lanctot et al., 2007; Mateos-Langerak et al., 2007). This process is conserved through vertebrate species. Cell type-specific signals induce non-random spatial proximity of transcription units, which has been demonstrated to be important for gene activation or regulation in different cell types and signaling pathways (Fraser, 2006; Fraser and Engel, 2006; Hu et al., 2008; Kosak et al., 2002; Spilianakis et al., 2005). Indeed, the eukaryotic genome forms an extensive and dynamic three-dimensional architecture, which modulates activation and repression of the transcription units in the emerging "interactome" (Chakalova and Fraser, 2010; Cope et al., 2010; Eskiw et al., 2010; Harismendy et al., 2011; Schoenfelder et al., 2010). The vigorous and complicated patterns of interaction between genome regions and the relocation of transcription units

Molecular Logic Underlying Chromosomal Translocations, Random or Non-Random? 259

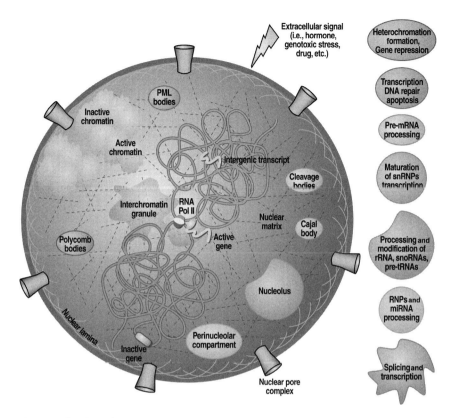

Fig. 5 The three-dimensional organization of an interphase nucleus. In the nucleus and subnulcear compartments, including PML bodies, cleavage bodies, Cajal bodies, the nucleolus, the perinucleolar compartment, and interchromatin granules are well organized. The gene-rich and transcriptionally active chromosomes locate relative to the nuclear center. For gene activation, chromatin loops are extended to the nuclear matrix or other subnuclear compartments for transcription. Extracellular signals could trigger interactions between potential translocation partners, which facilitate the formation of chromosomal translocation. (See Page 8 in Color Section at the back of the book.)

between subnuclear architecture have been demonstrated to be important for gene regulation. However, the patterns of interaction and mechanisms underlying the rearrangement of nuclear architecture are far from being understood.

The role of spatial proximity between genome regions has been speculated to be involved in chromosome translocation (Misteli, 2004; Parada et al., 2004) (Fig. 5). The physical closeness of translocation partners is remarkably correlated with the frequency of chromosomal translocations in cancer, especially in promyelocytic leukemia, acute myelocytic leukemia, Burkitt's lymphoma, and thyroid lymphoma (Misteli, 2008; Parada et al., 2004; Takizawa et al., 2008). Even in mouse lymphocytes, chromosomes 12, 14, and 15 that form a spatial-cluster in high-incidence undergo translocations (Parada *et al.*, 2004). In human leukemia, translocation partners including *MYC* and *IgH*, *BCR* and *Abl* gene loci are spatially proximal in a tissue- and cell type-specific pattern, which contribute to the correlated translocations observed in patients.

These experiments provide evidence to support the "contact-first" model of chromosome translocation, suggesting that the translocation partners are localized in a spatial proximal pattern. The generation of free chromatin from DNA DSBs has a higher chance to be ligated with close translocation partners to form a chimeric chromosome. Cell type-specific signaling induces precise interactions between gene loci, which may contribute to chromosomal translocations. For example, upon AR signaling, the *TMPRSS2* locus and the *ERG* or *ETV1* loci were found to interact with each other, which facilitate the *TMPRSS2–ERG* and *TMPRSS2–ETV1* chromosomal translocations (Haffner *et al.*, 2010; Lin *et al.*, 2009; Mani *et al.*, 2009). The signal-induced nuclear architecture rearrangement and transcription unit interaction are usually involved in the formation of chromatin loops, which have often been considered as being outside of their own chromosome territory invading neighboring chromosome territory or interchromatin compartments. Therefore, the extended chromatin fibers are susceptible to DNA damage and radicals generated from cell metabolism due to their decondensed nature.

Growing evidence supports the theory that the genome is organized in an extensive and dynamic three-dimensional architecture in the interphase nucleus and this organization is correlated with transcription regulation and genome stability. Transcription activation or DNA damage repair could induce chromosome territory rearrangement, thereby facilitating physical proximity of translocation partners (Chiarle *et al.*, 2011; Klein *et al.*, 2011; Lin *et al.*, 2009; Mani *et al.*, 2009). Understanding the mechanisms of relocation of gene loci between chromosome territory and subnuclear compartments will likely shed light on the mechanisms underlying chromosomal translocation.

E. Ligation

The first clue to understand how DSBs are ligated for generating chimeric chromosomes comes from inherited syndromes, which are associated with genomic instability, chromosome translocation, and hematologic malignancies. Patients with a mutation of the ATM gene suffer from ataxia-telangiectasia causing a deficiency in DNA damage repair, a hypersensitivity to X-rays, and a predisposition to lymphomas (Lavin, 2008). ATM knockout mice develop chromosomal translocations harboring antigen receptor genes and pre-T-lymphoblastic leukemia/lymphoma (Gapud *et al.*, 2011; Ramiro *et al.*, 2006). In yeast, DNA damage is suppressed by an ATM homologue to prevent formation of chromosomal translocations (Lee *et al.*, 2008). Similarly, individuals with a genetic disorder impairing DNA damage repair, including Bloom's syndrome, Fanconi's anemia, and Li-Fraumeni syndrome, have a higher chance of developing chromosomal translocation and hematologic malignancies (Gaymes *et al.*, 2002; Kuramoto *et al.*, 2002).

It is generally proposed that generation of chromosomal translocations involves misrepair of broken chromatin ends. The major DNA repair pathways include Homologous Recombination (HR) and Non-homologous End-Joining (NHEJ) (Shrivastav *et al.*, 2008). In lymphatic cells, illegitimate AID/RAG-mediated V(D)J or switch recombination plays a significant role in mediating translocations harboring antigen receptor genes (Ramsden *et al.*, 2010). Mice that are deficient in either HR or NHEJ pathways exhibit genome instability and higher incidences of chromosomal translocations (Ferguson and Alt, 2001). HR utilizes sister chromatids as a template to repair missing genetic information to avoid creating deletions/mutations. However, the NHEJ machinery directly ligates broken DNA ends for repair, which may cause deletions surrounding the ligation site (Karanjawala *et al.*, 1999; Lieber *et al.*, 2003). Recently, the base excision repair strategy has also been linked to chromosomal translocations (Nemec *et al.*, 2010).

During the generation of chromosomal translocations, the ligation of DNA ends is misrepaired to form genomic rearrangements. Depending on the cell stage and types of genomic rearrangements, different repair pathways may play critical roles in chromosomal translocations. Because HR uses sister chromatid template sequences for repair, this repair pathway is actively functional during S and G2 phases. The nature of the error-proofing property of this pathway prevents the formation of chromosomal translocations. During meiosis, homologous recombination produces new recombinations of genetic materials, promoting genetic variations during evolution. Defects of HR leads to improper chromosome copies, numbers, and structures such as Down's syndrome with an extra copy of chromosome 21 (Lamb *et al.*, 2005); or malignant chromosomal translocation when RecQ helicase is deficient in Bloom's syndrome (Modesti and Kanaar, 2001); or cancer formation when *BRCA1* or *BRCA2* are mutated in breast and ovarian cancer (Powell and Kachnic, 2003).

Unlike the HR, the NHEJ pathway is functional throughout the cell cycle, which makes this pathway more commonly used for DNA damage repair. The NHEJ pathway has been suggested to suppress translocation that is independent of sequence homologue structure (Boboila *et al.*, 2010; Simsek and Jasin, 2010). Missing one or two key players of the NHEJ pathway leads to the accumulation of chromosomal translocations (Boboila *et al.*, 2010; Simsek and Jasin, 2010). The microhomology-dependent alternative NHEJ pathway is more error-prone and less efficient, which may play a relatively important role in generating chromosomal translocations (Lieber et al., 2010). The genome-wide identification of chromosomal translocations during IgH class switching indicates that micro-homologue sequences are detected in the majority of translocation breakpoints (Chiarle et al., 2011; Klein et al., 2011). These studies provide solid evidence to support the critical role of the microhomology-dependent alternative NHEJ pathway as the major machinery involved in chromosomal translocation formation.

III. THE ROLE OF EPIGENETICS IN CHROMOSOMAL TRANSLOCATIONS

Epigenetic changes including DNA methylation/demethylation and histone modifications are molecular steps intrinsically involved in the regulation of chromatin structure and gene expression (Chi et al., 2010). Accumulating evidence indicates that remodeling of chromatin structures are crucial for establishing stable epigenetic states that restrict or permit chromosome rearrangements in a number of diseases such as cancer and other syndromes involving chromosomal instability (Bartova et al., 2008; Slotkin and Martienssen, 2007).

A. DNA Methylation and Genome Instability

DNA methylation has been suggested as an important factor in maintaining genomic stability. This notion is based on studies in tumor cell lines and animal models indicating that genome-wide DNA demethylation may cause genomic instability and hence facilitate or accelerate tumor progression (Jones and Gonzalgo, 1997). For example, in human leukemia and lymphomas, DNA demethylation induced by 5-azacytidine promotes chromosome rearrangements in specific genomic regions including the heterochromatic; juxtacentromeric regions of chromosomes 1, 9, and 16; and distal heterochromatic segment of the long arm of the Y chromosome (Vilain et al., 2000; Xu et al., 1999). Several studies have also shown that DNA hypomethylation can lead to chromosomal instability in *Dnmt1−/−* embryonic stem cells and human colon cancers (Lengauer et al., 1997). Although these observations clearly suggest that DNA methylation plays a pivotal role in maintaining genome stability, the mechanisms by which DNA methylation protects chromosomes remains elusive. Recent studies have demonstrated that CpG dinucleotide, which is the major DNA methylation motif in mammalian cells, accounts for 40–70% of breakpoints at pro-B/pre-B stage translocation regions in leukemia and lymphomas, suggesting a functional relationship between DNA methylation and DSBs generation required for chromosomal rearrangements (Tsai *et al.*, 2008). Interestingly, studies in prostate cancer cells have shown that the simultaneous knock down of MeCP2 and overexpression of Dot1L, a histone H3K79 methyltransferase, synergistically enhanced androgen, and genotoxic stress-induced chromosomal translocations between *TMPRSS2* and *ERG/ETV1* (Lin *et al.*, 2009). These observations suggest that DNA methylation may play a role in maintaining genomic stability that is mediated by methyl-binding proteins whose functional roles are to antagonize histone H3K79 methylation at genomic DSB regions (Fig. 6).

B. Chromatin Modifications and Genome Instability

Since chromosomal rearrangements arise when DSBs occur at two separate regions of the genome and that the resulting DNA ends are aberrantly joined into a new configuration, it is speculated that chromatin modifications, which generally modulate DSB repair, may influence the chromosomal

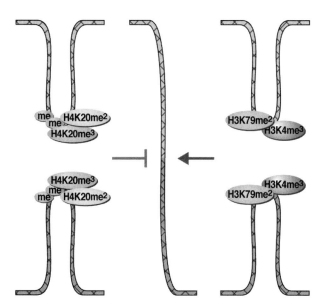

Fig. 6 The role of epigenetics on chromosomal translocation. DNA methylation and histone modifications are suggested to affect the formation of chromosomal translocation and genome integrity. DNA methylation, H4K20 di- and trimethyl are suggested to potentially protect chromatin from genome instability and chromosomal rearrangement. On the contrary, preliminary data suggest that H3K79 dimethyl and H3K4 trimethyl modifications might facilitate chromosomal translocation to occur.

rearrangements. Histone H3K79 methylation, a histone modification involved in DNA damage repair, enhances gross chromosomal rearrangements in yeast (Myung et al., 2001). Consistently, overexpression of the H3K79 methyltransferase Dot1L in prostate cancer cells significantly increases androgen- and genotoxic stress-induced *TMPRESS2–ETS* chromosomal translocation (Lin et al., 2009). Moreover, androgen-induced enrichment of H3K79me2 levels at *TMPRESS2* and *ERG/ETV1* DSB regions has been observed (Lin et al., 2009). Several histone modifications are also involved in programmed DNA rearrangements during B- and T-cell maturation. For example, Suv420 h1/h2 double knockout cells, which cause the genome-wide transition to a H4K20me1 state, show increased sensitivity to damaging stress and susceptibility to chromosomal aberrations (Schotta et al., 2008). Notably, Suv420 h1/h2 double knockout B cells are defective in immunoglobulin CSR (Schotta et al., 2008). Recently, it has also been shown that sequence-independent DNA binding of the RAG complex is mediated by H3K4me3, and that H3K4me3 facilitates the catalytic turnover of the RAG complex (Ji et al., 2010; Shimazaki et al., 2009) (Fig. 6).

Here, we have reviewed evidence pointing to the modulation of chromatin structure as a potential driving force of chromosomal rearrangements. While these observations are mostly correlative, recent technical advancements allowing genome-wide approaches have permitted the comprehensive interrogation of specific epigenetic changes at genomic regions susceptible to rearrangements. This knowledge provides the backdrop for developing new "epigenomic" drugs for diseases that occur when chromosomes go awry, such as leukemia, schizophrenia, additional cancers, and growth defects in children.

IV. CONCLUSION

Since identification of the Philadelphia chromosome, chromosomal translocations have been a central question in human genomic studies and cancer research. The direct link between chromosomal translocations and leukemia, lymphoma, and many other types of cancer mandates extensive studies of the mechanisms underlying chromosomal translocation formation and the regulation of chromosomal translocations.

Considering recent progress on the spatial organization of nuclear architecture, ligand-induced DNA DSBs and initiation of chromosomal translocation, a general model for chromosomal translocation-formation is suggested (Fig. 7). The genomic regions, especially the potential translocation partners, are non-randomly positioned in the interphase nuclei, probably due to cell- and tissue type-specific signaling. In other words, cell- or

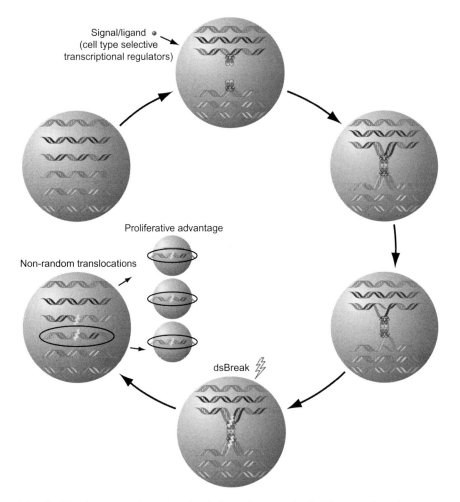

Fig. 7 The chromosomal translocation is formed non-randomly. The genomic regions containing potential translocation partners are non-randomly positioned in the interphase nuclei. Cell- and tissue-specific signals induce the reorganization of nuclear architecture leading to spatial proximity of translocation partners. In the transcription active sites, the formation of R-loop make non-template DNA strand susceptible to DNA damage. The misrepair between site-specific DSBs of spatial proximal translocation partners leads to chromosomal translocations.

tissue type-specific signals induce the reorganization of nuclear architecture, leading to spatial proximity of translocation partners, which are more prone to be fused as chromosomal translocations. Because the decondensed chromatin loop is triggered by signals, the extended chromatin fibers are susceptible to mutagens and DNA damage induced through cell metabolism. The binding of transcription factors facilitates the formation of RNA:DNA hybrids (R-loop), which are very fragile. If the accumulated single-stranded DNA lesions and double-stranded DNA breaks are not properly repaired by faithful DNA damage-repair machinery, chromosomal translocations will emerge as a consequence.

Mechanisms that protect genetic material from genomic instability and chromosomal translocations are critical for survival. DNA methylation and certain histone modifications act to condense the chromatin fibers and silence repetitive elements to inhibit their retrotransposon activity. These epigenetic strategies are also important to modulate the organization of nuclear architecture and transcriptional activity. Therefore, the formation of chromosomal translocation is not a single, isolated event. In this context, the generation of chromosomal translocation is linked to a myriad of biological activities of a living cell and may be the consequence of dysregulation of these indispensable cellular activities.

Tissue type-specific chromosomal translocation and rearrangement could serve as useful diagnostic markers for many types of cancers. Application of genetic tools detecting the presence of fusion gene or fusion proteins by *in situ* hybridization (ISH) or FISH can be applied to biopsy specimens or cancer cells in blood or urine sediments samples to provide molecular diagnostic information (Mao *et al.*, 2008; Tognon *et al.*, 2002). PCR-based technologies are also sensitive to detect fusion-positive cells (Soda *et al.*, 2007). The fusion proteins or dysregulated factors resulted form chromosomal translocation facilitate tumor cell proliferation or migration, therefore the presence of gene fusion is likely to be linked to poor prognosis. In certain tumor types, such as sarcomas, it is difficult to determine the tumor subtype based on routine cytogenetic examination. Identification of tumor-specific chromosomal translocation provides useful information both for diagnosis and prognosis.

Because the fusion gene and fusion proteins are tumor cell specific and the presence of fusion proteins are mostly facilitate tumor cells grow and proliferation. Therefore, targeting chromosomal translocation to inhibit the fusion gene expression or function represents novel therapeutic tool. To inhibit the fusion protein expression level could be achieved through delivery of anti-sense DNA oligo or siRNA via viral or nonviral based technology. Generation of specific antibody targeting fusion proteins is another plausible proposes. Delivering peptides containing fusion sites sequences to stimulate immune system for generation of specific antibodies against translocation-derived fusion proteins is under intensive investigation (Aryee *et al.*, 2006; Fukukawa *et al.*, 2005).

Developing small molecules inhibiting the kinase activity of fusion proteins have been successfully applied to leukemia and sarcoma. Imatinib, a small molecule compound that inhibit the tyrosine kinase activity of BCR–ABL in leukemia or PDGFB–COLIA1 in sarcoma has been extensively studied (Gambacorti-Passerini, 2008). Small molecules inhibiting ERG transcription activity is also under developing. Enormous effort has also been directed in developing clinically useful drugs to target hormones, kinases, and epigenetic regulators in order to block chromosomal translocations that lead to human malignancies. New techniques are developing to "cut and paste" the translocated chromosomes and restore a normal version of the chromosome. The most important future direction is to prevent chromosomal translocation events, which will preclude a large fraction of inherited genetic disorders, treatment-related leukemia, and progression of many types of cancer. However, to achieve this preventative approach has been elusive. Enhancing the maintenance of proper epigenetic status of the genome, evoking hormonal protective strategies, as well as preventing exposure to mutagens and free radicals, and faithfully preserving the DNA damage-repair machinery are the first groups of targets to explore.

ACKNOWLEDGMENTS

We apologize to our colleagues for our inability to cite their important publications because of length constraints. We thank J. Hightower for artwork; D. Benson and R. Pardee for assistance with the chapter. NIH, NIC, DoD supported work cited from our laboratories, as well as a Prostate Cancer Foundation grants to M. Rosenfeld. Dr. Michael G. Rosenfeld is a Howard Hughes Medical Institute Investigator; L. Yang is the recipient of a DoD Era of Hope Postdoctoral Award (W81XWH-08-0554); and C. Lin is the recipient of a Susan G. Komen for the Cure Postdoctoral Fellowship (KG080247).

REFERENCES

Aghib, D. F., Bishop, J. M., Ottolenghi, S., Guerrasio, A., Serra, A., and Saglio, G. (1990). A 3′ truncation of MYC caused by chromosomal translocation in a human T-cell leukemia increases mRNA stability. *Oncogene* **5**, 707–711.
Aguilera, A. (2002). The connection between transcription and genomic instability. *EMBO J.* **21**, 195–201.
Aguilera, A. (2005). mRNA processing and genomic instability. *Nat. Struct. Mol. Biol.* **12**, 737–738.
Aplan, P. D. (2006). Causes of oncogenic chromosomal translocation. *Trends Genet.* **22**, 46–55.
Aryee, D. N., Kreppel, M., Bachmaier, R., Uren, A., Muehlbacher, K., Wagner, S., Breiteneder, H., Ban, J., Toretsky, J. A., and Kovar, H. (2006). Single-chain antibodies to the EWS NH(2) terminus structurally discriminate between intact and chimeric EWS in Ewing's sarcoma and interfere with the transcriptional activity of EWS *in vivo*. *Cancer Res.* **66**, 9862–9869.
Asakawa, J., Kuick, R., Kodaira, M., Nakamura, N., Katayama, H., Pierce, D., Funamoto, S., Preston, D., Satoh, C., Neel, J. V., *et al.* (2004). A genome scanning approach to assess the genetic effects of radiation in mice and humans. *Radiat. Res.* **161**, 380–390.

Asatiani, E., Huang, W. X., Wang, A., Rodriguez Ortner, E., Cavalli, L. R., Haddad, B. R., and Gelmann, E. P. (2005). Deletion, methylation, and expression of the NKX3.1 suppressor gene in primary human prostate cancer. *Cancer Res.* **65,** 1164–1173.

Asmann, Y. W., Hossain, A., Necela, B. M., Middha, S., Kalari, K. R., Sun, Z., Chai, H. S., Williamson, D. W., Radisky, D., Schroth, G. P., et al. (2011). A novel bioinformatics pipeline for identification and characterization of fusion transcripts in breast cancer and normal cell lines. *Nucleic Acids Res.* **39,** e100.

Attard, G., Clark, J., Ambroisine, L., Mills, I. G., Fisher, G., Flohr, P., Reid, A., Edwards, S., Kovacs, G., Berney, D., et al. (2008). Heterogeneity and clinical significance of ETV1 translocations in human prostate cancer. *Br. J. Cancer* **99,** 314–320.

Bartek, J., Hamerlik, P., and Lukas, J. (2010). On the origin of prostate fusion oncogenes. *Nat. Genet.* **42,** 647–648.

Bartova, E., Krejci, J., Harnicarova, A., Galiova, G., and Kozubek, S. (2008). Histone modifications and nuclear architecture: a review. *J. Histochem. Cytochem.* **56,** 711–721.

Barwick, B. G., Abramovitz, M., Kodani, M., Moreno, C. S., Nam, R., Tang, W., Bouzyk, M., Seth, A., and Leyland-Jones, B. (2010). Prostate cancer genes associated with TMPRSS2-ERG gene fusion and prognostic of biochemical recurrence in multiple cohorts. *Br. J. Cancer* **102,** 570–576.

Bastus, N. C., Boyd, L. K., Mao, X., Stankiewicz, E., Kudahetti, S. C., Oliver, R. T., Berney, D. M., and Lu, Y. J. (2010). Androgen-induced TMPRSS2:ERG fusion in nonmalignant prostate epithelial cells. *Cancer Res.* **70,** 9544–9548.

Batra, V., Kesavan, V., and Mishra, K. P. (2004). Modulation of enzymes involved in folate dependent one-carbon metabolism by gamma-radiation stress in mice. *J. Radiat. Res. (Tokyo)* **45,** 527–533.

Belancio, V. P., Deininger, P. L., and Roy-Engel, A. M. (2009). LINE dancing in the human genome: transposable elements and disease. *Genome Med.* **1,** 97.

Belancio, V. P., Roy-Engel, A. M., Pochampally, R. R., and Deininger, P. (2010). Somatic expression of LINE-1 elements in human tissues. *Nucleic Acids Res.* **38,** 3909–3922.

Beletskii, A., and Bhagwat, A. S. (1996). Transcription-induced mutations: increase in C to T mutations in the nontranscribed strand during transcription in Escherichia coli. *Proc. Natl. Acad. Sci. USA* **93,** 13919–13924.

Boboila, C., Jankovic, M., Yan, C. T., Wang, J. H., Wesemann, D. R., Zhang, T., Fazeli, A., Feldman, L., Nussenzweig, A., Nussenzweig, M., et al. (2010). Alternative end-joining catalyzes robust IgH locus deletions and translocations in the combined absence of ligase 4 and Ku70. *Proc. Natl. Acad. Sci. USA* **107,** 3034–3039.

Boehm, T., Mengle-Gaw, L., Kees, U. R., Spurr, N., Lavenir, I., Forster, A., and Rabbitts, T. H. (1989). Alternating purine-pyrimidine tracts may promote chromosomal translocations seen in a variety of human lymphoid tumours. *EMBO J.* **8,** 2621–2631.

Branzei, D., and Foiani, M. (2005). The DNA damage response during DNA replication. *Curr. Opin. Cell Biol.* **17,** 568–575.

Brill, S. J., DiNardo, S., Voelkel-Meiman, K., and Sternglanz, R. (1987). DNA topoisomerase activity is required as a swivel for DNA replication and for ribosomal RNA transcription. *NCI Monogr.* (4), 11–15.

Brock, R. D. (1971). Differential mutation of the beta-galactosidase gene of Escherichia coli. *Mutat. Res.* **11,** 181–186.

Burhans, W. C., and Weinberger, M. (2007). DNA replication stress, genome instability and aging. *Nucleic Acids Res.* **35,** 7545–7556.

Cairns, P., Polascik, T. J., Eby, Y., Tokino, K., Califano, J., Merlo, A., Mao, L., Herath, J., Jenkins, R., Westra, W., et al. (1995). Frequency of homozygous deletion at p16/CDKN2 in primary human tumours. *Nat. Genet.* **11,** 210–212.

Carter, M. T., Barrowman, N. J., St Pierre, S. A., Emanuel, B. S., and Boycott, K. M. (2010). Risk of breast cancer not increased in translocation 11;22 carriers: analysis of 80 pedigrees. *Am. J. Med. Genet. A* **152A**, 212–214.

Caudill, C. M., Zhu, Z., Ciampi, R., Stringer, J. R., and Nikiforov, Y. E. (2005). Dose-dependent generation of RET/PTC in human thyroid cells after *in vitro* exposure to gamma-radiation: a model of carcinogenic chromosomal rearrangement induced by ionizing radiation. *J. Clin. Endocrinol. Metab.* **90**, 2364–2369.

Chakalova, L., and Fraser, P. (2010). Organization of transcription. *Cold Spring Harb. Perspect. Biol.* **2**, a000729.

Chaudhuri, J., and Alt, F. W. (2004). Class-switch recombination: interplay of transcription, DNA deamination and DNA repair. *Nat. Rev. Immunol.* **4**, 541–552.

Chaudhuri, J., Khuong, C., and Alt, F. W. (2004). Replication protein A interacts with AID to promote deamination of somatic hypermutation targets. *Nature* **430**, 992–998.

Chi, P., Allis, C. D., and Wang, G. G. (2010). Covalent histone modifications—miswritten, misinterpreted and mis-erased in human cancers. *Nat. Rev. Cancer* **10**, 457–469.

Chiarle, R., Zhang, Y., Frock, R. L., Lewis, S. M., Molinie, B., Ho, Y. J., Myers, D. R., Choi, V. W., Compagno, M., Malkin, D. J., *et al.* (2011). Genome-wide translocation sequencing reveals mechanisms of chromosome breaks and rearrangements in B cells. *Cell* **147**, 107–119.

Christman, M. F., Dietrich, F. S., and Fink, G. R. (1988). Mitotic recombination in the rDNA of S. cerevisiae is suppressed by the combined action of DNA topoisomerases I and II. *Cell* **55**, 413–425.

Collier, L. S., and Largaespada, D. A. (2007). Transposable elements and the dynamic somatic genome. *Genome Biol.* **8**(Suppl. 1), S5.

Cooke, M. S., Evans, M. D., Dizdaroglu, M., and Lunec, J. (2003). Oxidative DNA damage: mechanisms, mutation, and disease. *FASEB J.* **17**, 1195–1214.

Cope, N. F., Fraser, P., and Eskiw, C. H. (2010). The yin and yang of chromatin spatial organization. *Genome Biol.* **11**, 204.

Cox, C., Bignell, G., Greenman, C., Stabenau, A., Warren, W., Stephens, P., Davies, H., Watt, S., Teague, J., Edkins, S., *et al.* (2005). A survey of homozygous deletions in human cancer genomes. *Proc. Natl. Acad. Sci. USA* **102**, 4542–4547.

Cremer, T., and Cremer, C. (2001). Chromosome territories, nuclear architecture and gene regulation in mammalian cells. *Nat. Rev. Genet.* **2**, 292–301.

Datta, A., and Jinks-Robertson, S. (1995). Association of increased spontaneous mutation rates with high levels of transcription in yeast. *Science* **268**, 1616–1619.

Deininger, M., Buchdunger, E., and Druker, B. J. (2005). The development of imatinib as a therapeutic agent for chronic myeloid leukemia. *Blood* **105**, 2640–2653.

Deininger, M. W., Bose, S., Gora-Tybor, J., Yan, X. H., Goldman, J. M., and Melo, J. V. (1998). Selective induction of leukemia-associated fusion genes by high-dose ionizing radiation. *Cancer Res.* **58**, 421–425.

Di Noia, J., and Neuberger, M. S. (2002). Altering the pathway of immunoglobulin hypermutation by inhibiting uracil-DNA glycosylase. *Nature* **419**, 43–48.

Epstein, C. J. (2006). Down's syndrome: critical genes in a critical region. *Nature* **441**, 582–583.

Ergun, S., Buschmann, C., Heukeshoven, J., Dammann, K., Schnieders, F., Lauke, H., Chalajour, F., Kilic, N., Stratling, W. H., and Schumann, G. G. (2004). Cell type-specific expression of LINE-1 open reading frames 1 and 2 in fetal and adult human tissues. *J. Biol. Chem.* **279**, 27753–27763.

Eskiw, C. H., Cope, N. F., Clay, I., Schoenfelder, S., Nagano, T., and Fraser, P. (2010). Transcription factories and nuclear organization of the genome. *Cold Spring Harb. Symp. Quant. Biol.* **75**, 501–506.

Fanning, E., Klimovich, V., and Nager, A. R. (2006). A dynamic model for replication protein A (RPA) function in DNA processing pathways. *Nucleic Acids Res.* **34**, 4126–4137.

Feng, Q., Moran, J. V., Kazazian, H. H., Jr., and Boeke, J. D. (1996). Human L1 retrotransposon encodes a conserved endonuclease required for retrotransposition. *Cell* **87**, 905–916.

Ferguson, D. O., and Alt, F. W. (2001). DNA double strand break repair and chromosomal translocation: lessons from animal models. *Oncogene* **20**, 5572–5579.

Fix, D. F., and Glickman, B. W. (1987). Asymmetric cytosine deamination revealed by spontaneous mutational specificity in an Ung- strain of Escherichia coli. *Mol. Gen. Genet.* **209**, 78–82.

Florl, A. R., Lower, R., Schmitz-Drager, B. J., and Schulz, W. A. (1999). DNA methylation and expression of LINE-1 and HERV-K provirus sequences in urothelial and renal cell carcinomas. *Br. J. Cancer* **80**, 1312–1321.

Fraser, P. (2006). Transcriptional control thrown for a loop. *Curr. Opin. Genet. Dev.* **16**, 490–495.

Fraser, P., and Engel, J. D. (2006). Constricting restricted transcription: the (actively?) shrinking web. *Genes Dev.* **20**, 1379–1383.

Frederico, L. A., Kunkel, T. A., and Shaw, B. R. (1990). A sensitive genetic assay for the detection of cytosine deamination: determination of rate constants and the activation energy. *Biochemistry* **29**, 2532–2537.

Fukukawa, C., Nakamura, Y., and Katagiri, T. (2005). Molecular target therapy for synovial sarcoma. *Future Oncol.* **1**, 805–812.

Gambacorti-Passerini, C. (2008). Part I: milestones in personalised medicine—imatinib. *Lancet Oncol.* **9**, 600.

Gapud, E. J., Dorsett, Y., Yin, B., Callen, E., Bredemeyer, A., Mahowald, G. K., Omi, K. Q., Walker, L. M., Bednarski, J. J., McKinnon, P. J., et al. (2011). Ataxia telangiectasia mutated (Atm) and DNA-PKcs kinases have overlapping activities during chromosomal signal joint formation. *Proc. Natl. Acad. Sci. USA* **108**, 2022–2027.

Gaymes, T. J., North, P. S., Brady, N., Hickson, I. D., Mufti, G. J., and Rassool, F. V. (2002). Increased error-prone non homologous DNA end-joining–a proposed mechanism of chromosomal instability in Bloom's syndrome. *Oncogene* **21**, 2525–2533.

Gnatt, A. L., Cramer, P., Fu, J., Bushnell, D. A., and Kornberg, R. D. (2001). Structural basis of transcription: an RNA polymerase II elongation complex at 3.3 Å resolution. *Science* **292**, 1876–1882.

Gostissa, M., Alt, F. W., and Chiarle, R. (2011). Mechanisms that promote and suppress chromosomal translocations in lymphocytes. *Annu. Rev. Immunol.* **29**, 319–350.

Graziano, F., Kawakami, K., Ruzzo, A., Watanabe, G., Santini, D., Pizzagalli, F., Bisonni, R., Mari, D., Floriani, I., Catalano, V., et al. (2006). Methylenetetrahydrofolate reductase 677C/T gene polymorphism, gastric cancer susceptibility and genomic DNA hypomethylation in an at-risk Italian population. *Int. J. Cancer* **118**, 628–632.

Haffner, M. C., Aryee, M. J., Toubaji, A., Esopi, D. M., Albadine, R., Gurel, B., Isaacs, W. B., Bova, G. S., Liu, W., Xu, J., et al. (2010). Androgen-induced TOP2B-mediated double-strand breaks and prostate cancer gene rearrangements. *Nat. Genet.* **42**, 668–675.

Hagan, C. R., Sheffield, R. F., and Rudin, C. M. (2003). Human Alu element retrotransposition induced by genotoxic stress. *Nat. Genet.* **35**, 219–220.

Handler, A. M., and Gomez, S. P. (1997). A new hobo, Ac, Tam3 transposable element, hopper, from Bactrocera dorsalis is distantly related to hobo and Ac. *Gene* **185**, 133–135.

Hansmann, I., Wiedeking, C., Grimm, T., and Gebauer, J. (1977). Reciprocal or nonreciprocal human chromosome translocations? The identification of reciprocal translocations by silver staining. *Hum. Genet.* **38**, 1–5.

Harismendy, O., Notani, D., Song, X., Rahim, N. G., Tanasa, B., Heintzman, N., Ren, B., Fu, X. D., Topol, E. J., Rosenfeld, M. G., et al. (2011). 9p21 DNA variants associated with coronary artery disease impair interferon-gamma signalling response. *Nature* **470**, 264–268.

Hata, K., and Sakaki, Y. (1997). Identification of critical CpG sites for repression of L1 transcription by DNA methylation. *Gene* **189**, 227–234.

Helgason, G. V., Karvela, M., and Holyoake, T. L. (2011). Kill one bird with two stones: potential efficacy of BCR-ABL and autophagy inhibition in CML. *Blood* **118**, 2035–2043.

Hollander, M. C., Blumenthal, G. M., and Dennis, P. A. (2011). PTEN loss in the continuum of common cancers, rare syndromes and mouse models. *Nat. Rev. Cancer* **11**, 289–301.

Hu, Q., Kwon, Y. S., Nunez, E., Cardamone, M. D., Hutt, K. R., Ohgi, K. A., Garcia-Bassets, I., Rose, D. W., Glass, C. K., Rosenfeld, M. G., et al. (2008). Enhancing nuclear receptor-induced transcription requires nuclear motor and LSD1-dependent gene networking in interchromatin granules. *Proc. Natl. Acad. Sci. USA* **105**, 19199–19204.

Huang, G. S., and Keil, R. L. (1995). Requirements for activity of the yeast mitotic recombination hotspot HOT1: RNA polymerase I and multiple cis-acting sequences. *Genetics* **141**, 845–855.

Huang, M. E., and Kolodner, R. D. (2005). A biological network in Saccharomyces cerevisiae prevents the deleterious effects of endogenous oxidative DNA damage. *Mol. Cell* **17**, 709–720.

Ichijima, Y., Yoshioka, K., Yoshioka, Y., Shinohe, K., Fujimori, H., Unno, J., Takagi, M., Goto, H., Inagaki, M., Mizutani, S., et al. (2010). DNA lesions induced by replication stress trigger mitotic aberration and tetraploidy development. *PLoS One* **5**, e8821.

Ikeda, H., and Matsumoto, T. (1979). Transcription promotes recA-independent recombination mediated by DNA-dependent RNA polymerase in Escherichia coli. *Proc. Natl. Acad. Sci. USA* **76**, 4571–4575.

Iljin, K., Wolf, M., Edgren, H., Gupta, S., Kilpinen, S., Skotheim, R. I., Peltola, M., Smit, F., Verhaegh, G., Schalken, J., et al. (2006). TMPRSS2 fusions with oncogenic ETS factors in prostate cancer involve unbalanced genomic rearrangements and are associated with HDAC1 and epigenetic reprogramming. *Cancer Res.* **66**, 10242–10246.

Izsvak, Z., and Ivics, Z. (2004). Sleeping beauty transposition: biology and applications for molecular therapy. *Mol. Ther.* **9**, 147–156.

Jankovic, M., Nussenzweig, A., and Nussenzweig, M. C. (2007). Antigen receptor diversification and chromosome translocations. *Nat. Immunol.* **8**, 801–808.

Ji, Y., Resch, W., Corbett, E., Yamane, A., Casellas, R., and Schatz, D. G. (2010). The *in vivo* pattern of binding of RAG1 and RAG2 to antigen receptor loci. *Cell* **141**, 419–431.

Jones, P. A., and Gonzalgo, M. L. (1997). Altered DNA methylation and genome instability: a new pathway to cancer? *Proc. Natl. Acad. Sci. USA* **94**, 2103–2105.

Ju, B. G., Lunyak, V. V., Perissi, V., Garcia-Bassets, I., Rose, D. W., Glass, C. K., and Rosenfeld, M. G. (2006). A topoisomerase IIbeta-mediated dsDNA break required for regulated transcription. *Science* **312**, 1798–1802.

Jung, D., and Alt, F. W. (2004). Unraveling V(D)J recombination; insights into gene regulation. *Cell* **116**, 299–311.

Kadesch, T. R., and Chamberlin, M. J. (1982). Studies of *in vitro* transcription by calf thymus RNA polymerase II using a novel duplex DNA template. *J. Biol. Chem.* **257**, 5286–5295.

Karanjawala, Z. E., Grawunder, U., Hsieh, C. L., and Lieber, M. R. (1999). The nonhomologous DNA end joining pathway is important for chromosome stability in primary fibroblasts. *Curr. Biol.* **9**, 1501–1504.

Kazazian, H. H., Jr. (2004). Mobile elements: drivers of genome evolution. *Science* **303**, 1626–1632.

Kearney, L., and Horsley, S. W. (2005). Molecular cytogenetics in haematological malignancy: current technology and future prospects. *Chromosoma* **114**, 286–294.

Kettenberger, H., Armache, K. J., and Cramer, P. (2004). Complete RNA polymerase II elongation complex structure and its interactions with NTP and TFIIS. *Mol. Cell* **16**, 955–965.

Klein, I. A., Resch, W., Jankovic, M., Oliveira, T., Yamane, A., Nakahashi, H., Di Virgilio, M., Bothmer, A., Nussenzweig, A., Robbiani, D. F., et al. (2011). Translocation-capture sequencing reveals the extent and nature of chromosomal rearrangements in B lymphocytes. *Cell* **147**, 95–106.

Koivisto, P., Kononen, J., Palmberg, C., Tammela, T., Hyytinen, E., Isola, J., Trapman, J., Cleutjens, K., Noordzij, A., Visakorpi, T., et al. (1997). Androgen receptor gene amplification: a possible molecular mechanism for androgen deprivation therapy failure in prostate cancer. *Cancer Res.* **57**, 314–319.

Konkel, M. K., and Batzer, M. A. (2010). A mobile threat to genome stability: the impact of non-LTR retrotransposons upon the human genome. *Semin. Cancer Biol.* **20**, 211–221.

Korsmeyer, S. J. (1992). Chromosomal translocations in lymphoid malignancies reveal novel proto-oncogenes. *Annu. Rev. Immunol.* **10**, 785–807.

Kosak, S. T., Skok, J. A., Medina, K. L., Riblet, R., Le Beau, M. M., Fisher, A. G., and Singh, H. (2002). Subnuclear compartmentalization of immunoglobulin loci during lymphocyte development. *Science* **296**, 158–162.

Kumar-Sinha, C., Tomlins, S. A., and Chinnaiyan, A. M. (2008). Recurrent gene fusions in prostate cancer. *Nat. Rev. Cancer* **8**, 497–511.

Kuppers, R. (2005). Mechanisms of B-cell lymphoma pathogenesis. *Nat. Rev. Cancer* **5**, 251–262.

Kuppers, R., and Dalla-Favera, R. (2001). Mechanisms of chromosomal translocations in B cell lymphomas. *Oncogene* **20**, 5580–5594.

Kuramoto, K., Ban, S., Oda, K., Tanaka, H., Kimura, A., and Suzuki, G. (2002). Chromosomal instability and radiosensitivity in myelodysplastic syndrome cells. *Leukemia* **16**, 2253–2258.

Lamb, N. E., Yu, K., Shaffer, J., Feingold, E., and Sherman, S. L. (2005). Association between maternal age and meiotic recombination for trisomy 21. *Am. J. Hum. Genet.* **76**, 91–99.

Lanctot, C., Cheutin, T., Cremer, M., Cavalli, G., and Cremer, T. (2007). Dynamic genome architecture in the nuclear space: regulation of gene expression in three dimensions. *Nat. Rev. Genet.* **8**, 104–115.

Lavin, M. F. (2008). Ataxia-telangiectasia: from a rare disorder to a paradigm for cell signalling and cancer. *Nat. Rev. Mol. Cell Biol.* **9**, 759–769.

Leder, P., Battey, J., Lenoir, G., Moulding, C., Murphy, W., Potter, H., Stewart, T., and Taub, R. (1983). Translocations among antibody genes in human cancer. *Science* **222**, 765–771.

Lee, K., Zhang, Y., and Lee, S. E. (2008). Saccharomyces cerevisiae ATM orthologue suppresses break-induced chromosome translocations. *Nature* **454**, 543–546.

Lengauer, C., Kinzler, K. W., and Vogelstein, B. (1997). DNA methylation and genetic instability in colorectal cancer cells. *Proc. Natl. Acad. Sci. USA* **94**, 2545–2550.

Li, C., Larsson, C., Futreal, A., Lancaster, J., Phelan, C., Aspenblad, U., Sundelin, B., Liu, Y., Ekman, P., Auer, G., et al. (1998). Identification of two distinct deleted regions on chromosome 13 in prostate cancer. *Oncogene* **16**, 481–487.

Li, Z., Tognon, C. E., Godinho, F. J., Yasaitis, L., Hock, H., Herschkowitz, J. I., Lannon, C. L., Cho, E., Kim, S. J., Bronson, R. T., et al. (2007). ETV6-NTRK3 fusion oncogene initiates breast cancer from committed mammary progenitors via activation of AP1 complex. *Cancer Cell* **12**, 542–558.

Lieber, M. R. (2010). The mechanism of double-strand DNA break repair by the nonhomologous DNA end-joining pathway. *Annu. Rev. Biochem.* **79**, 181–211.

Lieber, M. R., Gu, J., Lu, H., Shimazaki, N., and Tsai, A. G. (2010). Nonhomologous DNA end joining (NHEJ) and chromosomal translocations in humans. *Subcell. Biochem.* **50**, 279–296.

Lieber, M. R., Ma, Y., Pannicke, U., and Schwarz, K. (2003). Mechanism and regulation of human non-homologous DNA end-joining. *Nat. Rev. Mol. Cell Biol.* **4**, 712–720.

Liggett, W. H., Jr., and Sidransky, D. (1998). Role of the p16 tumor suppressor gene in cancer. *J. Clin. Oncol.* **16**, 1197–1206.

Lin, C., Yang, L., Tanasa, B., Hutt, K., Ju, B. G., Ohgi, K., Zhang, J., Rose, D. W., Fu, X. D., Glass, C. K., et al. (2009). Nuclear receptor-induced chromosomal proximity and DNA breaks underlie specific translocations in cancer. *Cell* **139**, 1069–1083.

Lindahl, T., and Nyberg, B. (1974). Heat-induced deamination of cytosine residues in deoxyribonucleic acid. *Biochemistry* **13**, 3405–3410.

Lindblom, A., Sandelin, K., Iselius, L., Dumanski, J., White, I., Nordenskjold, M., and Larsson, C. (1994). Predisposition for breast cancer in carriers of constitutional translocation 11q;22q. *Am. J. Hum. Genet.* **54**, 871–876.

Longerich, S., Basu, U., Alt, F., and Storb, U. (2006). AID in somatic hypermutation and class switch recombination. *Curr. Opin. Immunol.* **18**, 164–174.

Lonnig, W. E., and Saedler, H. (2002). Chromosome rearrangements and transposable elements. *Annu. Rev. Genet.* **36**, 389–410.

Mackie Ogilvie, C., and Scriven, P. N. (2002). Meiotic outcomes in reciprocal translocation carriers ascertained in 3-day human embryos. *Eur. J. Hum. Genet.* **10**, 801–806.

Magi-Galluzzi, C., Tsusuki, T., Elson, P., Simmerman, K., LaFargue, C., Esgueva, R., Klein, E., Rubin, M. A., and Zhou, M. (2011). TMPRSS2-ERG gene fusion prevalence and class are significantly different in prostate cancer of Caucasian, African-American and Japanese patients. *Prostate* **71**, 489–497.

Maher, C. A., Palanisamy, N., Brenner, J. C., Cao, X., Kalyana-Sundaram, S., Luo, S., Khrebtukova, I., Barrette, T. R., Grasso, C., Yu, J., et al. (2009). Chimeric transcript discovery by paired-end transcriptome sequencing. *Proc. Natl. Acad. Sci. USA* **106**, 12353–12358.

Mani, R. S., and Chinnaiyan, A. M. (2010). Triggers for genomic rearrangements: insights into genomic, cellular and environmental influences. *Nat. Rev. Genet.* **11**, 819–829.

Mani, R. S., Iyer, M. K., Cao, Q., Brenner, J. C., Wang, L., Ghosh, A., Cao, X., Lonigro, R. J., Tomlins, S. A., Varambally, S., et al. (2011). TMPRSS2-ERG-mediated feed-forward regulation of wild-type ERG in human prostate cancers. *Cancer Res.* **71**, 5387–5392.

Mani, R. S., Tomlins, S. A., Callahan, K., Ghosh, A., Nyati, M. K., Varambally, S., Palanisamy, N., and Chinnaiyan, A. M. (2009). Induced chromosomal proximity and gene fusions in prostate cancer. *Science* **326**, 1230.

Mao, X., Shaw, G., James, S. Y., Purkis, P., Kudahetti, S. C., Tsigani, T., Kia, S., Young, B. D., Oliver, R. T., Berney, D., et al. (2008). Detection of TMPRSS2:ERG fusion gene in circulating prostate cancer cells. *Asian J. Androl.* **10**, 467–473.

Marians, K. J. (2000). Replication and recombination intersect. *Curr. Opin. Genet. Dev.* **10**, 151–156.

Martin, S. L., and Bushman, F. D. (2001). Nucleic acid chaperone activity of the ORF1 protein from the mouse LINE-1 retrotransposon. *Mol. Cell. Biol.* **21**, 467–475.

Masse, E., and Drolet, M. (1999). R-loop-dependent hypernegative supercoiling in Escherichia coli topA mutants preferentially occurs at low temperatures and correlates with growth inhibition. *J. Mol. Biol.* **294**, 321–332.

Mateos-Langerak, J., Goetze, S., Leonhardt, H., Cremer, T., van Driel, R., and Lanctot, C. (2007). Nuclear architecture: is it important for genome function and can we prove it? *J. Cell. Biochem.* **102**, 1067–1075.

Mathias, S. L., Scott, A. F., Kazazian, H. H., Jr., Boeke, J. D., and Gabriel, A. (1991). Reverse transcriptase encoded by a human transposable element. *Science* **254**, 1808–1810.

McCord, J. M., and Fridovich, I. (1969). The utility of superoxide dismutase in studying free radical reactions. I. Radicals generated by the interaction of sulfite, dimethyl sulfoxide, and oxygen. *J. Biol. Chem.* **244**, 6056–6063.

Meaburn, K. J., and Misteli, T. (2007). Cell biology: chromosome territories. *Nature* **445**, 379–781.

Meaburn, K. J., Misteli, T., and Soutoglou, E. (2007). Spatial genome organization in the formation of chromosomal translocations. *Semin. Cancer Biol.* **17**, 80–90.

Medema, R. H., and Macurek, L. (2011). Checkpoint control and cancer. *Oncogene*.
Mehra, R., Han, B., Tomlins, S. A., Wang, L., Menon, A., Wasco, M. J., Shen, R., Montie, J. E., Chinnaiyan, A. M., and Shah, R. B. (2007). Heterogeneity of TMPRSS2 gene rearrangements in multifocal prostate adenocarcinoma: molecular evidence for an independent group of diseases. *Cancer Res.* **67**, 7991–7995.
Misteli, T. (2004). Spatial positioning: a new dimension in genome function. *Cell* **119**, 153–156.
Misteli, T. (2005). Concepts in nuclear architecture. *Bioessays* **27**, 477–487.
Misteli, T. (2008). Cell biology: nuclear order out of chaos. *Nature* **456**, 333–334.
Misteli, T. (2009). Self-organization in the genome. *Proc. Natl. Acad. Sci. USA* **106**, 6885–6886.
Misteli, T., and Soutoglou, E. (2009). The emerging role of nuclear architecture in DNA repair and genome maintenance. *Nat. Rev. Mol. Cell Biol.* **10**, 243–254.
Mistry, A. R., Felix, C. A., Whitmarsh, R. J., Mason, A., Reiter, A., Cassinat, B., Parry, A., Walz, C., Wiemels, J. L., Segal, M. R., et al. (2005). DNA topoisomerase II in therapy-related acute promyelocytic leukemia. *N. Engl. J. Med.* **352**, 1529–1538.
Mitelman, F., Johansson, B., and Mertens, F. (2007). The impact of translocations and gene fusions on cancer causation. *Nat. Rev. Cancer* **7**, 233–245.
Modesti, M., and Kanaar, R. (2001). Homologous recombination: from model organisms to human disease. *Genome Biol.* **2**, REVIEWS1014.
Morrish, T. A., Gilbert, N., Myers, J. S., Vincent, B. J., Stamato, T. D., Taccioli, G. E., Batzer, M. A., and Moran, J. V. (2002). DNA repair mediated by endonuclease-independent LINE-1 retrotransposition. *Nat. Genet.* **31**, 159–165.
Muramatsu, M., Kinoshita, K., Fagarasan, S., Yamada, S., Shinkai, Y., and Honjo, T. (2000). Class switch recombination and hypermutation require activation-induced cytidine deaminase (AID), a potential RNA editing enzyme. *Cell* **102**, 553–563.
Muramatsu, M., Sankaranand, V. S., Anant, S., Sugai, M., Kinoshita, K., Davidson, N. O., and Honjo, T. (1999). Specific expression of activation-induced cytidine deaminase (AID), a novel member of the RNA-editing deaminase family in germinal center B cells. *J. Biol. Chem.* **274**, 18470–18476.
Murmann, A. E., Gao, J., Encinosa, M., Gautier, M., Peter, M. E., Eils, R., Lichter, P., and Rowley, J. D. (2005). Local gene density predicts the spatial position of genetic loci in the interphase nucleus. *Exp. Cell Res.* **311**, 14–26.
Myung, K., Chen, C., and Kolodner, R. D. (2001). Multiple pathways cooperate in the suppression of genome instability in Saccharomyces cerevisiae. *Nature* **411**, 1073–1076.
Nambiar, M., and Raghavan, S. C. (2011). How does DNA break during chromosomal translocations? *Nucleic Acids Res.* **39**, 5813–5825.
Nelson, D. L., and Gibbs, R. A. (2004). Genetics. The critical region in trisomy 21. *Science* **306**, 619–621.
Nemec, A. A., Wallace, S. S., and Sweasy, J. B. (2010). Variant base excision repair proteins: contributors to genomic instability. *Semin. Cancer Biol.* **20**, 320–328.
Nowell, P. C. (2007). Discovery of the Philadelphia chromosome: a personal perspective. *J. Clin. Invest.* **117**, 2033–2035.
Nussenzweig, A., and Nussenzweig, M. C. (2010). Origin of chromosomal translocations in lymphoid cancer. *Cell* **141**, 27–38.
Ohka, F., Natsume, A., Motomura, K., Kishida, Y., Kondo, Y., Abe, T., Nakasu, Y., Namba, H., Wakai, K., Fukui, T., et al. (2011). The global DNA methylation surrogate LINE-1 methylation is correlated with MGMT promoter methylation and is a better prognostic factor for glioma. *PLoS One* **6**, e23332.
Oliver-Bonet, M., Navarro, J., Carrera, M., Egozcue, J., and Benet, J. (2002). Aneuploid and unbalanced sperm in two translocation carriers: evaluation of the genetic risk. *Mol. Hum. Reprod.* **8**, 958–963.

Ostertag, E. M., and Kazazian, H. H., Jr. (2001). Biology of mammalian L1 retrotransposons. *Annu. Rev. Genet.* **35**, 501–538.

Palanisamy, N., Ateeq, B., Kalyana-Sundaram, S., Pflueger, D., Ramnarayanan, K., Shankar, S., Han, B., Cao, Q., Cao, X., Suleman, K., et al. (2010). Rearrangements of the RAF kinase pathway in prostate cancer, gastric cancer and melanoma. *Nat. Med.* **16**, 793–798.

Parada, L. A., McQueen, P. G., and Misteli, T. (2004). Tissue-specific spatial organization of genomes. *Genome Biol.* **5**, R44.

Petersen-Mahrt, S. K., Harris, R. S., and Neuberger, M. S. (2002). AID mutates E. coli suggesting a DNA deamination mechanism for antibody diversification. *Nature* **418**, 99–103.

Pierotti, M. A., Santoro, M., Jenkins, R. B., Sozzi, G., Bongarzone, I., Grieco, M., Monzini, N., Miozzo, M., Herrmann, M. A., Fusco, A., et al. (1992). Characterization of an inversion on the long arm of chromosome 10 juxtaposing D10S170 and RET and creating the oncogenic sequence RET/PTC. *Proc. Natl. Acad. Sci. USA* **89**, 1616–1620.

Pogribny, I., Raiche, J., Slovack, M., and Kovalchuk, O. (2004). Dose-dependence, sex- and tissue-specificity, and persistence of radiation-induced genomic DNA methylation changes. *Biochem. Biophys. Res. Commun.* **320**, 1253–1261.

Pollack, J. R., Perou, C. M., Alizadeh, A. A., Eisen, M. B., Pergamenschikov, A., Williams, C. F., Jeffrey, S. S., Botstein, D., and Brown, P. O. (1999). Genome-wide analysis of DNA copy-number changes using cDNA microarrays. *Nat. Genet.* **23**, 41–46.

Powell, S. N., and Kachnic, L. A. (2003). Roles of BRCA1 and BRCA2 in homologous recombination, DNA replication fidelity and the cellular response to ionizing radiation. *Oncogene* **22**, 5784–5791.

Prado, F., and Aguilera, A. (2005). Impairment of replication fork progression mediates RNA polII transcription-associated recombination. *EMBO J.* **24**, 1267–1276.

Rabbitts, T. H. (1994). Chromosomal translocations in human cancer. *Nature* **372**, 143–149.

Rada, C., Williams, G. T., Nilsen, H., Barnes, D. E., Lindahl, T., and Neuberger, M. S. (2002). Immunoglobulin isotype switching is inhibited and somatic hypermutation perturbed in UNG-deficient mice. *Curr. Biol.* **12**, 1748–1755.

Ragu, S., Faye, G., Iraqui, I., Masurel-Heneman, A., Kolodner, R. D., and Huang, M. E. (2007). Oxygen metabolism and reactive oxygen species cause chromosomal rearrangements and cell death. *Proc. Natl. Acad. Sci. USA* **104**, 9747–9752.

Ramiro, A. R., Nussenzweig, M. C., and Nussenzweig, A. (2006). Switching on chromosomal translocations. *Cancer Res.* **66**, 7837–7839.

Ramiro, A. R., Stavropoulos, P., Jankovic, M., and Nussenzweig, M. C. (2003). Transcription enhances AID-mediated cytidine deamination by exposing single-stranded DNA on the nontemplate strand. *Nat. Immunol.* **4**, 452–456.

Ramsden, D. A., Weed, B. D., and Reddy, Y. V. (2010). V(D)J recombination: born to be wild. *Semin. Cancer Biol.* **20**, 254–260.

Rao, P. H., Cigudosa, J. C., Ning, Y., Calasanz, M. J., Iida, S., Tagawa, S., Michaeli, J., Klein, B., Dalla-Favera, R., Jhanwar, S. C., et al. (1998). Multicolor spectral karyotyping identifies new recurring breakpoints and translocations in multiple myeloma. *Blood* **92**, 1743–1748.

Rassool, F. V., Gaymes, T. J., Omidvar, N., Brady, N., Beurlet, S., Pla, M., Reboul, M., Lea, N., Chomienne, C., Thomas, N. S., et al. (2007). Reactive oxygen species, DNA damage, and error-prone repair: a model for genomic instability with progression in myeloid leukemia? *Cancer Res.* **67**, 8762–8771.

Ren, R. (2005). Mechanisms of BCR-ABL in the pathogenesis of chronic myelogenous leukaemia. *Nat. Rev. Cancer* **5**, 172–183.

Robbiani, D. F., Bunting, S., Feldhahn, N., Bothmer, A., Camps, J., Deroubaix, S., McBride, K. M., Klein, I. A., Stone, G., Eisenreich, T. R., et al. (2009). AID produces DNA double-strand breaks in non-Ig genes and mature B cell lymphomas with reciprocal chromosome translocations. *Mol. Cell* **36**, 631–641.

Rosti, G., Castagnetti, F., Gugliotta, G., Palandri, F., and Baccarani, M. (2011). Second-generation BCR-ABL inhibitors for frontline treatment of chronic myeloid leukemia in chronic phase. *Crit. Rev. Oncol. Hematol.*

Rowley, J. D. (1973). Chromosomal patterns in myelocytic leukemia. *N. Engl. J. Med.* **289**, 220–221.

Rowley, J. D. (2001). Chromosome translocations: dangerous liaisons revisited. *Nat. Rev. Cancer* **1**, 245–250.

Rudin, C. M., and Thompson, C. B. (2001). Transcriptional activation of short interspersed elements by DNA-damaging agents. *Genes Chromosomes Cancer* **30**, 64–71.

Sandberg, A. A., and Yamada, K. (1966). Chromosomes and causation of human cancer and leukemia. I. Karyotypic diversity in a single cancer. *Cancer* **19**, 1869–1878.

Schmitz, K. J., Otterbach, F., Callies, R., Levkau, B., Holscher, M., Hoffmann, O., Grabellus, F., Kimmig, R., Schmid, K. W., and Baba, H. A. (2004). Prognostic relevance of activated Akt kinase in node-negative breast cancer: a clinicopathological study of 99 cases. *Mod. Pathol.* **17**, 15–21.

Schoenfelder, S., Clay, I., and Fraser, P. (2010). The transcriptional interactome: gene expression in 3D. *Curr. Opin. Genet. Dev.* **20**, 127–133.

Schotta, G., Sengupta, R., Kubicek, S., Malin, S., Kauer, M., Callen, E., Celeste, A., Pagani, M., Opravil, S., De La Rosa-Velazquez, I. A., *et al.* (2008). A chromatin-wide transition to H4K20 monomethylation impairs genome integrity and programmed DNA rearrangements in the mouse. *Genes Dev.* **22**, 2048–2061.

Schrock, E., du Manoir, S., Veldman, T., Schoell, B., Wienberg, J., Ferguson-Smith, M. A., Ning, Y., Ledbetter, D. H., Bar-Am, I., Soenksen, D., *et al.* (1996). Multicolor spectral karyotyping of human chromosomes. *Science* **273**, 494–497.

Shaffer, D. R., and Pandolfi, P. P. (2006). Breaking the rules of cancer. *Nat. Med.* **12**, 14–15.

Shedge, V., Davila, J., Arrieta-Montiel, M. P., Mohammed, S., and Mackenzie, S. A. (2010). Extensive rearrangement of the Arabidopsis mitochondrial genome elicits cellular conditions for thermotolerance. *Plant Physiol.* **152**, 1960–1970.

Shimazaki, N., Tsai, A. G., and Lieber, M. R. (2009). H3K4me3 stimulates the V(D)J RAG complex for both nicking and hairpinning in trans in addition to tethering in cis: implications for translocations. *Mol. Cell* **34**, 535–544.

Shrivastav, M., De Haro, L. P., and Nickoloff, J. A. (2008). Regulation of DNA double-strand break repair pathway choice. *Cell Res.* **18**, 134–147.

Simsek, D., and Jasin, M. (2010). Alternative end-joining is suppressed by the canonical NHEJ component Xrcc4-ligase IV during chromosomal translocation formation. *Nat. Struct. Mol. Biol.* **17**, 410–416.

Sjoblom, T., Jones, S., Wood, L. D., Parsons, D. W., Lin, J., Barber, T. D., Mandelker, D., Leary, R. J., Ptak, J., Silliman, N., *et al.* (2006). The consensus coding sequences of human breast and colorectal cancers. *Science* **314**, 268–274.

Skandalis, A., Ford, B. N., and Glickman, B. W. (1994). Strand bias in mutation involving 5-methylcytosine deamination in the human HPRT gene. *Mutat. Res.* **314**, 21–26.

Skok, J. A., Brown, K. E., Azuara, V., Caparros, M. L., Baxter, J., Takacs, K., Dillon, N., Gray, D., Perry, R. P., Merkenschlager, M., *et al.* (2001). Nonequivalent nuclear location of immunoglobulin alleles in B lymphocytes. *Nat. Immunol.* **2**, 848–854.

Slade, I., Stephens, P., Douglas, J., Barker, K., Stebbings, L., Abbaszadeh, F., Pritchard-Jones, K., Cole, R., Pizer, B., Stiller, C., *et al.* (2010). Constitutional translocation breakpoint mapping by genome-wide paired-end sequencing identifies HACE1 as a putative Wilms tumour susceptibility gene. *J. Med. Genet.* **47**, 342–347.

Slotkin, R. K., and Martienssen, R. (2007). Transposable elements and the epigenetic regulation of the genome. *Nat. Rev. Genet.* **8**, 272–285.

Soda, M., Choi, Y. L., Enomoto, M., Takada, S., Yamashita, Y., Ishikawa, S., Fujiwara, S., Watanabe, H., Kurashina, K., Hatanaka, H., et al. (2007). Identification of the transforming EML4-ALK fusion gene in non-small-cell lung cancer. *Nature* **448**, 561–566.

Sohail, A., Klapacz, J., Samaranayake, M., Ullah, A., and Bhagwat, A. S. (2003). Human activation-induced cytidine deaminase causes transcription-dependent, strand-biased C to U deaminations. *Nucleic Acids Res.* **31**, 2990–2994.

Speicher, M. R., and Carter, N. P. (2005). The new cytogenetics: blurring the boundaries with molecular biology. *Nat. Rev. Genet.* **6**, 782–792.

Spilianakis, C. G., Lalioti, M. D., Town, T., Lee, G. R., and Flavell, R. A. (2005). Interchromosomal associations between alternatively expressed loci. *Nature* **435**, 637–645.

Srivatsan, A., Tehranchi, A., MacAlpine, D. M., and Wang, J. D. (2010). Co-orientation of replication and transcription preserves genome integrity. *PLoS Genet.* **6**, e1000810.

Stavnezer, J., Guikema, J. E., and Schrader, C. E. (2008). Mechanism and regulation of class switch recombination. *Annu. Rev. Immunol.* **26**, 261–292.

Stern, C. (1950). Boveri and the early days of genetics. *Nature* **166**, 446.

Sunami, E., de Maat, M., Vu, A., Turner, R. R., and Hoon, D. S. (2011). LINE-1 hypomethylation during primary colon cancer progression. *PLoS One* **6**, e18884.

Takizawa, T., Meaburn, K. J., and Misteli, T. (2008). The meaning of gene positioning. *Cell* **135**, 9–13.

Taub, R., Kirsch, I., Morton, C., Lenoir, G., Swan, D., Tronick, S., Aaronson, S., and Leder, P. (1982). Translocation of the c-myc gene into the immunoglobulin heavy chain locus in human Burkitt lymphoma and murine plasmacytoma cells. *Proc. Natl. Acad. Sci. USA* **79**, 7837–7841.

Tehranchi, A. K., Blankschien, M. D., Zhang, Y., Halliday, J. A., Srivatsan, A., Peng, J., Herman, C., and Wang, J. D. (2010). The transcription factor DksA prevents conflicts between DNA replication and transcription machinery. *Cell* **141**, 595–605.

Tognon, C., Knezevich, S. R., Huntsman, D., Roskelley, C. D., Melnyk, N., Mathers, J. A., Becker, L., Carneiro, F., MacPherson, N., Horsman, D., et al. (2002). Expression of the ETV6-NTRK3 gene fusion as a primary event in human secretory breast carcinoma. *Cancer Cell* **2**, 367–376.

Tomlins, S. A., Laxman, B., Dhanasekaran, S. M., Helgeson, B. E., Cao, X., Morris, D. S., Menon, A., Jing, X., Cao, Q., Han, B., et al. (2007). Distinct classes of chromosomal rearrangements create oncogenic ETS gene fusions in prostate cancer. *Nature* **448**, 595–599.

Tomlins, S. A., Rhodes, D. R., Perner, S., Dhanasekaran, S. M., Mehra, R., Sun, X. W., Varambally, S., Cao, X., Tchinda, J., Kuefer, R., et al. (2005). Recurrent fusion of TMPRSS2 and ETS transcription factor genes in prostate cancer. *Science* **310**, 644–648.

Trautinger, B. W., Jaktaji, R. P., Rusakova, E., and Lloyd, R. G. (2005). RNA polymerase modulators and DNA repair activities resolve conflicts between DNA replication and transcription. *Mol. Cell* **19**, 247–258.

Trigueros, S., and Roca, J. (2001). Circular minichromosomes become highly recombinogenic in topoisomerase-deficient yeast cells. *J. Biol. Chem.* **276**, 2243–2248.

Tsai, A. G., Lu, H., Raghavan, S. C., Muschen, M., Hsieh, C. L., and Lieber, M. R. (2008). Human chromosomal translocations at CpG sites and a theoretical basis for their lineage and stage specificity. *Cell* **135**, 1130–1142.

Turner, D. P., and Watson, D. K. (2008). ETS transcription factors: oncogenes and tumor suppressor genes as therapeutic targets for prostate cancer. *Expert Rev. Anticancer Ther.* **8**, 33–42.

van der Burg, M., Poulsen, T. S., Hunger, S. P., Beverloo, H. B., Smit, E. M., Vang-Nielsen, K., Langerak, A. W., and van Dongen, J. J. (2004). Split-signal FISH for detection of chromosome aberrations in acute lymphoblastic leukemia. *Leukemia* **18**, 895–908.

Varella-Garcia, M. (2003). Molecular cytogenetics in solid tumors: laboratory tool for diagnosis, prognosis, and therapy. *Oncologist* **8**, 45–58.

Vilain, A., Bernardino, J., Gerbault-Seureau, M., Vogt, N., Niveleau, A., Lefrancois, D., Malfoy, B., and Dutrillaux, B. (2000). DNA methylation and chromosome instability in lymphoblastoid cell lines. *Cytogenet. Cell Genet.* **90**, 93–101.

Vilette, D., Ehrlich, S. D., and Michel, B. (1995). Transcription-induced deletions in Escherichia coli plasmids. *Mol. Microbiol.* **17**, 493–504.

Voelkel-Meiman, K., Keil, R. L., and Roeder, G. S. (1987). Recombination-stimulating sequences in yeast ribosomal DNA correspond to sequences regulating transcription by RNA polymerase I. *Cell* **48**, 1071–1079.

Wai, D. H., Knezevich, S. R., Lucas, T., Jansen, B., Kay, R. J., and Sorensen, P. H. (2000). The ETV6-NTRK3 gene fusion encodes a chimeric protein tyrosine kinase that transforms NIH3T3 cells. *Oncogene* **19**, 906–915.

Walsh, C. P., Chaillet, J. R., and Bestor, T. H. (1998). Transcription of IAP endogenous retroviruses is constrained by cytosine methylation. *Nat. Genet.* **20**, 116–117.

Wang, J., Cai, Y., Ren, C., and Ittmann, M. (2006). Expression of variant TMPRSS2/ERG fusion messenger RNAs is associated with aggressive prostate cancer. *Cancer Res.* **66**, 8347–8351.

Washburn, R. S., and Gottesman, M. E. (2011). Transcription termination maintains chromosome integrity. *Proc. Natl. Acad. Sci. USA* **108**, 792–797.

Weinstock, D. M., Brunet, E., and Jasin, M. (2008). Induction of chromosomal translocations in mouse and human cells using site-specific endonucleases. *J. Natl. Cancer Inst. Monogr.* (39), 20–24.

Westover, K. D., Bushnell, D. A., and Kornberg, R. D. (2004). Structural basis of transcription: separation of RNA from DNA by RNA polymerase II. *Science* **303**, 1014–1016.

Wieland, I., Muschke, P., Volleth, M., Ropke, A., Pelz, A. F., Stumm, M., and Wieacker, P. (2006). High incidence of familial breast cancer segregates with constitutional t(11;22)(q23;q11). *Genes Chromosomes Cancer* **45**, 945–949.

Wilhelm, C. S., Kelsey, K. T., Butler, R., Plaza, S., Gagne, L., Zens, M. S., Andrew, A. S., Morris, S., Nelson, H. H., Schned, A. R., et al. (2010). Implications of LINE1 methylation for bladder cancer risk in women. *Clin. Cancer Res.* **16**, 1682–1689.

Williamson, M. P., Elder, P. A., Shaw, M. E., Devlin, J., and Knowles, M. A. (1995). p16 (CDKN2) is a major deletion target at 9p21 in bladder cancer. *Hum. Mol. Genet.* **4**, 1569–1577.

Wong, S., and Witte, O. N. (2004). The BCR-ABL story: bench to bedside and back. *Annu. Rev. Immunol.* **22**, 247–306.

Wu, H. Y., Shyy, S. H., Wang, J. C., and Liu, L. F. (1988). Transcription generates positively and negatively supercoiled domains in the template. *Cell* **53**, 433–440.

Xu, G. L., Bestor, T. H., Bourc'his, D., Hsieh, C. L., Tommerup, N., Bugge, M., Hulten, M., Qu, X., Russo, J. J., and Viegas-Pequignot, E. (1999). Chromosome instability and immunodeficiency syndrome caused by mutations in a DNA methyltransferase gene. *Nature* **402**, 187–191.

Yoder, J. A., Walsh, C. P., and Bestor, T. H. (1997). Cytosine methylation and the ecology of intragenomic parasites. *Trends Genet.* **13**, 335–340.

Yoshimoto, M., Joshua, A. M., Chilton-Macneill, S., Bayani, J., Selvarajah, S., Evans, A. J., Zielenska, M., and Squire, J. A. (2006). Three-color FISH analysis of TMPRSS2/ERG fusions in prostate cancer indicates that genomic microdeletion of chromosome 21 is associated with rearrangement. *Neoplasia* **8**, 465–469.

Yoshimoto, M., Joshua, A. M., Cunha, I. W., Coudry, R. A., Fonseca, F. P., Ludkovski, O., Zielenska, M., Soares, F. A., and Squire, J. A. (2008). Absence of TMPRSS2:ERG fusions and PTEN losses in prostate cancer is associated with a favorable outcome. *Mod. Pathol.* **21**, 1451–1460.

Zhang, F., Carvalho, C. M., and Lupski, J. R. (2009). Complex human chromosomal and genomic rearrangements. *Trends Genet.* **25,** 298–307.

Zhang, Y., Gostissa, M., Hildebrand, D. G., Becker, M. S., Boboila, C., Chiarle, R., Lewis, S., and Alt, F. W. (2010). The role of mechanistic factors in promoting chromosomal translocations found in lymphoid and other cancers. *Adv. Immunol.* **106,** 93–133.

Zhao, R., Bodnar, M. S., and Spector, D. L. (2009). Nuclear neighborhoods and gene expression. *Curr. Opin. Genet. Dev.* **19,** 172–179.

Index

Note: Page numbers followed by "*f*" indicate figures, and "*t*" indicate tables.

A

Activation-induced cytidine deaminase (AID)
 adaptive immunity, defined, 168
 chromosome translocations
  depiction, reciprocal/nonreciprocal, 168–170, 170*f*
  rearrangements, 168–170
  RSSs, 170–171
  schematic representation, 168–170, 169*f*
  translocation, reciprocal/nonreciprocal, 168–170
 cloning, 3
 CODA, 31
 CSR and SHM, 3
 CSR-specific cofactors, 181–182
 C-terminal mutations phenotypes
  DNA cleavage, 7
  NES motif, 7
 description, 171
 DNA cleavage mechanism (*see* DNA cleavage mechanism, AID)
 genome instability
  CSR and meiotic recombination homology, 22–23
  Top1, transcription-related, 24
 genome wide damage
  c-myc locus, 180
  DSBs, 179–180
  Ig variable regions, 179
  interchromosomal translocations, 181
  mutational analysis, 179
  partner ligation, DSB, 180
 genomic damage and tumorigenesis, 171–172
 haploinsufficiency/hyperactivity, 172
 influences, CSR, 182
 miR-155 and miR-181b binding sites, 172
 N-terminal mutations phenotypes
  DNA cleavage activity, 6
  G23S knock-in-animals, 6–7
 overexpression, 3–4
 pathogen-induced expression
  HTLV-1 oncogenes, 28–30
  human malignant tumors, 29*t*
 phenotypes, deficiency
  CSR and SHM activities, 11*t*
  HIGM II patients, 4–5
  immune cells activation, 4–5
  structure and function, 5*f*
 phosphorylation, 173
 PKA, 173
 protein modification, 172–173
 regulation, 30–31
 schematic representation, 168, 169*f*
 SHM and CSR, 168
 Spt5 and stalled Pol II
  ChIP-Seq analysis, 176
  description, 175
  DSIF and NELF, 175–176
  model, 176, 177*f*
 subcellular localization, 172
 template strand, exosomes
  description, 176–178
  RNA–DNA hybrids, 176–178
  stalled Pol II/Spt5 elongation complex, 176–178, 178*f*
  stalling model, 176–178
 transcription
  defined, 173–174
  mutator factor, 173–174
  RNA–DNA hybrid R loops, 174–175
  structure, ssDNA, 174
 tumorigenesis
  GC-derived B-cell lymphoma, 25–26
  human B-cell malignancy, 27–28
  murine tumors, 24–25

Activation-induced cytidine deaminase (AID) (continued)
  non-GC-derived B-cell lymphoma, 26–27
  VDJ recombination, 2
Adult T-cell leukemia/lymphoma (ATL)
  expression, lymphokines and lymphokine receptors, 93
  HTLV-I, 87
  leukemogenesis, 87–88
  vs. novel therapies, 106
  pathology, transgenic mice, 86–87
Adult T cell leukemia virus (ATLV), 49
AID. *See* Activation-induced cytidine deaminase (AID)
Ang/Tie pathway
  cytokines, 200–201
  description, 200–201
  dimerization, c-Met, 201
  HGF, 201
Antiangiogenetic therapy, tumor angiogenesis
  hypoxia pathway, 217
  mTOR pathway, 217
  novel clinical evaluation, 217–218
  resistance, mechanism
    angiogenic factors, 220–221
    hypoxia and c-Met, 219–220
    sprouting, 221
  VEGF pathway
    agents, target, 215–216, 215*f*
    monoclonal antibodies, 216–217
AP endonuclease (APE), 10–16
ATL. *See* Adult T-cell leukemia/lymphoma (ATL)

C

Cellular miRNA expression, HTLV-1 infected cell lines
  ATLL and hematological malignancies, 61*t*
  ATLL cells, 56
  CXCR4, 58
  miR-21, 59
  miR-24, 59–60
  miR-155, 60
  miR-223, 60–63
  miR-146a
    AICD, 58
    CXCR4, 58
    solid cancers, 58–59
    upregulation, 58
  quantitative RT-PCR, 56–57
  upregulation, 56–57
Cellular stress
  genomic instability and DNA repair, 252
  reactive oxygen species (ROS), 252
  replication, 252
Chromatin modifications
  DSB repair, 263–264
  "epigenomic" drugs, 264
  H3K79 methylation, 263–264
Chromosomal translocations
  vs. cancer
    benign and malignant neoplastic disorders, 246
    deletions/duplications, 247
    dysregulation, 248
    FISH, 247
    genome instability and initiation, 246
    molecular mechanisms, 248
    oncogene, 247–248
    paired-end sequencing technology, 247
    SKY, 246–247
  cancer progression, 241–242
  cell-based models
    human B-cell lymphoma, 249
    in vivo, 248–249
    insertion, I-Sec sites, 249
    prostate cancer, 249
    treatment, LNCaP, 249
  chromatin fibers and silence repetitive elements, 266
  coding regions, 243
  definition, 242–243
  DNA oligo/siRNA, 266
  Down syndrome, 244
  DSBs generation
    cellular stress, 252–253
    genotoxic stress, 251–252
  interphase nuclei, 264–266, 265*f*
  ISH/FISH, 266
  ligation, 260–262
    IgH class, 261–262
    inherited syndromes, 260–261
    misrepair, broken chromatin ends, 261
    NHEJ pathway, 261–262
    sister chromatid template sequences, 261
  non-homologous chromosomes, 242–243, 242*f*
  nonreciprocal, 243
  PDGFB-COLIA1, sarcoma, 266–267

reciprocal/non-homologous chromosomes, 242–243, 242f
recombination, coding region, 243
recombination, noncoding region, 243–244
role, epigenetics
   chromatin modifications and genome instability, 263–264
   DNA methylation and genome instability, 262–263
selection, growth pressure, 250
spatial proximity
   cell cycle progression and signalling, 258–259
   chromatin territory, 258
   "contact-first" model, 260
   interactome, 258–259
   interphase nucleus, 259–260, 259f
   nuclear architecture, 257–258
   transcription regulation and genome stability, 260
TMPRESS2–ERG model, 250
transcription and genome instability
   B-cell specific factor, 254–256
   collision, DNA replication, 253
   DNA damage and repair, 254–256
   DNA mutations and recombination, 253
   DNMT1 and DNMT3L knockout mice, 257
   genome integrity, 254
   hypomethylation, LINE-1, 257
   R-loop, 254, 255f
   supercoiled DNA, 253–254
transcription factors/bubbles, 264–266
transposable elements and endonuclease
   deregulation and abnormal gene expression, 257
   LINEs, 256, 256f
   move/transpose genomes, 256
types
   AR gene and FISH technology, 244–245, 245f
   breast cancer, 245
   cytogenetics, 246
   leukemia and lymphoma, 244
Class-switch recombination (CSR)
activities, mutant AID, 11t
AID-dependent, 168, 169f
and chromosome translocations, 179–180
cofactors, 181–182
C-terminal-specific AID activity, 7
description, 3

genome integrity, 168–170
influences, 182
inhibition, HSP90, 172
meiotic recombination homology
   H3K4me3 histone modification, 22–23
   molecules and mechanisms, 23t
NHEJ pathway, 171
nonlymphoid cells, 3
phosphorylation, T140, 172–173
and SHM, mechanistic role, 174
Spt5, 175–176
transcription, 173–174
UNG involvement
   functions, 10
   gene conversion (GC), 9
   IgM hybridomas, 9–10
   rescue, CSR, 15t
   structural features, mouse UNG2, 16f
   U accumulation, 10
   WXXF motif, 10, 15t
Colorectal cancer (CRC)
Amsterdam criteria, 123–125, 124t
decision tree, 148f
deletion carriers, EPCAM, 138
description, 122
genetic and epigenetic alterations, 145
HNPCC, 122–123
IBD, 142
LS, 126t
MLH1 promoter methylation, 135–136
sporadic, 128–130
Turcot syndrome, 142
Constitutional MMR-deficiency syndrome, 144–145
CRC. See Colorectal cancer (CRC)
CSR. See Class-switch recombination (CSR)

D

Delta-like ligand 4 (DLL4)/notch pathway, 205–206
De novo mutations, 143
DNA cleavage mechanism, AID
deamination hypothesis
   APE requirement, 10–16
   C-to-T mutations, *E. coli*, 8–9
   in vitro reaction, 8
   UNG involvement, CSR, 9–10
RNA-editing model (*see* RNA-editing model)
target specificity

DNA cleavage mechanism, AID (continued)
  markers, 21–22
  non-IG targets, 19–21
  tumorigenesis, 7
DNA deamination hypothesis
  activity
    C-to-T mutations, E. coli, 8–9
    in vitro reaction, 8
  AP endonuclease requirement, 10–16
  UNG involvement, CSR
    functions, 10
    gene conversion (GC), 9
    IgM hybridomas, 9–10
    rescue, CSR, 15t
    structural features, mouse UNG2, 16f
    U accumulation, 10
    WXXF motif, 10
DNA methylation
  CpG dinucleotide, 262–263
  H3K79, DSB region, 262–263, 263f
  tumor progression, 262–263
DNA mismatch repair (MMR) genes, detection, 132–133
Double-stranded DNA breaks (DSBs)
  aberrant, 170–171
  AID-dependent CSR and SHM, 169f
  chromosome positioning and proximal DNA damage response, 180
  chromosome translocations, 182–183
  c-myc locus, 180
  NHEJ pathway, 171
  physiological, 179–180
DRB sensitivity inducing factor (DSIF) complex, 175–176, 178f
DSBs. *See* Double-stranded DNA breaks (DSBs)

F

False negative, LS
  cases, MSI, 141
  conventional DNA sequencing, 141
  germline mutation, 140–141
  Knudson two-hit hypothesis, 140–141
  promoter methylation, 141
False positive, LS
  CRCs, 142
  IBDs, 141–142
  microsatellite markers, 142
Familial adenomatous polyposis (FAP)
  description, 122

HNPCC, 123–125
FAP. *See* Familial adenomatous polyposis (FAP)
FGFRs. *See* Fibroblast growth factor receptors (FGFRs)
FGFs. *See* Fibroblast growth factors (FGFs)
Fibroblast growth factor receptors (FGFRs), 204–205
Fibroblast growth factors (FGFs), 204–205

G

Genome instability, AID
  CSR and meiotic recombination homology
    H3K4me3 histone modification, 22–23
    molecules and mechanisms, 23t
  Top1, transcription-related
    2-5 base deletion, 24
    TAM, 24
Genotoxic stress
  nucleotide structure, DNA, 251
  purine and pyrimidine, 251
  site-specific DNA breaks, 251
  TOPII, 251–252

H

HAM. *See* HTLV-I-associated myelopathy (HAM)
Hereditary nonpolyposis colorectal cancer (HNPCC)
  Amsterdam I/II and Bethesda I/II cases, 123–125, 124t
  Bethesda criteria, 125
  complex algorithms and multivariate models, 128
  consecutive unselected CRC series, 123–125, 126t
  CRC, 122
  defect, MMR, 132–133
  definition, 123–125
  description, 122–123
  FAP, 122
  germline mutation, DNA MMR gene, 123
  "G" family, 122–123
  LS (*see* Lynch syndrome (LS))
  MMR germline mutation, 123–125
  MSI (*see* Microsatellite instability (MSI))
  sensitivity and specificity, Amsterdam criteria, 125, 129t
  tumors, LS patients, 123, 124f

Index **285**

HNPCC. *See* Hereditary nonpolyposis colorectal cancer (HNPCC)
HTLV-1. *See* Human T cell leukemia virus type 1
HTLV-I-associated myelopathy (HAM), 87–89
Human T cell leukemia virus type 1 (HTLV-1)
 diseases
  ATLL, 51–52
  flower cells, 52
  HAM/TSP, 52
  HAU, 52
 genome organization and expression
  RXRE, 50
  "simple" and "complex" retroviruses, 50
  Tax and Rex, 50
 transmission and persistence, 50–51
Hypoxia pathway and HIF-1
 cyclic AMP and NF-κB, 203
 deletion and disruption, 202–203
 description, 202
 physiological regulator, 202
 transcription factor, 202
 types, 202

I

IBDs. *See* Inflammatory bowel diseases (IBDs)
I-kappa B kinase (IKK), 90–91, 92, 93, 94*f*, 95
IKK. *See* I-kappa B kinase (IKK)
Inflammatory bowel diseases (IBDs), 141–142

K

Knudson two-hit hypothesis, 140–141

L

LS. *See* Lynch syndrome (LS)
Lynch syndrome (LS)
 decision tree, 147–148, 148*f*, 149
 diagnosis, 146
 genetic alterations
  chromosomal abnormalities, 145–146
  and epigenetic, 145
  gastrointestinal and endometrial, MSI, 146
  "instabilotyping", 145–146
  noncoding regulatory sequences, 146
 IHC, 147
 not Lynch, HNPCC, 127*t*, 133–134
 predictive models, 146–147
 sporadic, MSI (*see* Microsatellite instability (MSI))
 surveillance, individuals
  analysis, germline mutations, 149
  annual gynecological surveillance, 149
  clinical management, 149
  upper digestive tract, endoscopy, 150
 unusual variants
  constitutional MMR-deficiency syndrome, 144–145
  MTS, 143
  Turcot syndrome, 143–144

M

Mammalian target of rapamycin (mTOR) pathway, 204
MicroRNA (miRNA) regulatory network
 biogenesis
  canonical and mirtron pathways, 47–48, 48*f*
  pre-miRNA, 47–48
  splicing, 47–48
 CD4+ T cells
  activated cells, 54–55
  development, 53–54
  tregs, 55–56
 cellular expression, HTLV-1 infected cell lines, 56–63
 description, 46
 host gene expression machinery, 46–47
 HTLV-1 and ATLL
  ATLV, 49
  genome organization and expression, 50
  HAM/TSP, 52
  HTLV-2, 49
  subtypes, 51–52
  transmission and persistence, 50
  Treg function, 52
 miR-155, 46–47
 miRNA–mRNA interactions
  base pairing, 46
  complementary nucleotides, 46
  description, 48–49
  translational inhibition, 49
 miR-31, repression, 67–70
 profiling, ATLL samples
  miR-93, 63–64

MicroRNA (miRNA) regulatory network (continued)
  miR-150, 66–67
  miR-125a, 66
  miR-181a, 66
  miR-130b, 64
  miR-142-5p/3p locus, 66
  quantitative RT-PCR, 65–66
  TP53INP1, 64–65, 65f
  up/down regulation, 61t, 63
Microsatellite instability (MSI)
  BAT-26, 130–131
  defined, 128–130
  dinucleotide amplification profiles, 131
  germline mutation, 128–130
  MSH2 allele, 131
  pentaplex assay, 131–132
  sporadic
    CIMP, 136
    classic Amsterdam criteria, 136–137
    clinical consequences, 139–140
    de novo mutations, 143
    diagnosis, clinical, 137–138
    DNA methylation, 135–136
    familial and clinical features, 134–135
    genetic testing, 139
    germline mutations, 137
    inactivation, MMR, 135
    MLH1 promoter, 135–136
    molecular analysis, 140–142
    molecular differences, 135, 135f
    mutations, BRAF, 136
    non-Mendelian transmission, 138–139
    prescreening methods, 136–137
    TACSTD1, 138
    UVs, 138
  subgroups, 130
miR-31 expression, ATLL
  genetic deletion and polycomb-directed epigenetic silencing
    PRC2 components, 69
    YY1-binding motifs, 69
  NF-kB activity
    oncogenic cascade, repressive factors, 70f
    PRC2 knockdown, 69, 70
    NIK identification, 68
MSI. See Microsatellite instability (MSI)
MTS. See Muir Torre syndrome (MTS)
Muir Torre syndrome (MTS), 143
Multifaceted oncoprotein tax
  NF-κB pathway, 91–97

oncogenic retrovirus HTLV-I
  cell-to-cell contact, 86
  description, 86
  epigenetic change, 87–88
  leukemogenesis, ATL, 88
  ORFs, 86–87
  pathology, transgenic mice, 86–87
  transmission, routes, 86
  TSP/HAM, 87
tax posttranslational modifications
  and intracellular localization, 103–106
  and NF-κB activation, 97–102
transactivator and deregulator, cellular machinery
  cyclin/CDK complex and CDK inhibitors, 90
  DNA damage response, 91
  effectors, transcriptional, 89–90
  expression, HTLV-I-infected cells, 90
  Fas and TRAIL, 90–91
  survival pathways, 90
viral oncoprotein tax
  central region, 89
  dual subcellular localization, 88–89
  formation, homodimers, 89
  PBM, 89
  structural and functional domains, 88–89, 88f

N

Negative elongation factor (NELF) complex, 175–176
NF-κB-inducing kinase (NIK), 68
NF-κB pathway
  ATL, 93
  canonical and noncanonical pathways, 92, 95
  description, 91–92
  DNA-binding and transcriptional activity, 96–97
  IκB-α, 95
  IKKs, 93
  members, 92
  molecular mechanisms, 93, 94f
  NEMO, 92–93, 93f
  Rel homology domain, 91–92
  TAX1BP1, 95–96
  and tax posttranslational modifications (see Tax posttranslational modifications)

P

PBM. *See* PDZ-domain binding motif (PBM)
PDGF/PDGF receptor pathways, 207
PDZ-domain binding motif (PBM), 89, 97–98
Polycomb repressive complex (PRC), 69

R

RECIST. *See* Response evaluation criteria in solid tumor (RECIST)
Recombination signal sequences (RSSs), 170–171
Response evaluation criteria in solid tumor (RECIST), 217–218
RNA-editing model
  CSR, C-terminus, 18–19
  DNA cleavage
    target transcription, AID-induced, 17f
    Top1 protein reduction, 18
  hypothesis, 16–17
RNA polymerases (RNAPs), 176
RNAPs. *See* RNA polymerases (RNAPs)
RSSs. *See* Recombination signal sequences (RSSs)

S

SHM. *See* Somatic hypermutation (SHM)
Signaling pathways, tumor angiogenesis
  Ang/Tie, 200–201
  basement membrane
    antiangiogenic therapy, 208
    MMPs, 208
    type IV collagen and laminin, 207
  DLL4, 205–206
  FGFs/FGFRs, 204–205
  hypoxia and HIF-1, 202–203
  inflammation
    IL-8, 209
    TAMs, 208
    TNF-a, 209
  molecules
    angiogenin, 206
    netrins, 206
    Robo4, 207
  mTOR, 204
  paradoxical effects, NO
    description, 203
    iNOS inhibition, 203
    Ras-Raf-MEK-ERK pathway, 203
  PDGF/PDGF receptor, 207

VEGF/VEGFR, 199–200
Somatic hypermutation (SHM)
  AID-dependent, 168, 169f
  genomic integrity, 173
  in vivo and siRNA knockdown, 176–178
  internal surveillance and protection pathways, 179–180
  molecular mechanism, 171
  physiological, 172
  positive regulators, 172–173
  transcription, 173–174

T

Tax posttranslational modifications
  acetylation, 98
  description, 97
  and intracellular localization
    adaptor molecule, NEMO, 105
    cytoplasm, 105
    domains, 103
    extranuclear compartments, 104–105
    intrachromatin granules, 103
    NES, 104
    nuclear bodies, 103–104
  phosphorylation, 97–98
  SUMOylation, 101–102
  ubiquitination, 99–101
Transcription-associated mutagenesis (TAM), 24
Tropical spastic paraparesis (TSP)
  description, 87
  dual subcellular localization, Tax, 88–89
  pathogenesis, 87–88
TSP. *See* Tropical spastic paraparesis (TSP)
Tumor angiogenesis
  antiangiogenic therapy, 215–217
  complexity and abnormality, 193
  defined, vasculogenesis, 192
  heterogeneity, vasculature
    abnormalities, 194, 194f
    diversity, 195
    endothelial cells, 193
    normal *vs.* tumor, 193
    pericytes, 195
    proangiogenic growth factors and leukocytes, 193–194
    sprouting, 195
    undifferentiated and differentiated, markers, 195
  microvessels, stem-like cells

Tumor angiogenesis (continued)
    endothelial, CD31+, 198
    malignancies, 198
    therapeutic strategy, 198
  novel clinical evaluation
    LD and PD, 217–218
    RECIST, 217–218
  resistance, antiangiogenic therapy,
      218–221
  revascularization
    bevacizumab therapy, 213
    mouse xenograft models, 212–213
    preclinical analysis, 213
    regrowth, vessels, 213
  signaling pathways
    Ang/Tie, 200–201
    basement membrane, 207–208
    DLL4, 205–206
    FGFs/FGFRs, 204–205
    hypoxia and HIF-1, 202–203
    inflammation, 208–209
    molecules, 206–207
    mTOR, 204
    paradoxical effects, NO, 203–204
    PDGF/PDGF receptor, 207
    VEGF/VEGFR, 199–200
  vascular normalization, 214–215
  vasculature (see Vasculature, tumor)
  vessel co-option, vasculature
    description, 196
    HEVs, 196–198
    in metastatic breast cancer, 196–198,
        197f
    sentinel lymph node metastasis, 196

Tumorigenesis, AID
  GC-derived B-cell lymphoma, 25–26
  human B-cell malignancy, 27–28
  murine tumors, 24–25
  non-GC-derived B-cell lymphoma, 26–27
Turcot syndrome, 142, 143–144

V

Vascular normalization, 214–215
Vasculature, tumor
  MVD
    description, 209
    drawbacks, 210
    types, 209–210
  prognostic value
    HCC, 210
    molecular classification, 210–211, 212f
    morphological classification, 210, 211f
    urothelial carcinoma, 211
VEGF/VEGFR signaling pathways
  adverse events, 200
  biological effects, 200
  description, 199
  fenestration structures, 200
  hypertension, 200
  neuropilins, 199
  proteinuria, 200
Virological synapse, 50–51

W

WXXF motif, 10, 15t

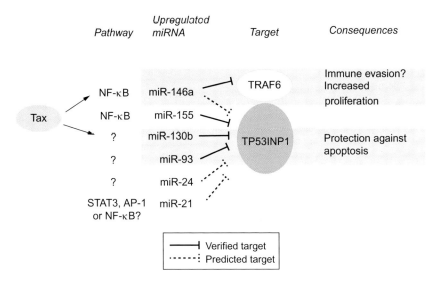

Fig. 3, Donna M. D'Agostino *et al.* (See Pages 65 of this volume.)

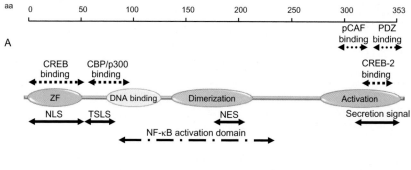

Fig. 1, Youmna Kfoury *et al.* (See Pages 88 of this volume.)

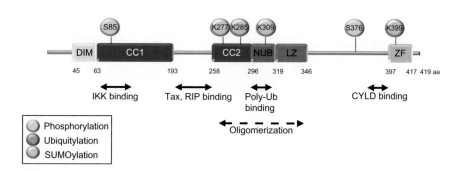

Fig. 2, Youmna Kfoury *et al.* (See Pages 93 of this volume.)

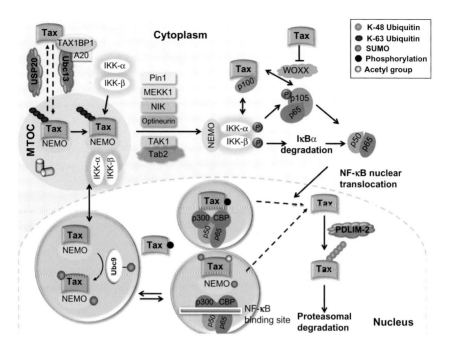

Fig. 3, Youmna Kfoury *et al.* (See Pages 94 of this volume.)

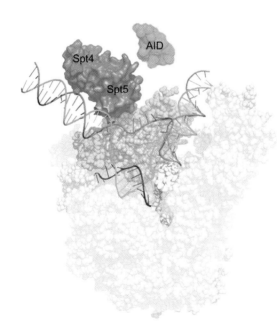

Fig. 3, Anna Gazumyan *et al.* (See Pages 177 of this volume.)

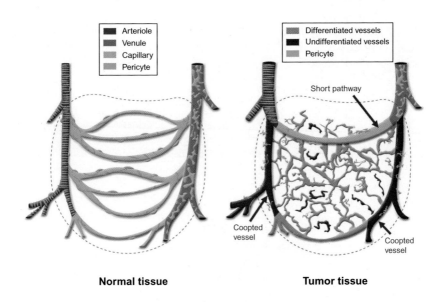

Fig. 1, Li Qin *et al.* (See Pages 194 of this volume.)

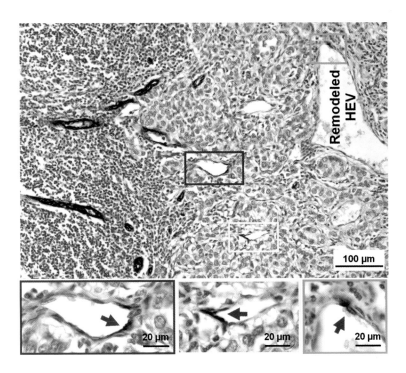

Fig. 2, Li Qin *et al.* (See Pages 197 of this volume.)

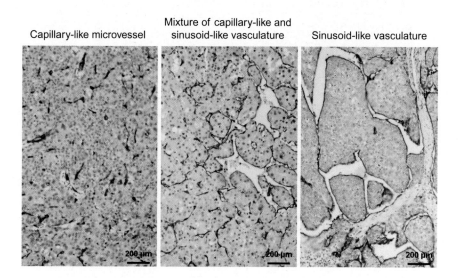

Fig. 3, Li Qin *et al.* (See Pages 211 of this volume.)

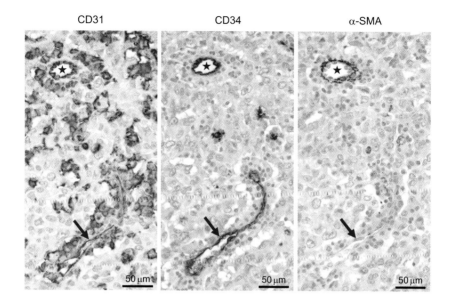

Fig. 4, Li Qin *et al.* (See Pages 212 of this volume.)

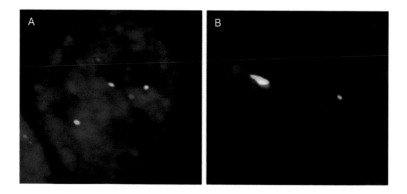

Fig. 2, Chunru Lin *et al.* (See Pages 245 of this volume.)

RT = Reverse transcriptase
EN = Endonuclease

Fig. 4, Chunru Lin *et al.* (See Pages 256 of this volume.)

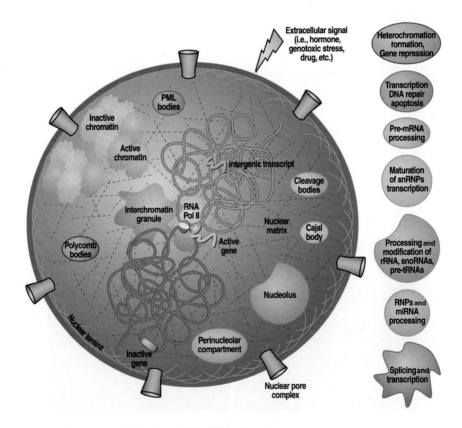

Fig. 5, Chunru Lin *et al.* (See Pages 259 of this volume.)